Conservation Science and Action

Edited by
William J. Sutherland

Professor of Biology at the
School of Biological Sciences
University of East Anglia

b

Blackwell
Science

© 1998 by
Blackwell Science Ltd
Editorial Offices:
Osney Mead, Oxford OX2 0EL
25 John Street, London WC1N 2BL
23 Ainslie Place, Edinburgh EH3 6AJ
350 Main Street, Malden
  MA 02148 5018, USA
54 University Street, Carlton
  Victoria 3053, Australia
10 rue Casimir Delavigne
  75006 Paris, France

Other Editorial Offices:
Blackwell Wissenschafts-Verlag GmbH
Kurfürstendamm 57
10707 Berlin, Germany

Blackwell Science KK
MG Kodenmacho Building
7–10 Kodenmacho Nihombashi
Chuo-ku, Tokyo 104, Japan

First published 1998

Set by Setrite Typesetters, Hong Kong
Printed and bound in Great Britain
by MPG Books Ltd, Bodmin, Cornwall

The Blackwell Science logo is a
trade mark of Blackwell Science Ltd,
registered at the United Kingdom
Trade Marks Registry

DISTRIBUTORS

  Marston Book Services Ltd
  PO Box 269
  Abingdon, Oxon OX14 4YN
  (*Orders*: Tel: 01235 465500
          Fax: 01235 465555)

USA
  Blackwell Science, Inc.
  Commerce Place
  350 Main Street
  Malden, MA 02148 5018
  (*Orders*: Tel: 800 759 6102
          781 388 8250
          Fax: 781 388 8255)

Canada
  Copp Clark Professional
  200 Adelaide St West, 3rd Floor
  Toronto, Ontario M5H 1W7
  (*Orders*: Tel: 416 597-1616
          800 815-9417
          Fax: 416 597-1617)

Australia
  Blackwell Science Pty Ltd
  54 University Street
  Carlton, Victoria 3053
  (*Orders*: Tel: 3 9347 0300
          Fax: 3 9347 5001)

A catalogue record for this title
is available from the British Library

ISBN 0-86542-762-3

Library of Congress
Cataloging-in-publication Data

Conservation science and action/edited by
William J. Sutherland.
        p.   cm.
Includes bibliographical references and
index.
ISBN 0-86542-762-3
1. Conservation biology.
I. Sutherland, William J.
QH75.C6815   1998
333.95' 16—dc21   97–20685
                                      CIP

Conservation
Science and Action

Contents

List of contributors

William M. Adams *Department of Geography, University of Cambridge, Cambridge CB2 3EN, UK*

Colin J. Bibby *BirdLife International, Wellbrook Court, Girton Road, Cambridge CB3 0NA, UK*

Michael J. Crawley *Department of Biology and NERC Centre for Population Biology, Imperial College at Silwood Park, Ascot SL5 7PY, UK*

Kevin J. Gaston *Department of Animal and Plant Sciences, University of Sheffield, Sheffield S10 2TN, UK*

H. Charles J. Godfray *Department of Biology and NERC Centre for Population Biology, Imperial College at Silwood Park, Ascot SL5 7PY, UK*

Nick Hanley *Institute of Ecology and Resource Management, University of Edinburgh, Edinburgh EH9 3JG, UK*

Susan Harrison *Division of Environmental Studies, University of California at Davis, Davis, California 95616, USA*

Martha F. Hoopes *Division of Environmental Studies, University of California at Davis, Davis, California 95616, USA*

Susan K. Jacobson *Department of Wildlife Ecology and Conservation, University of Florida, Gainsville, Florida 32611, USA*

Mallory D. McDuff *Department of Wildlife Ecology and Conservation, University of Florida, Gainsville, Florida 32611, USA*

Ian Newton *Institute of Terrestrial Ecology, Monks Wood, Huntingdon PE17 2LS, UK*

Stuart L. Pimm *Ecology and Evolutionary Biology, University of Tennessee, Knoxville, Tennessee 37996, USA*

Oliver Rackham *Corpus Christi College, Cambridge CB2 1RH, UK*

John D. Reynolds *School of Biological Sciences, University of East Anglia, Norwich NR4 7TJ, UK*

Daniel Simberloff *Ecology and Evolutionary Biology, University of Tennessee, Knoxville, Tennessee 37996, USA*

William J. Sutherland *School of Biological Sciences, University of East Anglia, Norwich NR4 7TJ, UK*

Graham Wynne *Royal Society for the Protection of Birds, The Lodge, Sandy SG19 2DL, UK*

Preface

Conservation science and its application are burgeoning subjects. Each year there are more conservation problems, more solutions and an ever increasing number of students wishing to study and do something useful with their newly acquired skills. This book is aimed at those who already have some knowledge of the basic concepts in conservation.

All the chapters in this volume have an applied component but I have arranged them so that those near the beginning tend to emphasize the science while those at the end concentrate more on practical application. The book leads with a discussion of biodiversity and the range of different meanings this term has acquired. Subsequent chapters describe the rate of extinctions and some of the main issues that bring these about—introduction of alien species, pollution and unsustainable exploitation. This is followed by a consideration of the important issues fundamental to current conservation practice. There has been a shift amongst conservationists from concentrating on the problems of small populations to considering more why populations have declined; the traditional emphasis of island biogeography in conservation biology is largely being replaced by a consideration of spatial processes; and many conservationists have yet to appreciate fully that an understanding of the history of an area or habitat is essential for correct management. The next two chapters discuss the very applied issues of how to select areas for conservation and the principles behind managing habitats and species. The book ends by tackling four subjects that are essential to long-term conservation: economics, education, politics and development, in the belief that these topics are too often omitted or underemphasized in conservation courses.

John Krebs and Nick Davies have produced a new edition of *Behavioural Ecology* every 6–7 years and they suggest that the best measure of the subject's continuing success will be the need for a fifth edition in another 7 years. By contrast, the success of conservation biology as a subject will be a reduced need for new students, research or textbooks.

Conservation is an applied subject. As well as reading textbooks and journals I hope that any student will also attempt to become involved in practical conservation either by starting projects or by working or volunteering for others.

I thank Ian Sherman at Blackwell Science for his enthusiastic and efficient support of this project.

William J. Sutherland
University of East Anglia

CHAPTER 1
Biodiversity

Kevin J. Gaston

1.1 What is biodiversity?

There is a touch of irony in the multiplicity and range of meanings which have come to be associated with the term 'biological diversity' and its contraction 'biodiversity' (hereafter used throughout). Upwards of a dozen different definitions have now been published which, although virtually all variants on a basic theme, differ in their detailed content (McAllister 1991; Gaston 1994b, 1996a) (Table 1.1). The range of ways in which the term 'biodiversity' is actually used is yet wider, perhaps as an inevitable combined consequence both of its frequency of use (the rate of increase in the number of science publications applying the term approaches an exponential function) (see Haila & Kouki 1994; Harper & Hawksworth 1994) and of the changing nuances it has acquired in its application in a growing number of spheres (e.g. academia, conservation, law, media, politics, science administration).

This lack of consistency has in some quarters been derided. It has led to the application of the term 'biodiversity' being widely regarded as something of a bandwagon, of more value in making the work of those who use it appear trendy and relevant than in identifying anything distinctive. The situation has not been helped because biodiversity has also become pseudo-cognate, in that many users tend to assume that everyone shares the same intuitive definition (Williams 1993). Unfortunately, they plainly do not.

In attempting to disentangle this arguably confused state of affairs, in order that it is understood if not resolved, it becomes apparent that much of the problem arises from some fundamental differences in the viewpoints about biodiversity which have developed (largely in different contexts). Or at least it arises from a failure to acknowledge the existence of these differences and to perceive their likely consequences. Previously I have distinguished three such viewpoints (Gaston 1996a). Here I revisit these, and add a fourth.

1.1.1 Biodiversity as a concept

In the main, explicit formal definitions of biodiversity outline a broad,

1

Table 1.1 Definitions of 'biological diversity' and 'biodiversity'.

'Biological diversity refers to the variety and variability among living organisms and the ecological complexes in which they occur. Diversity can be defined as the number of different items and their relative frequency. For biological diversity, these items are organized at many levels, ranging from complete ecosystems to the chemical structures that are the molecular basis of heredity. Thus, the term encompasses different ecosystems, species, genes and their relative abundance.'   (OTA 1987)

'Biodiversity is the variety of the world's organisms, including their genetic diversity and the assemblages they form. It is the blanket term for the natural biological wealth that undergirds human life and well-being. The breadth of the concept reflects the interrelatedness of genes, species and ecosystems.'   (Reid & Miller 1989)

'"Biological diversity" encompasses all species of plants, animals and microorganisms and the ecosystems and ecological processes of which they are parts. It is an umbrella term for the degree of nature's variety, including both the number and frequency of ecosystems, species or genes in a given assemblage.'   (McNeely *et al.* 1990)

'Biodiversity is the genetic, taxonomic and ecosystem variety in living organisms of a given area, environment, ecosystem or the whole planet.'   (McAllister 1991)

'Biodiversity is the total variety of life on earth. It includes all genes, species and ecosystems and the ecological processes of which they are part.'   (ICBP 1992)

'Biological diversity (= biodiversity). Full range of variety and variability within and among living organisms, their associations, and habitat-oriented ecological complexes. Term encompasses ecosystem, species, and landscape as well as intraspecific (genetic) levels of diversity.'   (Fielder & Jain 1992)

'Biodiversity. The variety of organisms considered at all levels, from genetic variants belonging to the same species through arrays of species to arrays of genera, families, and still higher taxonomic levels; includes the variety of ecosystems, which comprise both the communities of organisms within particular habitats and the physical conditions under which they live.'   (Wilson 1992)

'"Biological diversity" means the variability among living organisms from all sources including, *inter alia*, terrestrial, marine and other aquatic ecosystems and the ecological complexes of which they are part; this includes diversity within species, between species and of ecosystems.'   The Convention on Biological Diversity (Johnson 1993)

'... biodiversity—the structural and functional variety of life forms at genetic, population, species, community and ecosystem levels ...'   (Sandlund *et al.* 1992)

'For the purposes of the Global Biodiversity Assessment, biodiversity is defined as the total diversity and variability of living things and of the systems of which they are a part. This covers the total range of variation in and variability among systems and organisms, at the bioregional, landscape, ecosystem and habitat levels, at the various organismal levels down to species, populations and individuals, and at the level of the population and genes. ... It also covers the complex sets of structural and functional relationships within and between these different levels of organization, including human action, and their origins and evolution in space and time.'   (Heywood 1995, p. 9)

---

sometimes rather vague, and essentially abstract, concept. They are embellishments or expansions to the idea that biodiversity is an expression of the 'variety (or variability) of life' (the distinction between variety and variability in this context is not readily apparent and I will treat them as synonymous) (Table 1.1). Indeed, this idea could perhaps be regarded as the sole common denominator of all usage of the term. It would also seem to accord with much early usage (for a discussion of the origins and history of the terms 'biological diversity' and 'biodiversity' see Wilson & Peter 1988; Shetler 1991; Harper & Hawksworth 1994; Norse 1994).

Major differences between definitions of the concept of biodiversity lie not in their core, but rather in their extent; what does and does not fall within the bounds of the concept? Particularly important distinctions include whether or not biodiversity is explicitly centred on human interests (e.g. the future of humankind), and embraces the abiotic components of the environments in which organisms live, or embraces process (e.g. disturbance, energy flow, nutrient cycling, succession). To the author's mind, all of these should lie firmly outside the bounds of the concept. The inclusion or exclusion of process is, however, especially contentious. Various authors have argued categorically for (e.g. Noss 1990; Western 1992; Mosquin 1996) or against (e.g. Angermeier & Karr 1994) its importance. Much of the debate has been associated with differing opinions on how close is the association between the concept of biodiversity and the goals of conservation, a point which will be returned to later. However, I concur with Angermeier & Karr (1994) in advocating differentiation between biodiversity, which concerns biotic elements, and biological integrity, which concerns both biotic elements and those processes which generate and maintain them. This concords with the implicit distinction between biodiversity and processes which many ecologists have drawn in exploring relationships between the two (e.g. Naeem *et al.* 1994; Tilman & Downing 1994; Risser 1995).

The debate over the relevance of process constitutes part of a disturbing trend in definitions of biodiversity towards greater and greater inclusiveness. Increasingly, definitions of the concept of biodiversity are becoming so broad as to effectively equate it with the whole of biology (Gaston 1996a). This undermines any claims that the concept may have to being at all useful, and should be strongly resisted. Nonetheless, one must recognize that the preferred limits to the concept will, perhaps inevitably, reflect the context in which biodiversity is being explored.

1.1.2 Biodiversity as a measurable entity or entities

In both pure and applied science, biodiversity is widely construed as measurable. That is, rather than simply remaining a broad concept, it can be made operational and reduced to measurable quantities. Such a viewpoint forms the basis for a variety of research questions which are currently in vogue. These range from the reasonably focused (such as the relationship between biodiversity and stability, and the extent of geographical congruence of hot spots of biodiversity across different taxonomic groups) to the more general (such as how much biodiversity there is and where it occurs).

Given the breadth of virtually any definition, it is inevitable that no single all-embracing measure of biodiversity can be derived. Instead, only measures of certain components, and appropriate for restricted purposes, can be obtained (Norton 1994) (Section 1.3). Whilst one may feel uncomfortable with this notion, it applies, though perhaps not so obviously, in making many other concepts operational. For example, the concept of body size is widely utilized in biology, and yet there is no such thing as *the* body size of an organism. Rather, size can be expressed in a variety of ways, none of which has any obvious logical precedence. Consider, for example, two individuals similar in mass, but differing in linear dimensions (McKinney 1990). Which is larger?

Closely associated with the idea that the concept of biodiversity can be made operational is that of surrogacy. A host of theoretical measures of biodiversity can undoubtedly be conceived. It has proven possible to apply some of these in certain circumstances, and it may become possible to apply others. However, numerous dimensions of biodiversity will be difficult to quantify in many situations (for reasons of practicality and cost), and it will be necessary to identify surrogates for them. For example, it is, and will probably remain, impossible to determine the diversity of characters (be they genetic, morphological or behavioural) represented in a large assemblage of organisms, but numbers of species and numbers of higher taxa may under some circumstances serve as suitable surrogate measures of these quantities (Williams & Humphries 1996).

1.1.3 Biodiversity as a field of study

The existence of professorial chairs for the study of, laboratories for the investigation of, courses for the teaching of, and conferences for the discussion of biodiversity bears witness to the fact that it is viewed by many as a field of study, on a par with botany, ecology and genetics. Indeed, some definitions of biodiversity, particularly those that seek to be very wide-ranging, seem to rest more easily as descriptors of the limits to a field of study than to a concept. That is, they concern the boundaries to an area of interest. Just as ecology has been defined as 'the scientific study of the interactions that determine the distribution and abundance of organisms' (Krebs 1972), biodiversity might perhaps succinctly be defined as the scientific study of the patterns in, and the determinants and consequences of, the variety of life.

Although biodiversity may be discriminated as a field of study, the questions which tend to be recognized as falling within its sphere (e.g. how many species are there?; what is the wider functional role of individual taxa or groups of taxa?) might equally fall within those of more traditional fields (e.g. ecology, genetics, systematics). The importance of biodiversity has been to give these questions a more central significance (where previously they have been of peripheral or marginal interest) and to recognize their fundamental nature and ramifications. As such, the field is very much at the 'what?' stage of its development, with much discussion of issues of quantification and pattern, and has yet to pass to the 'how?' and 'why?' stages and a wider discussion of mechanisms (Wiegert 1988). Furthermore, the study of biodiversity has served to combine elements of pure research in evolutionary biology and ecology with the practical applications of such research for applied biology and public policy, especially in the spheres of conservation, medicine, forestry and agriculture (Ehrlich & Wilson 1991).

1.1.4 Biodiversity as a social or political construct

For many, the term 'biodiversity' is value laden. It carries with it connotations that biodiversity is *per se* a good thing, that its loss is bad, and that something should be done to maintain it. This has reached the point where the discussion of biodiversity may often be taken as implying acceptance of these premises, and they have effectively been incorporated into the working definitions of biodiversity which many people use. This position has been associated with the development, as the touchstone of much of present thinking in conservation biology, of the idea that it is desirable to maintain the variety of life (e.g. ICBP 1992; Forey *et al.* 1994; Frankel *et al.* 1995; Hunter 1996; and virtually any recent issue of the journal *Conservation Biology*). Implicit are notions that this end is not appropriately achieved through preserving this variety *ex situ* in zoo- logical and botanical gardens (which would anyway prove prohibitively expensive), nor through introductions of non-native species to areas (Angermeier 1994). This idea has stimulated an important shift away from an obsession with conserving a few, usually large-bodied, endangered species. As yet, this shift is perhaps more noticeable in policy than in management practice, possibly reflecting the greater difficulties in identifying management approaches directed at a more diffuse target than particular species.

A consequence of this conservation value-laden view of biodiversity has been, as previously noted, a move to broaden the definitions of biodiversity to embrace everything that conservationists believe it is important to conserve. Thus, for example, one repeatedly finds statements to the effect that biodiversity has to embrace processes because their conservation is crucial to maintaining genes, species and ecosystems (e.g. Noss 1990; Western 1992). The definition of biodiversity has in this way been used to identify conservation targets. This is unhelpful. Though some might deny it, the variety of life poses many interesting questions, whether its existence is threatened or not. As a concept, biodiversity has far broader implications than those which pertain to conservation alone, and such constraints serve to deny this.

1.1.5 Where next?

Recognizing that biodiversity means different things to different people does not constitute a resolution of this situation. Indeed, it is highly doubtful whether such a resolution is achievable, and arguable whether it is entirely desirable. Attempts prescriptively to standardize the usage of terminology are seldom effective, and the variety of viewpoints on what biodiversity actually is may each have significant roles to play. What is perhaps more important is a general recognition that your 'biodiversity' may not be the same as mine, and that we should clarify the ways in which we are using the term if we are to pre-empt confusion and misunderstanding. Such clarification may facilitate improved dialogue between those with differing viewpoints, potentially to the benefit of all concerned.

1.2 A pragmatic approach

Having recognized the complexity of the patterns of usage of the term 'biodiversity', I want for present purposes to put this to one side. Whatever the sources and possible solutions to the confusion, the fact remains that the natural world is infinitely diverse (or as good as). To borrow from the closing paragraph of *Origin of Species* (Darwin 1888):

> 'It is interesting to contemplate a tangled bank, clothed with many plants of many kinds, with birds singing on the bushes, with various insects flitting about, and with worms crawling through the damp earth, and to reflect that these elaborately constructed forms, so different from each other, and dependent upon each other in so complex a manner, have all been produced by laws acting around us. … There is grandeur in this view of life, with its several powers, having been originally breathed by the Creator into a few forms or into one; and that, whilst this planet has gone cycling on according to the fixed law of gravity, from so simple a beginning endless forms most beautiful and most wonderful have been, and are being, evolved.'

Defined in the most straightforward way as 'the variety (or variability) of life' (in all its manifestations), biodiversity captures something that is tangible and important. That is, notwithstanding characterizations of such a definition as popular (as opposed to scientific) and simplistic. The fact that the term also carries connotations of struggles to characterize, measure and conserve this variety is perhaps best regarded as a reflection of some harsh realities. As such, though it may often prove frustrating, this may be no bad thing.

The rest of this chapter, taking as a starting point a definition of biodiversity as 'the variety of life', addresses a number of very basic questions: how do we quantify this variety? (Section 1.3), how much is there? (Section 1.4), where is it? (Section 1.5) and why does it matter?

(Section 1.6). Entire books have been written on one or more of these topics (e.g. Norton 1987; Wilson & Peter 1988; Groombridge 1992; Wilson 1992; Hawksworth 1995; Heywood 1995; Perrings *et al.* 1995a; Gaston 1996b). The aim here is not to review comprehensively the available information which impinges on the answer to each question, but rather to identify some of the primary issues, themes and directions of research associated with them.

1.3 How is biodiversity quantified?

1.3.1 Elements and hierarchies

The variety of life is expressed in a multiplicity of ways. To capture this variety thus necessitates a multiplicity of measures. Phrasing it in another way, Haila & Kouki (1994) characterize the concept of biodiversity as descriptively complex, in that it can be divided into parts using alternative criteria and the result looks different depending on which is adopted. We can recognize some general ways of creating these divisions.

First, we can distinguish a number of elements of biodiversity (elsewhere these have variously been termed units, components and levels). These include genes, populations, species, genera, tribes, families, phyla, kingdoms, habitats, ecosystems, biomes and so forth. Some maintain that one particular element is fundamental, others that this is not so (e.g. Noss 1990; Solbrig 1991; Wilson 1992). In practice, species are treated in many quarters as the fundamental elements of biodiversity, species richness as the fundamental expression of biodiversity, and the high level of species extinction as the primary manifestation of the biodiversity crisis. Arguably, this is more a reflection of practical considerations than theoretical niceties. Whatever one's view, to focus quantification on a single element is arbitrarily to ignore most of biodiversity (Angermeier & Karr 1994). Moreover, in the context of conservation, it is important to remember that concentration on a particular element of biodiversity essentially places differential value on that facet of the variety of life (Williams *et al.* 1994a).

The elements of biodiversity can usefully be viewed as nested components of one or more organizational hierarchies. The range of different sets of such hierarchies that have been proposed is broad, though there are substantial overlaps between most schemes. The most widely accepted distinguishes between hierarchies of genetic diversity, organismal diversity and ecological diversity (the names are not always consistent) (Angermeier & Karr 1994; Harper & Hawksworth 1994; Heywood 1995). Under one interpretation, genetic diversity comprises nucleotides, genes, chromosomes, individuals and populations, organismal diversity comprises individuals, populations, subspecies, species, genera, families, phyla and kingdoms, and ecological diversity comprises populations, niches, habitats, ecosystems, landscapes, bioregions and biomes (Heywood 1995).

Biodiversity is typically quantified in terms of an element within a hierarchy. For elements such as genes, populations, species and families, it is comparatively easy to see how this might be done. For other elements, such as ecosystems and landscapes, it is more obscure how this might usefully be achieved, at least at scales less than the entire globe. In practice, for these elements, biodiversity is seldom expressed in terms of their diversity, but rather in terms of the diversity of some other element (e.g. species), physically nested within them. Species diversity and ecosystem diversity, for example, may thus in some instances confusingly equate to the same thing.

1.3.2 Number and difference

Elsewhere, I have described biodiversity as a biology of numbers and difference (Gaston 1996b). This distinguishes the two major components of any measure of diversity, the number of entities (in this case, the number of elements) and the difference between these entities. Indeed, different measures of diversity can, in principle, be placed on a scale reflecting the relative emphasis which they give to these components—from those such as species richness, which essentially ignore any difference component (e.g. weighting all species equally), to those such as many measures of ecological (or species) diversity, which combine richness and the evenness of the relative abundances of species (relative abundance being used as a metric of difference).

In biology, the theory of measures of diversity is probably best developed in the context of ecological diversity (Peet 1974; Magurran 1988; Kvalseth 1991; Balle 1994; Brewer & Williamson 1994), although its roots lie in information theory. However, application of these measures has widely fallen into disrepute, because they may often fail to distinguish between different forms of variation; the same diversity value may be obtained for assemblages of quite different structure (e.g. numbers of species and distributions of abundances amongst them). It remains an area of contention how serious a problem this really is; it certainly generalizes far more widely than simply measures of ecological diversity. Its significance will depend critically upon the question being addressed.

In measuring biodiversity, almost by definition the breadth of ways in which differences can be expressed is potentially infinite. Think, for example, of the ways in which one could discriminate between two species. These might include facets of their biochemistry, biogeography, ecology, genetics, morphology and physiology.

Weitzman (1995) suggests two general criteria which distinguish a good diversity function. First, it should a priori be sensible, in that it embodies an intuitively plausible formulation that does not admit of seriously damaging counter examples. Second, there should be some special case, forming the central paradigm, for which the particular formulation is exactly the right answer to a rigorously well-posed problem. The majority of recent research into diversity functions has focused on a suite of possible problems characterized, in an oft-repeated scenario, as how best to capture the diversity differences between assemblages of identical species richness but comprising species of differing degrees of relatedness. For example, one might envisage three assemblages comprising, respectively, say, three species of mouse, two species of mouse and a mustelid, and one species each of a mouse, a mustelid and an ungulate. One rationale for focusing on this question is that it offers a potential method of addressing the great variety of ways in which species might differ. A large proportion of this variance might, however, reasonably be expected to be captured by measuring phylogenetic diversity (or disparity), reflecting the fact that, on average, the more distantly related two species are, the more different their biologies are likely to be (this is not to say that aspects of the biologies of some very distantly related species may not be very similar, marsupial and eutherian mice for example).

I do not wish to dwell on the various solutions which have been offered for the measurement of phylogenetic diversity (for a discussion see Vane-Wright *et al.* 1991; Humphries *et al.* 1995; Williams & Humphries 1996; and references therein). However, it has served to highlight considerations of more general import. First, it is helpful to distinguish between those measures of biodiversity which are theoretically ideal and those which can in practice be applied to data which are available or can be sufficiently readily obtained. Much of the discussion which has surrounded some biodiversity measures (e.g. those based directly on genetic distances between species) is, in an applied sense, largely irrelevant, because the data demands such measures impose are too high, particularly in the face of requirements for rapid solutions to pressing environmental problems. Pragmatism is necessary, returning us to earlier comments on the importance of surrogacy and the indirect measurement of particular dimensions of biodiversity (Section 1.1.2). Second, one can distinguish between measures which express the biodiversity of an assemblage in terms of the richness of the measure of difference between elements (e.g. the richness of character differences between species), or in terms of the richness of different combinations of the measure of difference between elements (e.g. the number of combinations of characters) (Williams 1993; Williams *et al.* 1994a). Third, levels of biodiversity as measured using many indices do not always closely reflect patterns of diversity to which people respond first and foremost and which are most directly amenable to the senses. If conservation priorities were to be assessed and implemented using such indices, a general understanding of this mismatch would seem desirable.

1.3.3 Integrating across levels

Rather than expressing biodiversity primarily in terms of one element, in principle it could be expressed in terms of some integration across elements at different levels within a hierarchy (Leemans 1996). For example, one might imagine integrating under a curve linking numbers of individuals, populations, species, families, orders, phyla and kingdoms. No accepted methodology exists to do this, and the extent to which it would fulfil any useful purpose remains questionable (Angermeier & Karr 1994).

1.4 How much biodiversity is there?

1.4.1 Species richness

Given the multiplicity of ways in which the variety of life is expressed and hence the multiplicity of ways in which it can be quantified, there can of course be no over-arching answer to the question of how much biodiversity there is. There can, however, be answers framed in terms of some of its elements.

At the global scale, most emphasis on estimating the magnitude of biodiversity has been placed on species richness (biodiversity and species richness are widely, if incorrectly, treated as synonymous). A working estimate of extant species numbers has been suggested at around 13.5 million, with lower and upper estimated numbers of about 3.5 and 111.5 million species, respectively (Heywood 1995; see also Hammond 1992, 1995) (Fig. 1.1). The upper boundary appears wildly improbable (if for no other reason than that it is not obvious where all these species are to be found), and evidence in support of the working estimate is becoming increasingly convincing (albeit, categorical demonstrations of its validity do not exist).

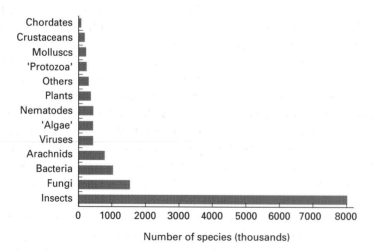

Fig. 1.1 Conservative 'working estimates' of the possible overall present species richness of different groups of organisms. The reliability of these estimates is likely to vary widely. (From data in Heywood 1995.)

The more rigorous estimates of global species numbers largely are arrived at, implicitly or explicitly, by summation of estimates from a variety of sources. Major uncertainties remain in figures for particular taxonomic groups (e.g. viruses, bacteria, nematodes, mites), functional groups (e.g. parasites), and habitats or biomes (e.g. soils, deep-ocean benthos). Indeed, the relative contributions of some groups and areas compared with others continue to be, sometimes vigorously, debated. For example, views on the relative magnitudes of marine and terrestrial richness are highly divergent, with claims that terrestrial systems are most diverse, that marine systems are most diverse, and that the diversities of terrestrial and marine systems may be broadly similar (Grassle & Maciolek 1992; May 1992, 1994a; Lambshead 1993; Angel 1994; Briggs 1994; Hammond 1995). It seems very probable that, at the species level, terrestrial systems are substantially more diverse than are marine systems, although the pattern is undoubtedly reversed at high taxonomic levels (e.g. many more animal phyla are known from marine systems than from terrestrial).

Encouragingly, these problem areas have narrowed somewhat from those one might have identified just a few years ago. Confidence in estimates of the probable numbers of extant species of insect, for example, has improved greatly (e.g. Gaston 1991a, 1996c; Hammond 1992; Gaston & Hudson 1994; Heywood 1995; Gaston et al. 1996). In major part this is because suggestions of astronomical numbers of insect species in moist tropical forest canopies have on closer examination found little firm foundation. Early impressions were strongly influenced by inadequate levels of sampling, which overestimated levels of spatial turnover in species identities, and assumptions that species are more host-specific and that canopy faunas are more discrete than is actually the case.

Periodically, when first subjected to serious scrutiny, various habitat types have been described as representing fresh frontiers of species richness, liable to lead to dramatic upward revisions in global species numbers (e.g. tropical forest canopies, soils, marine benthos). In practice, and particularly following more extensive and intensive sampling, these dramatic revisions are seldom sustained. It seems unlikely that any further entirely unexpected sources of substantial additional richness remain to be discovered. None-theless, some habitats may prove more diverse than suspected on superficial inspection. Usher (1994) provides three examples from the British Isles: acidic soils, for which the taxonomy of many occupants remains unresolved; upland heaths, which are misleadingly perceived as uniform stands of vegetation; and farm woodlands, which may be overlooked because frequently they are planted and not especially old.

Alongside questions of the numbers of species in particular taxonomic or functional groups, habitats and biomes, some more general issues in estimating contemporary global species numbers remain outstanding (e.g. Gaston 1991a, 1991b; Hammond 1994; May 1994a, 1994b). Together with some accompanying remarks, these include:

1 Can we reasonably talk of global numbers of species, when that total must include groups of organisms for which those species concepts which are employed (rather than simply espoused) differ widely (May 1990; Gaston 1991a, 1996c; Groombridge 1992)? This problem has repeatedly been reiterated, though its ultimate severity remains poorly understood. Perhaps the greatest divide lies between those groups (predominantly vertebrate, or of economic or medical importance) which have been the subject of intensive investigations of the cytological, genetic, behavioural and other differences between putative taxa, and those groups which have attracted little attention and where taxa have been discriminated exclusively on morphological traits. Should we shy away from adding apples and oranges, or simply accept that they are different forms of fruit and can usefully be summed?

2 How do we, with confidence, scale up known species numbers at small spatial scales (which we can realistically survey) to predict numbers at very large ones (which we cannot survey)? This is more than an issue simply of species–area relationships, because such divergent-sized areas embrace several different forms of the species–area relationship, and for any overall pattern the variance in species richness explained by area can become quite small (Rosenzweig 1995).

3 In rigorous taxonomic revisions (involving, where necessary, fresh collecting efforts), or in samples from previously little explored areas, what proportions of species are previously taxonomically undescribed? Explorations to date suggest that these proportions are inadequate to support very high estimates of global species richness (May 1994a, 1994b; Gaston *et al.* 1996). For example, examination of recent taxonomic revisions of hymenopteran taxa (a notoriously poorly known insect order, but constituting a sizeable proportion of all insect species) in South America reveals that, in all cases, of the total number of species recognized substantially more than 10% had previously been described (Gaston *et al.* 1996). Accepting that these revisions predominantly concern groups with larger-bodied species and may therefore possibly present a biased picture, such figures support suggestions that the Hymenoptera as a whole may number perhaps only a million species.

The gulf between numbers of described species and the total number of extant species is predominantly determined by small-bodied invertebrate taxa. However, it should not be forgotten that, in some areas, some vertebrate taxa are still comparatively poorly known (e.g. < 50% of African freshwater fishes are estimated to have been described; Ribbink 1994), and that new species in many vertebrate taxa continue to be discovered (e.g. Vuilleumier *et al.* 1992; Pine 1994). Sixteen new living species of large mammals alone have been discovered during the period 1937 to the present, about three per decade (Pine 1994); these are two porpoises (*Lagenodelphis hosei*, *Phocoena sinus*), four beaked whales (*Tasmacetus shepherdi, Mesoplodon ginkgodens, M. carlhubbsi, M. peruvianus*), a wild pig (*Sus heureni*), a peccary

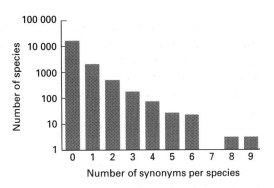

Fig. 1.2 The distribution of the number of synonyms per valid species name for described geometrid moths. Most valid names have no or few synonyms, whilst a few valid names have a large number of synonyms. (Redrawn from Gaston *et al.* 1995b.)

(*Catagonus wagneri*), four deer (*Mazama chunyi, Moschus fuscus, Muntiacus atherodes, M. gongshanensis*), the kouprey (*Bos sauveli*), a gazelle (*Gazella bilkis*), a wild sheep (*Pseudois schaeferi*) and a 'bovid' (*Pseudoryx nghetinhensis*).

4 What proportion of formal species names are synonyms (additional names given to species previously described under other names)? Evidence for insect taxa suggests that typically 20% of species group names are known to be synonyms and a substantially higher proportion may actually be so (because many remain unrecognized) (Gaston 1991a; Gaston & Mound 1993; Solow *et al.* 1995) (Fig. 1.2). The consequences of such levels for our view of described species richness and its relation to real richness may be profound. For example, the existence of so many synonyms may markedly reduce estimates of the proportions of all species which have actually been described (because some described species have been at least double or triple counted).

5 What is the relationship between the set of species which has been described thus far and the global set of extant species? On the basis of analyses of only a few taxa, it has become apparent that species which are described earlier on average are larger-bodied, more abundant (locally or regionally), more widely distributed, occupy a larger number of habitats or life zones, and derive disproportionately from temperate zones (Gaston 1991c, 1993, 1994b; André *et al.* 1994; Gaston & Blackburn 1994; Patterson 1994; Blackburn & Gaston 1995; Gaston *et al.* 1995a). Such biases narrowly prescribe our knowledge of the natural world and our ability accurately to generalize about it.

1.4.2 Other measures

Beyond species numbers, at the global scale, we might imagine expressing the overall magnitude of biodiversity in a variety of other ways. Remarkably, most have at best received no more than passing consideration. The following are some of the simpler to consider.

Number of genes. Most discussion of the genetic dimension to biodiversity is concerned with the levels of genetic diversity within individual species, and differences in these levels between species with differing taxonomic affinities and exhibiting differing traits (e.g. Solbrig 1991; Heywood 1995; Baur & Schmidt 1996; Mallet 1996). In principle, more broad-ranging issues could be addressed, such as the number of different genes summed across all species.

Number of individual organisms. Although probably a more tractable issue than the number of species, the global number of individual organisms has received remarkably little attention. A few disparate estimates of components of this total suggest that it must be enormous. Thus, amongst soil fauna, Wallwork (1976) collates various estimates of numbers per square metre of 16 million testate Protozoa, 4–20 million Nematoda, 35 thousand Collembola and 176–410 thousand Cryptostigmata (Acari), whilst Boucher & Lambshead (1995) estimate numbers of marine nematodes, in some instances of more than $5000/10 \, cm^2$. At a global scale it has been estimated that there are 200–400 billion birds (Gaston & Blackburn 1997) and 10^{18} insects (Williams 1960).

Number of functional species. If some species fulfil similar ecological roles or functions, then the number of different 'functional' species may be considerably smaller than the number of species. That is, there would be functional redundancy (Walker 1992; Lawton & Brown 1993). Determining its level (i.e. the difference between the numbers of species and functional species) is complicated by the difficulty of objectively defining function, and of discriminating whether functional equivalence under one set of circumstances (environmental conditions, abundance levels, etc.) would be maintained under others (Martinez 1996). Whilst on the one hand it can be argued that every species is, by definition, different and hence fulfils a different function, on the other hand it can be argued that the differences are often so minor that, to all intents and purposes, many species are functional equivalents. Just as Lawton (1992) has argued that there are not 10 million kinds of population dynamics, there are not 10 million kinds of function. One can recognize guilds, functional types and other groupings of species (e.g. Hawkins & MacMahon 1989; Jaksic & Medel 1990; Simberloff & Dayan 1991). Unfortunately, the conflation of the concept of biodiversity with conservation priorities (Section 1.1) has served mistakenly to confuse objective debate of this issue with concerns that ideas of functional redundancy constitute suggestions that it is not necessary, useful or desirable to conserve some or all species (Walker 1992, 1995; Martinez 1996).

Number of species combinations. Across areas of a particular size, what is the total number of different combinations of species which occur? Such questions have thus far only explicitly been addressed in terms of small numbers of very closely related species (e.g. Brown & Kurzius 1987), although the importance of conserving the variance in species combinations (i.e. the diversity of species–species interactions) has been highlighted (e.g. Janzen 1986). Depending on the extent to which one's view of communities departs from an individualistic concept (in which there is continuous variation in community composition, and no degree of integration) and converges on the view that communities exist as repeated units with a considerable degree of integration, the more one's view of the likely number of combinations will depart from the number of areas. Although categorical general statements are always difficult when dealing with continua, communities are probably best characterized as individualistic (Clements 1916; Gleason 1926; Huntley 1991; Buzas & Culver 1994; McIntosh 1995; McKinney *et al.* 1996).

1.4.3 Extinctions

Levels of biodiversity, almost however quantified, exhibit temporal variation. The present trajectory for global biodiversity is predominantly downward (though this is superimposed on a general pattern of long-term increase through geological time). This is perhaps most starkly illustrated by rates of species extinction, with estimates of impending rates orders of magnitude faster than the background rates seen in the fossil record (May *et al.* 1995; Pimm *et al.* 1995). Loss of other dimensions of biodiversity (such as population losses) may, however, be greater still, as well as more insidious (e.g. Ehrlich & Daily 1993; Ehrlich 1994). Such observations raise the question of whether, in attempting to determine how much biodiversity there is, we are chasing a rapidly moving target. This seems particularly likely for fine-grained measures of biodiversity (e.g. numbers of species, numbers of populations), if not for coarse-grained ones (e.g. numbers of phyla, numbers of biomes).

1.5 Where is biodiversity?

In few, if any, senses can the biodiversity of the Earth be seen as evenly distributed across its surface, or through the media (e.g. air, water) which blanket it. We remain far away, however, from being able to produce a set of maps, of even moderately fine spatial resolution, which illustrate the highs and the lows of biodiversity across the planet. That is, we cannot provide an atlas of biodiversity. Nonetheless, it is a high priority.

Without wishing to misrepresent the very real obstacles which exist, this pessimism should, however, be tempered. We do have many of the pieces of information which we would require to generate such maps. Indeed, we know more than we are frequently prepared to admit (Rosenzweig 1995). For example, many basic patterns in the broad-scale spatial distribution of diversity have been documented (albeit with some, often important, exceptions), and we have a general grasp of the mechanisms underlying some of them (for general reviews see Myers & Giller 1988; Ricklefs & Schluter 1993; Huston 1994; Heywood 1995; Rosenzweig 1995; Gaston & Williams 1996). These patterns include: increase in diversity with increasing area (Connor & McCoy 1979; Anderson & Marcus 1993; Palmer & White 1994), increase in endemism with increasing area (Diamond 1984; Major 1988; Anderson 1994), increase in diversity towards lower latitudes (Pianka 1966; Stevens 1989; Rohde 1992; Rosenzweig 1992), increase in endemism towards lower latitudes (Rapoport 1982; Stevens 1989), peaks of diversity at mid-elevations and intermediate depths (Rex 1981; Rahbek 1995), and relationships between diversity and various environmental parameters (e.g. species richness increases with primary productivity at large spatial scales and shows a hump-shaped relationship with productivity at small to moderate scales) (Adams & Woodward 1989; Currie 1991; Rosenzweig & Abramsky 1993; Wright *et al.* 1993). Indeed, studies of many of these patterns continue to accrue at a respectable rate (e.g. Brener & Ruggiero 1994; Eggleton 1994; Kouki *et al.* 1994; McAllister *et al.* 1994; Scheiner & Rey-Benayas 1994; Kaufman 1995; Petanidou *et al.* 1995; Poulin 1995; Quicke & Kruft 1995; Blackburn & Gaston 1996). Our understanding of some is founded solely on species richness, but others have been explored in broader terms and similar results seem to pertain. For example, gradients of declining diversity with increasing latitude are observed at the levels of orders, families and genera, as well as species (e.g. Stehli *et al.* 1967; Stehli 1968; Cook 1969; Fleming 1973; Wilson 1974; Rabinovich & Rapoport 1975; Taylor & Taylor 1977; Eggleton *et al.* 1994; Gaston & Blackburn 1995; Gaston *et al.* 1995b; Kaufman 1995).

Over and above these general trends, information is available on many other facets of spatial variation in the biodiversity of particular taxonomic groups. This includes estimates of absolute or relative numbers of species in different biogeographic and geopolitical units (e.g. Groombridge 1992; Davis *et al.* 1994, 1995) and study plots (e.g. Gentry 1990), identification of centres of endemism (e.g. ICBP 1992; Davis *et al.* 1994, 1995; Williams *et al.* 1994b) and, increasingly, broad-scale but low-resolution maps of richness (e.g. Williams 1993; Eggleton *et al.* 1994; McAllister *et al.* 1994; Williams *et al.* 1994b; Gaston *et al.* 1995b). These data are highly non-random with respect to their taxonomic and geographical coverage. Perhaps above all, they serve to illuminate how patterns of biodiversity are strongly contingent upon spatial scale. Peaks and troughs may move in space when biodiversity is calculated at high and low spatial resolutions.

How do we go about putting these different pieces of the jigsaw puzzle together? This brings us to perhaps the most fundamental issue in developing an atlas of biodiversity, and whose resolution is rapidly becoming paramount in biodiversity research. What is the extent of congruence in the patterns of spatial variation in the diversities of different taxonomic groups of organisms? To what extent can we assume that patterns observed for one or more groups and measures also hold for others? The rapidity with which we can generate an atlas will be enhanced by greater congruence. Unfortunately, to date, analyses of patterns of congruence have not been greatly encouraging in this regard (Prendergast *et al.* 1993; Gaston 1996d). Evidence is poor that there is strong coincidence in the geographical hot spots in the biodiversity of different groups, or that there are strong correlations in the biodiversity of different groups across multiple areas. For example, Lombard (1995) finds low levels of coincidence of hot spots of richness, endemism and rarity within taxa, and greatly variable proportional overlaps of species-richness hot spots (0–72%) in comparisons of six vertebrate taxa (freshwater fish, frogs, tortoises and terrapins, snakes, birds and various mammal orders) in South Africa. Although statistically significant relationships have been documented in some studies, these often provide inadequate foundations for extrapolation. Interpretation of the generality of these conclusions continues to be hampered by the small number of studies which have been performed and the heterogeneity of the spatial scales they concern. For example, it is evident that congruence in species-richness hot spots tends to be greater at coarser levels of spatial resolution (Curnutt *et al.* 1994). Nonetheless, it appears increasingly unlikely that spatial patterns in the biodiversity of a few well-known groups can either be extrapolated to other groups or to biodiversity more broadly.

High levels of congruence have been identified as facilitating not only the production of an atlas of biodiversity but also the prioritization of areas for conservation, on the grounds that they would give some confidence that areas protected because they contained high diversities of one taxon would also contain high diversities of others. High diversity does not, however, necessarily equal high value for biodiversity protection. The potential value of an area depends on which taxa it contains, not how many. This means that what is required is more than simply maps of the distributions of the highs and lows of diversity, but also detailed information on the occurrences of individual taxa. This is a yet more daunting proposition than even an atlas, but serves again to sharpen the focus on some fundamental questions for research, such as the forms of spatial variation in beta diversity for different taxa.

1.6 Why does biodiversity matter?

As previously noted, in many circles the term 'biodiversity' carries with it connotations of the importance of the variety of life, of the crisis represented by its loss, and of the need for conservation action. But why does the maintenance of biodiversity matter? This is a topic which has received a great deal of attention, although a wide gulf continues to exist between public understanding of what biodiversity is (let alone why it is important) and deliberations on the importance of biodiversity which have been carried out largely amongst circles of academics. Personal experience suggests that, amongst the general public, otherwise well-informed individuals continue to profess ignorance of what biodiversity is, despite the publicity surrounding the Rio Convention and the existence of well-written popular books on the topic (e.g. Wilson 1992).

Following Kunin & Lawton (1996), three groups of reasons can be identified for conserving biodiversity. First, biodiversity should be conserved because it provides sources of marketable commodities. The scale of exploitation of biodiversity in economic terms remains to be fully evaluated, but it is enormous and varied. It embraces not only the use of components of biodiversity for food (from wild and cultivated sources) and as a source of medicinal compounds, but also for biological control, as sources of (or templates for) an enormous variety of materials of industrial value (not least of which is wood), and as an initial or ongoing basis for recreational harvesting (hunting, fishing) and culturing (gardening).

Second, biodiversity should be conserved because it provides non-market goods and services. These include environmental modulation and ecosystem functions (the interactions of organisms with their, and our, environment), and ecological roles (the interactions of organisms with one another). These are valuable functions which are served by wild organisms in some sense at no cost to humankind, although the price we might pay for their loss could be immense. These goods and services also embrace knowledge (embodied in the biology of organisms), aesthetic values (e.g. the pleasure derived from the experience of wild organisms) and existence values (the values people place on organisms, even though they may never personally encounter them). Conservation for all these reasons is stimulated by a mix of risk aversion (and hence a desire to maintain biodiversity because of its potential significance to our own life support systems) and recognition that the natural world has beauty and enriches our lives.

Third, biodiversity should be conserved because it has intrinsic value, and therefore humankind has moral and ethical responsibilities towards it. This justification is the least seldom enunciated in broad debate over the significance of biodiversity.

The appropriate weighting to give these different reasons is not readily judged, and will inevitably ultimately rest on the value systems employed by the individual or society concerned. The economic benefits of biodiversity have been given particularly strong emphasis by many conservationists, apparently in the belief that to politicians they will prove the most convincing argument for its preservation. Whilst these benefits are undoubtedly very great, it would seem impossible, wholly undesirable and potentially hazardous to reduce everything to economic values. Other imperatives may to some minds appear less readily quantifiable, at least in some common currency, but remain important nonetheless.

Whatever the justification for its conservation, a number of things militate against rapid and widespread acknowledgement of the importance of biodiversity. For example:

1 Much of its loss is presently slow and insidious compared with the

timescales of changes of which people are readily aware (albeit exceedingly rapid in biological terms).

2 Typically, people have poor perceptions of real patterns of risk (Stewart 1990), and the risks associated with the loss of biodiversity are often not readily calculated.

3 A substantial and growing proportion of people are to a high degree disconnected, directly if not indirectly, on a day-to-day basis from the natural world.

4 Most people in western nations, which have the most marked per capita impact on biodiversity, have not encountered pristine habitats (e.g. Gavin & Sherman 1995) and are separated geographically from areas in which biodiversity is presently declining most rapidly.

5 For many individuals, there are no obvious immediate benefits that result if they halt those of their own actions which contribute to the loss of biodiversity.

These and other factors sum to the, perhaps somewhat depressing, observation that most of the loss of biodiversity results from the independent decisions of many millions of individual users of environmental resources, and its underlying cause is to be found in the constraints on, and other determinants of, those decisions (Perrings *et al.* 1995b). Such determinants include the goals in decision making, the preferences that decisions express, the opportunities for decision making (including those associated with cost), and the restrictions placed on the behaviour of individuals by the cultural, religious, institutional and legal frameworks in which they operate.

Consideration of the variety of life has given rise to a number of novel scientific questions, and some bold attempts to resolve them. If that variety is to be preserved we will likewise have to employ approaches that are both novel and bold.

Acknowledgements

K.J.G. is a Royal Society University Research Fellow. I am grateful to Paul Williams for much discussion of these topics over the years. Andrew Balmford, Sian Gaston, Phil Warren and Paul Williams kindly commented on the manuscript.

CHAPTER 2

Extinction

Stuart L. Pimm

2.1 The human future and extinctions: crisis or conspiracy?

A joint statement of the US National Academy of Sciences and Britain's Royal Society in May 1992 concluded: 'If current predictions of population growth prove accurate ... the future of our planet is in the balance' (Press & Atiyah 1992). How could an argument follow such an unequivocal statement from the world's two most influential scientific academies? Yet the debates about the consequences of human population growth are neither new nor resolved. Malthus (1798) was concerned with increasingly rapid population growth and its effects on 'human happiness'. Our numbers have increased dramatically since his time. So, too, has our technology: who would want to live in an age before antibiotics and electricity?

Consider the (live) debate held in New York in October 1992 between the Maryland marketing professor Julian Simon and the Oxford environmentalist Norman Myers, reported in the appropriately titled book *Scarcity or Abundance?* (Myers & Simon 1994). Simon writes, 'We now have in our hands ... the technology to feed, clothe, and supply energy to an ever-growing population for the next 7 billion years'. To which Myers replies '... as many as 200 000 generations to come will be impoverished because of what we are doing [now]'.

For Simon, the ultimate resource is people. The more we have, the better off we shall be, for the more readily will our collective ingenuity solve the problems we create. Ingenuity can substitute an electric light bulb for a whale-oil lamp, but it cannot replace the whales we may hunt to extinction. Extinction is irreversible. Species have economical, cultural and spiritual values (Ehrlich & Ehrlich 1981; NRC 1996). Consequently, the speed at which our increasing population drives a species to extinction is a part of the answer to what our future will hold.

Predictably, Simon finds high estimates of current and future extinction rates to be 'statistical flummery' and contends that it is the 'facts, not the species' that are endangered (Simon & Wildavsky 1984, 1993). He is not alone. Mann & Plummer (1995) consider the estimates to be 'strident, inconsistent, and data-free'. Budiansky (1993) considers them to be 'doomsday myths'. He dismisses one of their proponents, E.O. Wilson, as 'a relentless popularizer' whose conclusions 'are something closer to politics than science' (Budiansky 1995). Easterbrook (1995) presents much the same criticisms.

In this chapter, I cannot delve into why the journalists Mann, Plummer, Budiansky and Easterbrook find fault with scientists' estimates of extinction and other global environmental changes. Nor can I speculate on why political bodies, such as the Senate of the USA, sought the opinions of these journalists in equal measure with those of scientists during the 1995 hearings on pending legislation on endangered species. One distinguishing feature of conservation issues is their ability to generate heated public debate in the face of near scientific consensus. I will, however, explain the science behind estimates of recent and future extinction rates and consider what uncertainties are involved in their calculation. We shall see that Wilson's (1988b, 1989) estimates of future rates are not 'alarmist' (S. Budiansky personal communication, 1995), but remarkably close to the mean of all other studies.

2.2 How many species are there?

Any *absolute* estimate of extinction rate, such as the number of extinctions per year, requires that we know how many species there are. We do not and the problems of estimating their numbers are formidable (May 1990; Pimm *et al.* 1995). Only about 10^6 species are described and fewer than 10^5—terrestrial vertebrates, some flowering plants and invertebrates with pretty shells or wings—are popular enough to be known well. Birds are exceptional in that differences in taxonomic opinion (estimates range between about 8500 and 9500 species) far exceed the annual descriptions of new species (*c.* 1) (Sibley & Monroe 1990, 1993).

In some groups we may have more names than the species they represent (Chapter 1). Those who describe species cannot always be certain that the specimen in hand has not been given a name by someone else in a different country and (sometimes) in a different century (May & Nee 1995). The more serious error is that in all potentially species-rich groups, the estimates of numbers of species far exceed the number of named species. Moreover, some potentially rich communities, such as the deep-sea benthos, have been sparsely sampled (Grassle & Maciolek 1992).

Consider Hawksworth's (1991) estimate of the number of fungi that exist. Britain has six times as many species of fungi as of plants. If this ratio applies worldwide, a world total of *c.* 250 000 plant species would predict *c.* 1.5 million fungi. Only about 70 000 species currently have names. The predicted total would be higher if the warm, wet places of the world (such as tropical rainforests) hold more fungi per plant species than temperate Britain.

For insects, there are $c.\ 10^6$ described species, yet estimates range from 10 to 100 times this number. Erwin (1982), Stork (1988) and Hodkinson & Casson (1990) produce startling estimates from small samples taken from tropical trees. A large sample of canopy-dwelling beetles from one species of tropical tree had 163 species specific to it. There are $c.\ 5 \times 10^4$ tree species, and so we predict $(163)(5 \times 10^4) \approx 8 \times 10^6$ species of canopy beetle. Since 40% of described insects are beetles, the total number of canopy insects could be 20×10^6. Adding half that number for arthropod species on the ground gives a grand total of 30×10^6. Another extrapolation assumes that only 20% of canopy insects are beetles, but that there are at least as many ground as canopy species. The grand total is then 80×10^6. Another calculation in this genre found that, of the 1690 insect species on approximately 500 Indonesian tree species, only 37% were previously recorded. If that proportion were typical, then the total of $c.\ 10^6$ described insect species suggests a total of $(100/37) \times 10^6 = 2.7 \times 10^6$ insects.

Two concerns follow. First, any absolute estimates of extinctions must be extrapolations from the $<10^5$ well-known species to the $c.\ 10^6$ described species, to the grand totals of 10^7–10^8 species. No wonder one reads such a wide range of estimates for the numbers of species lost per year or, dramatically, per day! Absolute estimates can vary a hundredfold because of our uncertainties about the total number of species.

We can derive *relative* estimates: the proportion of well-known species that go extinct in a given interval. Estimating such proportions consumes the rest of this chapter and raises a second concern. Are the proportions typical of the great majority of species for which we lack names? They are likely to be so if extinction rates in widely different species groups and regions are broadly similar.

There is another way in which estimates of extinctions must be made relative. Extinctions have always been a part of Earth's history. Any claims of massive future extinction must, therefore, be scaled relative to what we would expect for the past.

2.3 Background extinction rates

What is the background rate of extinction—that is, how fast did species disappear in the absence of humanity? Studies of marine fossils show that species last from 10^6 to 10^7 years (May *et al.* 1995). Let us suppose that extinctions happen independently and not simultaneously—as they did at the end of the Cretaceous, when the dinosaurs disappeared. Then, with a sample of 10^6 species, we would expect to see one extinction every 1–10 years. With a sample of 10^4 species (roughly the number of modern bird species), we should see an extinction every 100–1000 years, and so on.

Such examples suggest that, for ease of comparison, we can use the number of extinctions (E) per million species years (MSY) or E/MSY. The values of E/MSY that correspond to the fossil species' lifetimes

Table 2.1 Average species life spans of different species groups and the extinctions per million species years (*E*/MSY) assuming extinctions are independent events. (From data in May *et al.* 1995.)

Group	Species' average life span (millions of years)	*E*/MSY
All fossil groups	0.5–5	2.0–0.2
Marine animals	4–5	0.25–0.2
All invertebrates	11	0.9
Marine invertebrates	5–10	0.2–0.1
Diatoms	8	0.12
Dinoflagellates	13	0.08
Planktonic foraminifera	7	0.13
Cenozoic bivalves	10	0.1
Echinoderms	6	0.16
Silurian graptolites	2	0.5
Mammals	1	1.0
Cenozoic mammals	1–2	1.0–0.5

(*c.* 0.1–1 *E*/MSY) are shown in Table 2.1. In this chapter, I emphasize terrestrial vertebrates. There are only two studies of such fossils and these suggest relatively high background rates (*c.* 1 *E*/MSY).

These estimates derive from the abundant and widespread species that dominate the fossil record. The species most prone to current extinction are rare and local, so perhaps these fossil data underestimate extinction rates. Importantly, we can supplement these estimates from our knowledge of the speciation rates of species that often are neither common nor widespread. These rates cannot be much less than the extinction rates or the groups would not be here for us to study.

Molecular phylogenies are now produced rapidly and extensively; they exist for 1700 bird species (Sibley & Ahlquist 1990). Using the relative time axis of molecular distances, we can elucidate the patterns of species formation. Models in which every lineage has the same, constant probability of giving birth to a new lineage (speciation) or going extinct (death) permit an estimation of the rate parameters (Nee *et al.* 1992, 1994). The rich details of this approach offer hope in testing for important factors controlling relative rates of background speciation and extinction. Obviously, absolute rates require accurately dated events, such as the first appearance of a species or genus in the fossil record. There are genetic distance and palaeontological estimates of divergence times for 72 carnivore and 14 primate species or subspecies (Wayne *et al.* 1989, 1991). These broadly support an average origin time for modern species of *c.* 1 million years ago—humans are a case in point—but we need more compilations. If extinction rates match origination rates, we would again estimate extinction rates at 1 *E*/MSY. This value thus provides a rough, but simple, benchmark against which to compare modern extinction rates—which is why I chose it, of course.

2.4 The past as a guide to the future

Have our species increased extinction rates beyond these background rates? I will now review five well-known cases of recent extinctions. From them, I will deduce some general features about recent extinctions that also provide clues to the future.

2.4.1 Five case histories

Pacific island birds

The Pacific islands are the obvious place to start. Polynesians reached these islands—the planet's last habitable areas—within the last 1000–4000 years. Human impact is freshest here and it provides unambiguous evidence of massive extinction (Pimm *et al.* 1994; Steadman 1995). The bones of many bird species are found into, but not through, archaeological zones showing the presence of humans. No species disappeared in the longer intervals before the first human contact. With only Stone Age technology, the Polynesians may have exterminated more than 2000 bird species, approximately some 15% of the world total. Locally, they often exterminated all the species they encountered.

Colonists ate the large, probably tame, and often flightless species. They may have been the perfect accompaniment to *poi*—a purplish paste made from taro root. They introduced pigs and rats to islands far too remote to have native land mammals. The rats, too, would have found the naive birds to be easy pickings.

In the Hawaiian islands, we know 43 species only from their bones (Olson & James 1982, 1991; James & Olson 1991). Yet, bird bones are fragile and easily destroyed. We may never find bones of all the now extinct species, so how many are missing? The bone record would be complete only if all the recent species—those collected or seen in the last two centuries—were also found as bones. Simply, the proportion of recent species *also* found as bones estimates how complete is the sample of species found *only* as bones. Across the Hawaiian islands, we estimate there are about 40 species missing from the record (Pimm *et al.* 1994). Add this number to the 43 species known to be extinct and the body count rises to 83.

James Cook found the islands in 1778. Trade and colonization followed within a generation. The new people cleared forests and introduced cattle and goats. Like pigs, these destroyed native plants as unprepared for large mammalian herbivores as the birds were for rats. Today, our only records of 18 species of bird are the specimens collected by nineteenth century naturalists. We know they missed some species. On Moloka'i, for example, early naturalists heard a rail, but never collected it. The body count rises to at least 101.

What remains today? A dozen species are so rare that there is little hope of saving them. If we cannot find these species, then they probably cannot find each other. A further dozen we can find, but in numbers so small that their future survival is uncertain. Of an estimated 136 species, only 11 survive in numbers that suggest a confident future.

I calculate that, as the Polynesians colonized the Pacific from New Zealand, north to Hawai'i and east to Easter Island, they exterminated 500–1000 species (Pimm *et al.* 1994). Steadman estimates a much higher number. Indeed, he argues convincingly that there could be 2000 species of rail alone that are missing from the Pacific (Steadman 1995).

Every Pacific island that Steadman and others have searched carefully has yielded at least one unique species of rail either living or known from bones. This includes Henderson, so remote it still has a common living rail (*Porzana atra*), and tiny atolls, like Lisianski, where the rail was seen but never collected. The larger volcanic islands had several species of rail. Steadman argues that probably every one of 800 Pacific islands had one or more rail. In the time it took humans to colonize the Pacific, one of the greatest evolutionary radiations of vertebrates—the Pacific island rail—has almost completely disappeared.

This is only the rail. Excluding the most remote islands, all the islands also had several species of pigeon and parrot. Nor does it count all the small species.

Birds were not the only victims, incidentally. Of 980 native Hawaiian plants, 84 are extinct and 133 have wild populations of less than 100 individuals (Sohmer 1994). Across the Pacific, a predatory snail, *Euglandina rosea*, introduced to control another introduced snail, *Achatina fulica*, ate to extinction hundreds of taxa of native *Achatinella* and *Partula* land snails (Hadfield 1986; WCMC 1992). (I use the term 'taxa' to include recognized geographically distinct populations. Taxonomic uncertainties often raise and sink their specific status. For those that are now extinct we may never resolve the issue.)

The Pacific islands are not unusual. In the last 300 years, Mauritius, Rodrigues and Réunion in the Indian Ocean lost 33 species of bird, including the dodo *Raphus cucullatus*, 30 species of land snail and 11 species of reptile. St Helena and Madeira in the Atlantic Ocean have lost 36 species of land snail (WCMC 1992). This raises the obvious question: Do we find evidence of massive extinctions *only* on islands?

Flowering plants

A distinct and unusual flora defines the Cape Floristic Region, which occupies a small area of the southern tip of Africa. It comprises several vegetational types of which the fynbos is dominant in area and contributes the most species. Of about 8500 species, 36 species have become extinct in the region in the last century, and some 618 species are *threatened*. (I will always use 'threatened' in a specific, technical sense to mean those species thought likely to become extinct within at most a few decades. Quite how long rare species are likely to last is a topic I discuss below.)

Cowling (1992) identifies invading alien plants—particularly Australian wattle trees—and the conversion of natural areas to agriculture as the two major causes of species extinction and endangerment.

Freshwater mussels and clams

Williams *et al.* (1992) assessed the Mississippi and St Lawrence river basins. Of the 297 North American taxa of the two families Unionidae and Margeritifidae, an estimated 21 have probably gone extinct since the end of the nineteenth century. Another 120 taxa are threatened.

Habitat modification has been the primary cause of extinction. Dynesius & Nilsson (1994) mapped the rivers of North America and Eurasia to show how few (and how remote) are these continents' remaining wild rivers. Mussel beds suffer from the channellization and impoundment of almost all the basins' major rivers. Poor farming practices have resulted in run-off and enormous increases in the nutrients and silt load of these waters. Also, the increase in sewage effluent and urban run-off has produced a long-term decline in water quality. Freshwater mussels are sessile, filter feeding organisms and so are extremely sensitive to such changes. Furthermore, in their juvenile stages, most mussels are parasitic on resident fishes, which are also adversely affected by declines in water quality. Another threat is harvesting for use in the cultured pearl trade. Finally, two freshwater mussel introductions, the Asian clam, *Corbicula fluminea*, and the zebra mussel, *Dreissena polymorpha*, are rapidly expanding their ranges in the region.

Freshwater fish

Miller *et al.* (1989) found that, of *c.* 950 taxa of freshwater fish in the USA, Canada and Mexico 40 have become extinct in the last 100 years. Northern lakes, southern streams, wetlands and desert springs are very different habitats, yet all have lost species.

The arid region of the southwestern North American continent has lost most taxa, mainly from physical habitat changes and introduced species. Most of the species were from springs. These small habitats are highly sensitive to disturbance. Currently, some 50 taxa of Cyprinidae are threatened, including 14 species that inhabit Nevada spring systems and 14 species in the Colorado River system. Impoundments, ground-water extraction, channellization and irrigation schemes are reported as contributory factors in 18 extinctions.

Introduced species—stocking of fish and frogs for food, fish for sport and accidental aquarium releases—are also factors in 18 extinctions. Introductions cause extinctions through competitive exclusion, predation or hybridization. Hybridization may seem a surprising cause of extinction, but populations that have evolved in isolation need not retain their morphological and genetic distinctiveness when mixed with similar species from elsewhere. Simberloff shows how pervasive and important is hybridization as a cause of extinction (Chapter 6).

In the southeast states of the USA, Etnier & Starnes (1993) counted 488 freshwater fish taxa in the region, of which four have become extinct and 80 are threatened. Increasing development and chemical alteration of Appalachian and Cumberland mountain streams pose serious threats to many species.

Australian mammals

Of the 60 species of recent mammal extinctions, 19 are from Caribbean islands (WCMC 1992). This repeats the pattern of high extinction rates on islands and I will not consider them further here. Interestingly, 18 more were in Australia (WCMC 1992), representing about 6% of its non-marine mammal species. The extinctions have been equally divided between the southern arid zone—a sparsely inhabited area of mostly spinifex desert and extensive pastoralism—and the wheat belt of the southern tip of Western Australia—where 95% of the natural woodland has been cleared (Short & Smith 1994). Another 43 species have been lost from over 50% of their former ranges or only survive on protected offshore islands (Burbidge & McKenzie 1989). Medium-sized ground dwellers weighing between 35 g (large murids) and 5.5 kg (small wallabies) have been hit hardest. Bats, arboreal species (possums and gliders) and those that use rock piles for shelter have fared better.

Short & Smith (1994) suggest three causes: the destruction and fragmentation of natural habitats; introduced species; and recent changes in the continent's fire regime. Domestic farm animals may have destroyed vegetation cover and caused extensive soil erosion and compaction. Introduced rabbits are competitors for already declining food resources. The predatory red fox *Vulpes vulpes* introduced in the 1860s (Burbidge & McKenzie 1989) may well have destroyed small mammal populations, even in remote areas. Foxes are absent from the areas of the continent with the fewest extinctions. Fox control programmes are successful in halting the decline of some small populations.

2.4.2 Relative rates of extinction

To calculate extinction rates, consider the freshwater molluscs as an example. Divide the number of extinctions (21) into the region's total (297) times the number of years over which the extinctions occurred (on the order of 100). This yields a regional rate of one species per 1.4×10^3 species per year or 714 *E*/MSY.

Obviously, I selected the above five studies because of their high regional rates of extinction. Interestingly, we can use these numbers to calculate an extremely conservative estimate of the *global* rate of extinction, by supposing these were the only extinctions worldwide. (They are far from it; many other areas are losing species.) The global rate divides the known extinctions per year by the worldwide total of species. For the two mollusc families this is *c.* 1000; this yields a rate of *c.* 200 *E*/MSY.

The recent extinction rates for these five case histories are 20–200 *E*/MSY (Fig. 2.1). This is a small range given the many uncertainties. Should we average rates over a century or a shorter interval that reflects more recent human impacts, for example? High rates are typical of both mainlands and islands, of arid lands and rivers, and both plants and animals. Although we know less about invertebrates, high rates characterize bivalves of continental rivers and island land snails. There is nothing intrinsic to these species' diverse life histories to predict their being unusually extinction-prone. Indeed, all five studies involve species with generally high reproductive rates. It is therefore not just the species that recover only slowly from the damage we inflict upon them that are in jeopardy.

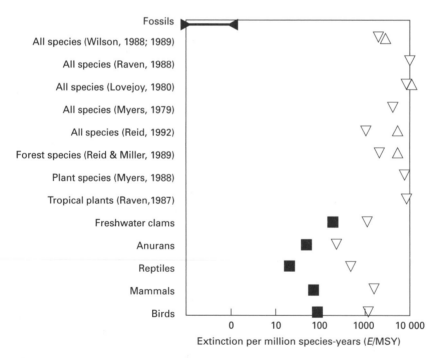

Fig. 2.1 Estimates of past (solid squares) and future (open triangles) extinction rates. Where an author suggests a lower bound (down-pointing triangle) and an upper bound (up-pointing triangle) for an estimate these are both shown. Data with no source are from Pimm *et al.* (1995) and are explained in the text.

2.4.3 What causes these extinctions?

Habitat loss is widely thought to be the predominant cause of extinction (WCMC 1992). Surprisingly, in the four continental studies, the introduction of alien species often ranks as the primary cause, as it always does for island floras and faunas (Pimm *et al.* 1994). These studies also demonstrate how much damage even small numbers of our species can cause, for none of them are from areas where human densities are particularly high. Introduced species, agriculture and dammed rivers cause extinctions well beyond the main mass of humanity. Elsewhere, quite small numbers of people are responsible for the current deforestation of tropical forests (see below), just as small numbers of Polynesians exterminated the Pacific's terrestrial birds.

People and the species we introduce are ubiquitous. So what are the features common to these extinction centres? First, we know the species and places well. Birds, flowers, land snails, freshwater bivalves, mammals and fish were collectors' items a century ago, so we are well informed of their fate.

Second, and most importantly, each area holds a high proportion of species restricted to the area. We call such species *endemics*. Remote islands are rich in endemics, but so are some continental areas (ICBP 1992). Endemics constituted 90% of Hawaiian plants (Sohmer 1994) and 100% of Hawaiian land birds (Pimm *et al.* 1994), but also *c.* 70% of fynbos plants (Cowling 1992), 74% of Australian mammals (WCMC 1992), over 90% of North American fish (Miller *et al.* 1989) and 'the great majority' of that continent's freshwater molluscs (Williams *et al.* 1992). In contrast, only about 1% of Britain's birds and plants are endemics (WCMC 1992).

Past extinctions are so concentrated in small, endemic-rich areas that the analysis of global extinction is effectively the study of extinctions in one or a few extinction centres (Nott *et al.* 1995; Pimm *et al.* 1995). Why should this be?

Consider some simple models of extinction. The simplest supposes only that some species groups are more vulnerable than others. This model does a poor job of predicting global patterns. First, the model predicts that the more species that are present, the more there will be to lose. Yet the number of species an area houses is not a good predictor of the number of extinctions. Relative to continents, islands house few species yet suffer many extinctions. Second, if island birds were intrinsically vulnerable to extinction, then Hawai'i and Britain—with roughly the same number of breeding land birds and both with widespread habitat modification—would have suffered equally. Hawai'i had over 100 extinctions, Britain only three (Parslow 1973).

All the Hawaiian species were restricted to the islands; none of the British species were. This suggests another model. Imagine a 'cookie cutter' model where some cause destroys ('cuts out') a randomly selected area

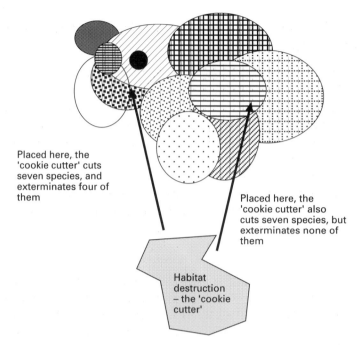

Placed here, the
'cookie cutter' cuts
seven species, and
exterminates four of
them

Placed here, the
'cookie cutter' also
cuts seven species, but
exterminates none of
them

Habitat
destruction
– the 'cookie
cutter'

Fig. 2.2 The 'cookie cutter' model. Habitat destruction may encompass a range of similar numbers of species in different areas (seven in each example). Only those species with ranges entirely encompassed by the destruction run the risk of extinction. By chance alone, extinctions are concentrated in areas with high numbers of species with small ranges.

(Fig. 2.2). Species also found elsewhere survive, for they can re-colonize. Only some of the endemics become extinct, the proportion depending on the extent of the destruction (see below). In this model, the number of extinctions correlates weakly with the area's total number of species, but strongly with the number of its endemics. By chance alone, small endemic-rich areas will contribute disproportionately to the total number of extinctions.

This model is consistent with known mechanisms of extinction. Habitat destruction 'cuts out' areas as the model implies. Introduced species also destroy species regionally. Species need not be entirely within the area destroyed to succumb: the populations outside may be too small to persist (see below). Moreover, across many kinds of organism, species with small ranges typically have lower local densities than widespread species (Brown 1984; Gaston 1994a) (several examples of this pattern are compiled in Gaston 1994a, pp. 66–71). Range-restricted species are not only more likely to be 'cut' in the first place, but their surviving populations will have smaller densities and therefore higher risks of extinction than widespread species.

This entirely self-evident model emphasizes the localization of endemics as the key variable in understanding global patterns of recent—and future—extinctions. These are what Myers (1988, 1990) calls 'hot spots'.

2.5 Predicting future rates of extinction from species currently threatened

Projecting past extinction rates into the future is absurd for no other reason than that the ultimate cause of these extinctions—our human population—is still increasing rapidly (Cohen 1995). So let us explore two other methods to predict future rates of extinction. In this section, I will consider how quickly species now threatened with extinction will persist before they expire. Then I will explore the relationship between habitat destruction and extinction.

For vertebrates, we have worldwide surveys of threatened species (WCMC 1992). For birds, 1100 of *c.* 10^4 species are threatened (Collar *et al.* 1994). Will *all* these threatened species be extinct in less than 100 years? If so, then future rates will be 200–1500 *E*/MSY (Fig. 2.1).

For 13 groups of species, there is an important compilation for the USA (TNC 1996). It counts extinct and probably extinct species. It also identifies those that are *critically imperilled* (thought to number < 1000 individuals) and those merely *imperilled* (1000–3000 individuals). The percentages of the US flora and fauna in both imperilled categories are: butterflies (4.4%), birds (6.2%), mammals (6.2%), tiger beetles (6.4%), reptiles (7.1%), dragonflies and damselflies (7.6%), ferns (9.8%), conifers (13.1%), flowering plants (15.9%), freshwater fish (20.6%), amphibians (23.2%), crayfish (37.4%) and freshwater mussels (43%).

Notice that birds are proportionately the second least endangered species. The detailed estimates of bird extinctions I present later must be viewed in this context. Extinction rates for other, less well-known species are likely to be much higher.

2.5.1 Many threatened species will not last another century

The data on threatened species beg the obvious question: how long will these rare species last? We have to decide into which of several categories they belong.

Inexorable declines

Some threatened species are declining rapidly and will go extinct quickly and predictably unless we intervene. The California condor, *Gymnogyps californianus*, for example, declined year after year for decades, before the few remaining individuals entered a captive breeding programme.

Very small populations

Other species are not so obviously doomed. Some have small numbers (< 100) and no possibility of achieving larger numbers because they survive

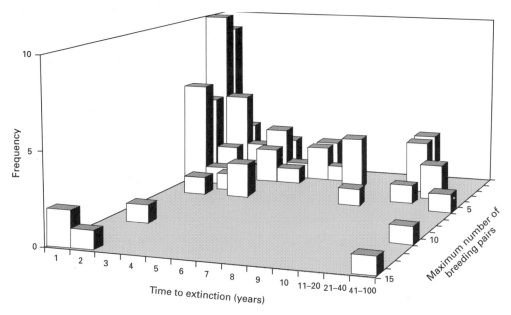

Fig. 2.3 Times to extinction of hawks, owls and crows on small islands off the coast of Britain. The times between colonizations and observed extinctions are 1, 2 ... to 10 years, 11–20 years, 21–40 years and 41–100 years. The maximum number of breeding pairs is the maximum number observed for a particular species on a particular island. Thus, one species achieved a maximum population size of 15 pairs before becoming extinct in 41–100 years; subsequent colonizations of this species lasted only 1–2 years. (This maximum is thus an estimate of the population's ceiling and not the largest population size achieved before extinction.) Notice that, on average, the smaller the ceiling, the shorter the time to extinction. Most populations with ceilings of < 10 pairs are extinct within 20 years.

only in small, protected areas. Such species risk the inevitable vagaries of sex (the difficulties of finding a suitable mate, most of the young of a generation being of the same sex, etc.) and death (most of the individuals dying before reproducing from independent causes). The history of British royalty (by decision, a small population) provides a detailed 1000-year record of such disasters. The fate of populations of birds on small islands off the coast of Britain provides a mere half-century record. It shows times to extinction of the order of decades (Pimm *et al.* 1988, 1993) (Fig. 2.3). Such results match those of mathematical models (Goodman 1987; Lande 1993) and provide crucial advice to those of us who manage threatened species (NRC 1996).

Large but still threatened populations

A broad range of environmental vagaries (cold winters, droughts, disease, food shortages, etc.) cause all population sizes to fluctuate considerably (Pimm 1991). Fluctuations, like death and taxes, are unavoidable. Butterfly numbers, for example, can drop 100-fold in 1 year. Thus, the 27 (of 610) US butterfly species that number less than 3000 individuals could disappear, if not overnight, then from one year to the next. Indeed, there are another 73 species (12% of the US fauna) that number less than 10 000 (TNC 1996). Insect densities can vary 10 000-fold over 20 years (Pimm 1991). Within a century or so, such fluctuations can obviously drive even the largest populations of currently threatened species to those low levels where recovery is unlikely or impossible.

Some threatened species will probably survive the twentieth century, some through good luck, others by our good management of them (Collar *et al.* 1994). The more serious error in calculating future extinction rates is that species not now threatened will become extinct.

2.5.2 Some species not now threatened will not last another century either

First, for birds—the one group for which we have detailed lists of the causes of threats—limited habitat is the most frequently cited factor. It is implicated in about 75% of threatened species (Collar *et al.* 1994). Increasingly well-documented studies (considered next below) show that habitat destruction is continuing and perhaps accelerating. Some now common species will lose their habitat within decades.

Second, accidentally or deliberately introduced species are blamed for only 6% of currently threatened birds (Collar *et al.* 1994). Yet introduced species, and the hybridization, predation, competition, disease and habitat modification they cause, are the most frequently cited factors in all the extinction centres I discussed above. Undoubtedly, many species will be lost to introduced species that we cannot now anticipate. For example, no one considered the birds on the island of Guam to be in danger 30 years ago. An introduced snake, *Boiga irregularis*, has eliminated all the island's birds since then (Pimm 1987; Savidge 1987). Were it to reach Hawai'i, all its birds would be at risk.

In sum, many of the species that are threatened now will be extinct within a century from now. The threatened species that do survive will likely be outnumbered by extinctions of species not now threatened. The next century's extinction rates will probably be at least 10 times greater than the current rate.

2.5.3 Estimating future species losses from habitat loss

So far, we have discussed well-known but disparate species whose high extinction rates are probably typical of the unknown majority. Now consider a typical mechanism of extinction: habitat loss. Can we predict species losses from habitat losses?

A rough calculation

Tropical moist forests may hold two-thirds of all species (Raven 1988)—more if one accepts the highest estimates of insect and fungal diversity (see above). Satellite imagery yields detailed and rapidly changing estimates showing their rapid depletion. The forests' global extent is variously estimated at 8×10^6 to $12.8 \times 10^6 \text{km}^2$ and their rate of clearing as 1.2×10^6 to $1.4 \times 10^6 \text{km}^2$ per decade (FAO 1992; Skole & Tucker 1993; Myers 1994). Suppose clearing continues to the last tree. Two-thirds of all species will disappear in from 67 years (= 8/1.2 decades) to 91 years (= 12.8/1.4 decades), yielding rates of 10 000 $E/$MSY and 7300 $E/$MSY, respectively (Fig. 2.1).

Surely some forests will remain. Saving 5% of them could save 50% of their species (see below), though saving this much will require massive conservation efforts. A few species will likely survive in the disturbed habitats and secondary forest that replaces the original forest. Because these values provide the highest numbers of future extinctions, can we specify when and where they will occur?

2.5.4 Calibrating species losses from habitat losses

Various small-scale studies illustrate that now isolated small forest patches contain fewer species than we would expect were they part of the once continuous and extensive habitat (e.g. Willis 1974; Karr 1982; Bierregaard *et al.* 1992); Diamond (1984) and Pimm (1991) provide reviews. Can we extrapolate from these studies that involve local extinctions (species A is missing from a given small area) to global extinctions (species A is missing from the planet)?

The function, $S = cA^z$, frequently provides an adequate description of the relationship between the size of an area, A, and the number of species, S, that it contains; c and z are constants. This suggests we can predict the reduction in numbers of species from $S_{original}$ to S_{now} as the habitat's area is reduced from $A_{original}$ to A_{now}. The proportion of species lost ($S_{now}/S_{original}$) should be $(A_{now}/A_{original})^z$. Thus S_{now} equals $S_{original}(A_{now}/A_{original})^z$ and, obviously, the number of extinctions $S_{extinct}$ equals $S_{original} - S_{now}$. Notice, that we need an estimate of the value of z but not of c. This algebra makes the critical assumption that the same function applies equally to cases where habitats are actively being destroyed as to the counts of species in areas of different sizes.

I pick two examples: the temperate forests of eastern North America and the tropical forests of insular South East Asia. In both places, about half the forest remains, fragmented into large habitat islands. Perhaps the best natural model of this are large islands within an archipelago, for which a value of z = 0.25 is typical (Rosenzweig 1995). How accurately can we predict these forests' extinctions?

Eastern North America

Extensive reductions in the forests of eastern North America occurred during the nineteenth century. Surprisingly, only four bird species have gone extinct: the passenger pigeon *Ectopistes migratorius*, Carolina parakeet *Conuropsis carolinensis*, ivory-billed woodpecker *Campehilus principalis* and Bachman's warbler *Vermivora bachmanii*. Birds are well known so we cannot plead ignorance of their extinctions, and critics have used this apparent discrepancy to claim that fears about massive global extinctions based on species–area predictions are 'simply wrong' (Budiansky 1994).

Some 48% of the area covered by the eastern forest at the time of European settlement (in 1620) was still wooded at the low point in 1872 (Pimm & Askins 1995). With $A_{1872}/A_{1620} = 0.48$ and $z = 0.25$, we predict that *c.* 17% of the region's 160 forest birds (27 species) should have become extinct. It is this prediction, some six times greater than the well-documented extinctions, that causes controversy.

Does this discrepancy cast doubt on the predictions of species losses from habitat reduction? It does not. Those who point to the small number of observed extinctions in the eastern forests mean *global extinctions*—species that are lost everywhere. The prediction of 27 extinctions is based on the total number of species. Most of these 160 would not become globally extinct even if *all* the eastern forests were cleared. For many species, their distribution across the boreal forests of Canada makes them invulnerable to forest losses in the USA.

Recall the 'cookie cutter' model of extinction presented above. It concludes that we should restrict our analysis to the endemics. It asks: How many species *could* become globally extinct if all the eastern forests were felled? That is, which species are found *only* in these forests. The answer is only 28. Now 17% of 28 is about 4.76. This prediction is three-quarters of a species higher than the number of extinctions observed. I will not push my luck to argue that the endangered red-cockaded woodpecker *Picoides borealis* is three-quarters of its way to extinction. The observed and predicted numbers are close enough to show that this case history is not the counter example critics claim it to be.

Insular South East Asia

The region comprises four archipelagos: the Philippines, the Greater Sundas (Java, Sumatra and Borneo), northern Wallacea (Sulawesi and the Moluccas) and the Lesser Sundas. Their forests hold 585 endemic species of bird—roughly 20 times that of America's eastern forest, in half the area. About 10% of the original area is cleared per decade. Most of this deforestation has occurred recently and approximately 60% of the original area is still forested (Collins *et al.* 1991). Unlike the previous example,

deforestation has not yet caused any confirmed bird extinctions in insular
South East Asia. Extinctions take time following habitat loss (see above)
(Diamond 1972).

Let us repeat an earlier calculation. Suppose that 60 years from now all
the forest has been cleared, save 5% in reserves. With $A_{now}/A_{original} = 0.05$,
we calculate $S_{extinct}/S_{original}$ to be roughly a half. The loss of c. 300 of the
region's endemic birds is interesting because only 120 are currently classified
as 'threatened' (Collar *et al.* 1994). As discussed above, the number of
species that may be soon doomed to extinction but are not now thought
to be threatened is probably quite large.

Again, we ask: Does the species–area recipe predict the details of where
these extinctions will occur? Across the region, some areas still have
most of their forests—Borneo has about 67%, for example. Other areas
have almost none—Cebu, in the Philippines, has less than 1% (Magsalay
et al. 1995). The region's structure of islands within archipelagos, and
archipelagos within the region, allows us to group species into four levels
of endemism: (i) the *'single-island' endemics*—those found on single islands;
(ii) the *'intra-archipelago' endemics*—species found on more than one island
within the archipelago, but not outside of it; (iii) the *'interarchipelago'*
endemics—species found in more than one of the four archipelagos, but
not outside of the region; and (iv) species that are not endemic to the
region.

Using the recipe, Brooks *et al.* (1997) sought to predict the numbers of
threatened species in each of these four groupings. They used a value of
z = 0.25 and values of deforestation that correspond to the four levels of
endemism: deforestation within the particular island, averaged across the
particular archipelago, averaged across the four archipelagos, and averaged
across both mainland and insular South East Asia.

Table 2.2 Number of species at different levels of endemism in insular South East Asia,
the number that are threatened with extinction, and the number predicted to go extinct
on the basis of current levels of deforestation. (From Brooks *et al.* 1997.)

Level of endemism	Species	Threatened	Predicted to go extinct from current deforestation
Single-island			
Philippines	103	58	31
Greater Sundas	84	15	12
Lesser Sundas	69	12	26
Northern Wallacea	146	16	14
Intra-archipelago			
Philippines	81	15	24
Greater Sundas	26	2	4
Lesser Sundas	17	0	6
Northern Wallacea	20	1	2
Interarchipelago	39	2	6
Non-endemic	500	17	72

For single-island endemics, the number of extinctions predicted by deforestation *underestimates* the number of threatened species in the Philippines (31 vs 58) but *overestimates* the number in the Lesser Sundas (26 vs 12). Very small forest fragments remain on some Philippine islands, such as Cebu (Magsalay *et al.* 1995). The small surviving populations encounter the rapid extinction discussed earlier. In general, the smallest forest fragments typically hold fewer species than the species–area model predicts using $z = 0.25$. Values of $z = 0.6$–1.0 provide a better fit to such data (Rosenzweig 1995). In contrast, the forests in the Lesser Sundas are regularly disturbed by typhoons. It is possible that their bird species are thus better adapted to open areas and secondary forest than species are elsewhere and so are less affected by deforestation.

Relative to what deforestation predicts, the four groups are threatened in inverse proportion to their ranges (Table 2.2). The non-endemics are threatened the least (17 vs 72 predicted), then interarchipelago endemics (two vs six), then intra-archipelago endemics (a total of 18 vs 36), and finally the single-island endemics which are the most vulnerable (a total of 101 vs 83). This is exactly what the 'cookie cutter' model expects, of course.

2.6 The future of biological diversity

If one values species, the information in this chapter starts with bad news and gets progressively worse. The extermination of 20% of the world's birds is not a fanciful future prediction, but a reasonable estimate of the effects of a Stone Age culture on an endemic-rich avifauna. Islands are often rich in endemics. So, too, are many continental areas that have lost large numbers of flowering plants, mammals, fish and other groups. Recent extinction rates are easily 100 times greater than background geological ones.

Assuming that these rates would continue into the future is bad enough, but two lines of argument suggest otherwise. The first is that the numbers of species threatened with extinction are 10 times the numbers of extinctions in the last century. Some of these threatened species will be extinct well before another century passes. While some may survive, they will surely be out-numbered by the extinctions of species not now considered to be threatened.

The second line of evidence comes from projecting current rates of habitat destruction—particularly the destruction of tropical forests. A simple function relates the number of species found in an area to its size. Reduce the extent of available habitat and the number of species should decline accordingly. Not all these species will become extinct, of course, because some will survive elsewhere. Common sense suggests we restrict the analysis to the species that cannot survive elsewhere—the endemics. The need for the details of global habitat losses is obvious, but these details are not sufficient. It matters where the 'cookie cutter' falls and we need to know which small areas hold many endemics. The fate of these endemic-rich areas dominates the calculations of future extinction rates.

So what do we know about the patterns of endemism? Many tropical areas are unusually rich in endemics. For example, 18 areas worldwide are so rich in endemic flowering plants as to encompass about 20% of the known species in a total area of $0.74 \times 10^6 km^2$ (Myers 1988, 1990). A larger area than this was cleared from the eastern American forests in the nineteenth century (see above). The destruction of 20% of the planet's flowering plants is thus well within our technical capabilities.

For birds, we have detailed maps of endemic-rich areas (ICBP 1992). (Indeed, some guides to tropical bird watching highlight the endemics one may encounter.) Again, small areas may hold many species. Insular South East Asia holds $c.$ 10% of the planet's species, and $c.$ 5% of its species are found nowhere else (see above). Unfortunately, we know the geographical ranges of only a small proportion of the already small proportion of the planet's species for which we have names. This raises a practical question: Can we set conservation priorities on the basis of groups, like birds, that we know very well and expect them to work for other species groups (see Chapter 1)?

What we know about the geographical patterns of species richness is not encouraging (Rosenzweig 1995). For example, across Australia, areas rich in lizard species house few bird species. Worse, in North America, areas rich in lizard species house relatively large numbers of bird species (Schall & Pianka 1978). Conserving bird-rich areas would help lizards in North America, but not Australia.

Even if the patterns of species richness matched consistently, the patterns of endemism need not mirror the patterns of species richness. And, indeed, they do not (Nott & Pimm 1997). Areas rich in species are not always rich in endemics. Curnutt *et al.* (1994) examining Australian birds and Prendergast *et al.* (1994) examining a range of British organisms both found that the areas that housed the greatest number of species were *not* the areas that housed the species with the smallest geographical ranges.

The arguments of the previous paragraph still leave open the possibility that areas rich in, say, endemic butterflies may coincide with those rich in endemic birds. There are some coincidences: the Philippines are rich in both endemic birds and flowering plants. There are counter examples too: the eastern USA holds very few endemic birds but more than 20% of the planet's salamanders (Conant & Collins 1991)! Apart from these anecdotes, the extent of the coincidence of endemism is an as yet unanswered problem. It is critically important. Without a proper understanding of endemism we cannot know the future of the planet's species with the precision needed to manage them.

CHAPTER 3

Introductions

H. Charles J. Godfray and Michael J. Crawley

3.1 Introduction

Over the last few thousand years, humans have had an enormous impact on virtually every ecological community on Earth. One of the most important ways in which natural communities have been altered by people is through the introduction of species from different geographical areas. Frequently, the introduction is accidental, as is the case for most agricultural weeds, but many deliberate introductions have also been carried out, for example rabbits for food in Britain and Australia, specific natural enemies of weeds and pests in biological control programmes, and the inspired lunacy of the nineteenth century Naturalization Society in New Zealand which sought, and largely succeeded, in recreating the ambience of the English countryside in the Antipodes by releasing common British songbirds. While a naturalization society would not be tolerated today, introductions are still occurring at a high rate. The ease of modern travel, particularly by air, and the continuing expansion of world trade result in a never-ending stream of new accidental introductions. The majority are unnoticed small plants and insects, occasionally supplemented by a spectacular introduction from a zoo or collection escape, such as the recent establishment of the Asian short-clawed otter *Amblonyx cinerea* in southern England. Horticulturists, always on the lookout for novelty, continue to bring new plants into cultivation, some of which escape into natural communities; and the release of biological control agents against pests has increased, in part due to growing problems with the evolution of insecticide resistance. The next 10 years will see a completely new type of introduction, of genetically engineered organisms carrying novel gene constructs.

Thus invasive alien species pose serious problems on an international scale, affecting health, agricultural potential and biodiversity, and these problems look set to become more acute. The Biodiversity Convention (Article 8(h)) includes sweeping requirements to 'prevent the introduction of, control or eradicate those alien species which threaten ecosystems, habitats or species', although there are few provisions for monitoring or enforcing the intent of these agreements. In this chapter we review some of the problems and the ecological theory available for tackling them. The next four sections deal with the steps involved in a successful invasion of a new environment by a non-indigenous species. First, the species must be *introduced* in sufficient numbers and into appropriate habitats. Second, the species must become *established* by forming self-replacing (typically breeding) populations. Third, the species will *spread* into new habitats beyond those into which it was originally introduced. Finally, we consider the *consequences* of establishment and spread. We finish by discussing some special problems raised by the introduction of genetically manipulated organisms.

3.2 Rates and types of introductions

What determines the rate at which new species are introduced into a habitat or locality, and what type of biological characters will these species possess? Consider first introduction through non-human agencies. As is clearly shown by studies of fossil and subfossil flora and fauna, biological communities are not static but change constantly over time (Davis 1994). All natural communities are subject to challenge by potential new invaders, although almost certainly at far lower rates than those experienced today.

An obvious determinant of the rate of species introduction is the distance from sources of colonization. The effect of distance can be seen most clearly in the colonization of islands, both actual islands and islands of particular habitats set in a sea of other vegetation. MacArthur & Wilson (1967), in their celebrated theory of island biogeography, argued that island biodiversity was set by a dynamic balance between colonization and extinction. Islands more distant from the mainland will have a lower rate of colonization and hence a lower equilibrium number of species. The evidence supports this prediction, with distant oceanic and habitat islands being particularly depauperate in both plants and animals (Diamond 1972; Williamson 1981). The pattern also occurs on a much more local scale: patches of plants situated at a distance from other patches of the same species often have fewer specialist herbivores (Strong *et al.* 1984). While confirming the importance of distance in setting the rate of introduction, it should be noted that these data do not necessarily support the classic island biogeography model. If islands accumulated new species at a rate proportional to the numbers of introductions, the same pattern would be observed.

Time since the creation of a new habitat also influences the number and type of introductions. Studies of post-glacial plant re-colonization provide an important model that may be relevant to vegetation change following climatic disruption. Detailed examination of the pollen record preserved in peat and in lake sediments shows that individual species moved northwards during post-glacial re-colonization at rates that were independent of one another, and which varied from one glacial period to another. Thus, the tree species that we observe growing together currently in North America are not survivors of some tightknit community that has marched southwards to its glacial refugium, then back northwards again to occupy its current range. The species composition and the relative abundance of species within communities have changed so dramatically that it makes little sense to think of the community as a structured (and hence closely regulated) assemblage (Davis 1994). There appears to be a great deal of serendipity in which species get to live together at any one period of geological time. A corollary is that the composition of many (perhaps even most) communities is determined by the history of introduction.

Do some species have features that make their natural introduction into new habitats more likely? Clearly, dispersal ability is important. Many plants possess adaptations that allow their seeds to be transported huge distances by sea currents—coconuts and mangroves, for example. Diamond (1975b), in his study of bird communities in the South Pacific, classified some species as 'super-tramps' because of their ability to colonize even remote islands. More generally, there is likely to be a correlation between dispersal ability and the likelihood a species is introduced to a new habitat or locality. Classic ecological theory, again originating with MacArthur & Wilson (1967), distinguishes two adaptive syndromes associated with species living in relatively stable habitats at fairly constant population densities (K-strategists) and species living in temporary habitats showing boom-and-bust population dynamics (r-strategists). Because of the temporary nature of their habitat, r-strategists are under strong selection for dispersal ability and so are likely to be over-represented among introduced species. However, life history strategies of plants and animals are too complex to be summarized along a single axis, and the power of r- and K-theory in predicting colonization ability is likely to be weak.

It is a useful exercise to consider the reasons why certain species are not present in a community. Logically, their absence must be due to one or both of two reasons: (i) failure to disperse; or (ii) failure to establish having successfully dispersed. If it were possible to show that a species was not limited by dispersal, but yet was absent from a community, then it might be considered a 'safe' species in the sense of being non-invasive.

While natural introductions provide important information about community structure and resilience, the majority of cases of introductions today occur through human agency. The chief predictor of likelihood of introduction is association with man. Consider plants—a large class of species have been transported by man because of their value as crops, forage or sources of fibre, timber, etc. (Pimentel 1995). Some of these species have escaped from cultivation and have invaded natural habitats, for example fruit trees like the guava and strawberry guava (*Psidium* spp., Myrtaceae) in Hawai'i (Vitousek 1988) and the introduced pine, *Pinus patula*, that has invaded high-altitude grassland on Mount Mulanje in Malawi (Davis *et al.* 1994). With useful plants come weeds. European settlers, travelling to the New World, carried stocks of seed corn that were contaminated by the seeds of countless weed species. They shared their ships with draft animals and domestic livestock; remnants of the hay used to feed the animals during the voyage were swept out onto the dock when the ship arrived at its New World destination, thereby introducing the seeds of an entire grassland flora in a single pulse. Shared agricultural implements and long-distance droving of livestock moved weed seeds from farm to farm. Ships returning to the home country took on sand and gravel as ballast for the return journey; this was full of seeds, some of which became established when the ballast was emptied out in European ports. Once thriving agricultural industries became established in the colonies, the trade in wool and hides formed a massive source of introduction of plants to the old country (for example, over 340 species of 'wool aliens' were found growing on the banks of the River Tweed, downstream from the mills of Galashiels in Scotland by Hayward & Druce in 1919). Finally, plants are introduced by humans for purely ascetic reasons. In Britain, the horticulturist's bible, *The Plant Finder* (Philip & Lord 1996), lists 65 000 species and cultivars of plants that can be purchased and planted, all potential colonists of native communities.

A similar classification can be drawn up for animals; species valued for food or fur, such as rabbits or mink, have escaped and invaded natural communities. Animals, particularly insects, but in the past vertebrates and increasingly microorganisms, have been introduced as biological control agents to control weeds or animal pests. Many ungulate species have been introduced onto islands to provide food for humans. However, the numbers of these deliberate introductions are dwarfed compared with the vast assemblage of pests that have been carried by man around the world. There are also some animal equivalents of the escape of ornamental plants: there are thriving cat populations on some sub-Antarctic islands, while the population of the mandarin duck in the UK is larger than in China, where it originated. Some animals have been deliberately introduced by man into natural ecosystems. For example, deer and many species of gamebird have been introduced into a number of countries for sport. A large variety of fish have been introduced into lakes and rivers to provide food or sport. The practice of pumping seawater into the holds of ships for ballast which is jettisoned at the destination has moved many planktonic and semiplanktonic species from ocean to ocean (Carlton & Geller 1993).

3.3 Establishment

One of the most fundamental concepts in population biology is the invasion criterion. In order for a species to establish itself, it must be able to increase when rare. The invasion criterion is not only important in considering the establishment of new species, but has a critical role in theories of species coexistence—for two species to coexist, each must have the ability to increase in numbers when rare (Tilman 1988). In epidemiology, the invasion criterion is normally written as $R_0 > 1$ where R_0 ('R nought') is the number of secondary infections caused by an initial infection. $R_0 > 1$ simply states that a pathogen must more than replace itself when rare (Anderson & May 1991).

What determines whether the invasion criterion is met? Perhaps the simplest case is an animal with a single synchronized generation per year; if the number in year t is called N_t, the invasion criterion requires

$$N_{t+1}/N_t = \lambda > 1.$$

If we were able to measure the average fecundity of the animal and all the mortality factors that impinged on the species, we could calculate λ

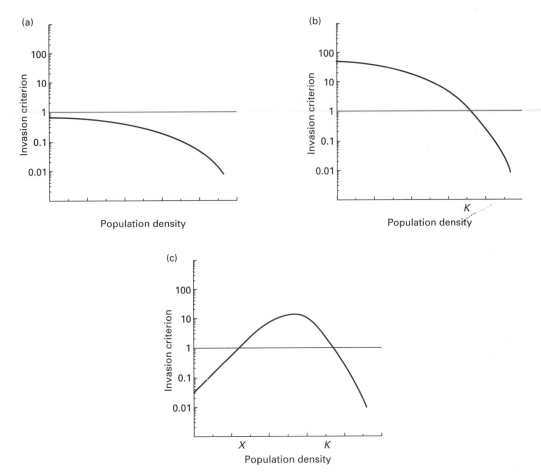

Fig. 3.1 The invasion criterion. (a) This species has $\lambda < 1$ at all densities, which means that it will not be invasive. (b) This species has $\lambda > 1$ at low densities, so it will invade and increase to its equilibrium population K; populations higher than K (e.g. as a result of mass emigration) will decline towards K. (c) Allee effects: this population is unable to increase when rare, and requires a threshold population size X before $\lambda > 1$; such populations are highly vulnerable to extinction if exploitation (e.g. harvesting) drives the population density below X.

and determine whether the invasion criterion was met (Fig. 3.1). Such an exercise is called population projection and is distinct from population prediction. Projection states that a population is currently changing at a geometric rate of λ; if fecundity and mortality remained constant, then this implies that in T generations' time the population density will be $N_t\lambda^T$.

This calculation is clearly simplistic as few species have such a simple life cycle, and both temporal and spatial variation are ignored. There is now a large body of theory, much developed with conservation in mind, that can be used to calculate the invasion criterion in more realistic situations. Consider first the complications that arise with overlapping generations. Techniques are available to calculate the equivalent of λ for populations with completely overlapping generations, or for populations with distinct age classes. The latter is the more simple as age-specific mortality and fecundity can be described in a matrix format (the Leslie matrix—technically, the invasion criterion is the dominant eigenvalue of the Leslie matrix). This formalism allows the relative contribution of mortality and fecundity at different ages to the population growth rate to be assessed (Caswell 1989; Tuljapurkar 1990).

Newly founded populations will not increase deterministically at a rate λ, but will be buffeted by different random factors. The theory of invasions in a stochastic environment is much harder than in a stable environment and is only just beginning to be applied to real examples. There are several ways in which stochasticity experienced by a population can be classified and here we distinguish between three: demographic, parametric and catastrophic (see also Chapter 6).

Demographic stochasticity describes the random effects of birth, death and mating in small populations. Consider a small population of annual insects or plants. By pure bad luck the population may go extinct because no individual reproduced in one year, or all the offspring were male. Similarly, by pure bad luck, random mortality might happen to hit all the members of a population in one year, or all the members may fail to find mates. The mathematics of demographic stochasticity is well understood (May 1973) and its effects are only significant at very small population densities. But very small population densities are exactly what occur in many cases immediately after invasion, and it is likely that many introduced species with a deterministic invasion criterion greater than one fail to establish because of demographic stochasticity.

There is some evidence suggesting a relationship between the probability of establishment and initial population size. In the biological control of insect pests using their insect natural enemies, the probability of successful establishment is related to the number of individuals released (e.g. Campbell 1976). There is contemporary documentation of introduction efforts for alien birds in New Zealand, which allows an unbiased comparison of successful and unsuccessful invaders (Veltman *et al.* 1996). From an analysis of 496 introductions of 79 species, it emerged that the likelihood of establishment increased with: (i) the total number of individuals released; (ii) the number of sites at which release was attempted; and (iii) the number of times that the release was repeated.

Fecundity and mortality vary from year to year depending on abiotic conditions as well as the local density of predators and competitors. Temporal variation means that we should write $N_{t+1}/N_t = \lambda_t$ emphasizing the fact that annual changes in population are not constant. The invasion criterion in the presence of this parametric stochasticity is that the long-term geometric average of the logarithm of the growth rate must be greater than zero; in symbols:

eqn 3.1

$$a = \lim_{T \to \infty} \left\{ \frac{1}{T} \sum_{t=1}^{T} \ln(\lambda_t) \right\} > 0.$$

Analogous quantities can be calculated for populations with overlapping generations (Caswell 1989; Tuljapurkar 1990). The most striking corollary of this equation is that the long-term logarithmic growth rate is undefined (implying invasion is impossible) if, in any one year, $\lambda_t = 0$. Because population change is multiplicative, any year in which the population falls to zero causes the geometric mean population size to be zero. In such cases, the only way the species can survive is if there are repeated invasions to re-establish the population after extinction. In the last section, we discussed the huge number of garden plants imported into the British Isles and the relatively small number that had established themselves. Many of these species, in most years, have $\lambda_t > 1$; but a number are frost intolerant and, in the occasional severe winters in the UK, suffer $\lambda_t = 0$ and extinction until the next introduction (see Fig. 3.1).

Our final category of random effects is catastrophic stochasticity, by which we mean extinction of a local population irrespective of its size and vigour. Catastrophic stochasticity occurs if a habitat is destroyed, naturally or by man; or if an epidemic or some other disaster sweeps through a population. By definition, this type of random effect results in the loss of a local population, although the invading species may still establish if it survives in other unaffected populations. To study the establishment of a species subject to catastrophic stochasticity, we need to consider assemblages of populations or metapopulations rather than individual populations (as well as possible new immigration). If all populations are roughly the same size, we can use the invasion criterion to decide whether a species can invade, but instead of counting the number of individuals in a population, we count the number of component populations in a metapopulation. Invasion occurs as long as each subpopulation, before its extinction, sends out sufficient colonists to found at least one new subpopulation. The extensions to overlapping generations and demographic and parametric stochasticity also follow. Of course, the populations that make up a metapopulation will vary in size, but even then we can make use of versions of the invasion criterion designed to incorporate age or size structure where the size of the population takes the place of the size of the individual. The analogy to models of individuals breaks down when the explicit spatial locations of the components of a metapopulation are important. In such cases, establishment may depend critically on the distance between populations, and the spatial covariance of catastrophic mortality. A final point is that much of what we have just discussed also applies to species that live in ephemeral habitats where population destruction is more deterministic than random.

So far we have discussed population projection of species that are sufficiently rare so that we can ignore the normal types of density-dependent mortality or reduction in fecundity that will eventually limit their abundance should they become established. However, there is another type of density dependence that can operate in very small populations. Some small populations will suffer from a failure to find mates, inbreeding depression, a breakdown in social structure or, in group-living species, a reduction in foraging efficiency. Such effects are not due to demographic stochasticity, but to an 'Allee effect', a lower than average fecundity, or higher than average mortality at low population densities (see Fig. 3.1c). Such species may only be able to establish ($\lambda > 1$) when population density exceeds a certain threshold; the same species will be more susceptible to parametric stochasticity because if densities drop below the threshold they will be unable to recover. The possibility of an Allee effect is an alternative explanation to demographic stochasticity for the relationship between establishment success and the number of organisms released.

Predicting whether the invasion criterion will be met is difficult as the parameters that determine the rate of population change should be measured in the species' potential new range, with its new competitors and natural enemies. The best way to obtain appropriate estimates of the demographic parameters of an invading species is to carry out the invasion. But this is often too risky a prospect—it is prudent to assume that introductions are irreversible once established. One solution is to conduct the experiments in the organism's original range, but this may severely underestimate its invasion potential. In the process of translocation, alien plants are often freed from their specialist pathogens and herbivores which are left behind in the country of origin. Alien plants often grow to larger size and live longer in their new environment than in their native countries (Crawley 1987). Two classic examples of herbivore release are the Monterey pine *Pinus radiata* and rubber *Hevea brasiliensis* (Euphorbiaceae). In its native home on the Monterey peninsula in California, *P. radiata* is often a scrubby and rather sickly little tree, but growing in plantations in New Zealand it is a straight-stemmed giant. Rubber simply cannot be grown as a commercial plantation crop in its native Brazil because of pest problems, but it flourishes under plantation conditions in South East Asia (Crawley 1997a).

With invading plants, the most likely cause of failure to establish is competition with established native vegetation (e.g. strongly asymmetric interspecific competition between seedlings of the invader and adult perennial natives). This kind of microsite limitation is critically dependent on the disturbance regime (e.g. the timing, intensity and frequency of destruction of plant biomass). Consumption of seeds, seedlings or mature plants by native vertebrate herbivores is the next most likely cause of failure. It is, however, relatively unusual for native invertebrate herbivores or pathogens to cause the exclusion of an invading plant species that has overcome the hurdles of plant competition and vertebrate herbivory (see Crawley 1990 for details).

Where it is not possible to estimate the invasion criterion directly, it may be possible to use comparative studies to identify potential invaders, and also the kinds of habitats that are likely to be most susceptible to invasion. Several predictable patterns have been identified: (i) pest species in one country are likely to become pests when introduced to another climatically matched country; (ii) invasive crop plants in one country will be invasive in other countries at similar latitudes; (iii) the rate of establishment of alien species will be proportional to the frequency and intensity of disturbance of the habitat (e.g. by alien ungulates); (iv) the higher the rate of introduction of propagules, and the greater the degree of matching of the ecological attributes of the source and target habitats, the greater the number of alien species is likely to be (Crawley *et al.*, 1996; Williamson 1996).

Species inhabiting disturbed areas associated with man are more likely to be transported to new regions, where they will tend to be deposited in very similar habitats to those from which they originated. It is not surprising, therefore, that alien plant species richness is positively correlated with proximity to human transport centres (docks, cities, railways, arterial roads) and negatively correlated with the degree of isolation of a habitat within the new, introduced range, e.g. alien plant species richness is low in mountain areas and on uninhabited islands (Fig. 3.2). Thus, the sample of species picked up is likely to be highly biased, and much of it is preadapted to human-disturbed habitats (Crawley 1997a).

While it is relatively straightforward to draw up inventories of introduced species, only in a few cases is it possible to study the probability of establishment because seldom is information available on the species that were introduced but did not become established. A total of somewhere between 20 000 and 200 000 alien plant species have been introduced into Britain. Of the 1169 naturalized alien species, about 70 have become sufficiently widespread in seminatural habitats that a visiting botanist might mistake them for natives. Only about 15 alien species are regarded as problem plants, but even amongst these species there is far from unanimous agreement as to their pest status (Williamson 1993). Thus, somewhere between 0.5% and 5% of introduced species have become naturalized,

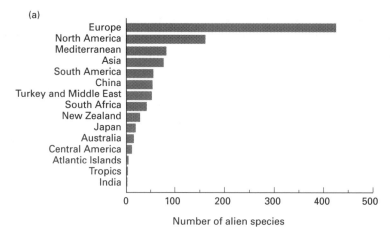

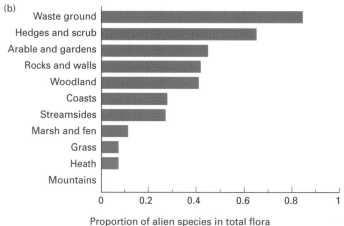

Fig. 3.2 The alien flora of the British Isles. (a) Geographic origin of alien species, showing the relative importance of proximity, trade and climatic matching (e.g. note the small number of alien species from tropical environments; these species generally lack the frost hardiness necessary to survive the British winter). (b) Invasible communities: note the large number of alien species in open, disturbed habitats close to human settlements, and the low number of species in isolated, relatively undisturbed communities.

and about 6% of naturalized species behave like natives. Even by the most generous estimate, well under 0.1% of introduced species have become pests.

What traits, if any, distinguish the native and alien species? In phylo-genetically controlled contrasts (Harvey & Pagel 1991) comparing native and alien plants in the British flora, the most clearcut difference was in plant height; alien plants were taller than their native counterparts, perhaps associated with greater competitive ability for light. The aliens also had significantly larger seeds than the natives (this is consistent with the positive correlation known to exist between plant height and seed size) (Rees 1996). The seed banks of natives and aliens were significantly different—aliens were less likely to show no seed dormancy and more likely to show protracted (> 20-year) seed dormancy than were native species. Flowering phenology was also different—alien plant species tended to flower either relatively early or relatively late in the year compared with native species, and were more likely to be pollinated by insects. Not surprisingly, recent introductions had more restricted geographical ranges within the British Isles than had long-established aliens (Crawley *et al.* 1996). The importance of climatic matching is demonstrated by the fact that the number of alien species originating from different regions declines monotonically with increasing difference in latitude between their native and alien ranges (e.g. lack of frost-hardiness in tropical and subtropical species is the most obvious cause of this pattern) (Crawley *et al.* 1996). Rejmánek & Richardson (1996) have recently used statistical methods to look for life history correlates of invasiveness within the genus *Pinus* (although the results have to be interpreted with caution as there is no control for phylogeny). Invasiveness is associated with early reproduction and frequent mast years (i.e. high intrinsic rate of increase) and also with small seed size (which may exert its influence through correlation with seed number, dispersal ability, speed of germination, etc.).

Certain communities appear to be more susceptible to colonization than others. For example, open disturbed plant communities are more invasible than closed, less disturbed communities (Crawley 1987). Animal communities on remote oceanic islands are much more invasible than equivalent communities in continental areas (Elton 1958). Vitousek (1988) suggests a number of possible reasons for this: (i) island species may exhibit reduced competitive ability as a result of founder effects and genetic bottlenecks; (ii) small populations on islands may be capable of maintaining little genetic diversity; (iii) this low genetic diversity could result in a relative lack of adaptability to change; (iv) there has often been a loss of resistance to predators, pathogens and herbivores (perhaps because many island plant species were never exposed to mammalian herbivores); (v) a loss of coevolved organisms or failure of potential mutualist ever to arrive; (vi) lack of disturbances such as fire in recent evolutionary history; (vii) intensive exploitation by humans; and, most importantly, (viii) the introduction of ungulates by ancient mariners as an insurance against shipwreck. Many oceanic islands of the tropics and subtropics support feral populations of goats, pigs, sheep, cattle and horses, with devastating impact on local endemic plant communities. Feral animals alter the disturbance regime in ways that favour the establishment of unwanted alien plants and hinder the natural regeneration of valued native species (Stone *et al.* 1992).

Changes in invasibility may result from altered land management, for example, winter salting of motorways facilitates the inland spread of coastal halophytes like *Cochlearia danica*; changing fashions in agricultural crops (e.g. the spectacular rise in popularity of oilseed rape *Brassica napus* ssp. *oleifera*); climate change (e.g. the escape of *Conyza sumatrensis* from London's 'heat island'); or changed grazing practice (e.g. the removal of cattle from upland pastures following change to agricultural subsidies, allowing the rapid spread of bracken *Pteridium aquilinum*) (Stewart *et al.* 1994; Pysek *et al.* 1995; Crawley 1997b).

3.4 Spread

Let us assume that a plant or animal has established itself in a new site and has increased in numbers sufficiently that its random extinction is now unlikely. What are the factors that determine the rate at which it increases its new range?

Clearly, the availability of suitable habitat will be a major determinant of range expansion. However, it is helpful to begin by considering the spread of an organism through a uniform habitat. To predict the increase in range, we need to know something about local population dynamics, and something about the dispersal behaviour of the animal. Let us assume that the population increases when rare at a rate governed by its intrinsic rate of increase, $r = \ln(\lambda)$, but that its growth rate declines with increasing population density so that the population reaches a stable equilibrium density, K, its carrying capacity (i.e. a logistic population model, Fig. 3.1b). Now we need to describe the pattern of dispersal. Consider a particular individual born at a certain spatial location and consider the spatial distribution of its progeny. If we were dealing with a sedentary organism such as a plant, we might ask where its seeds will germinate, while if we were dealing with a mobile animal, we might ask where its young will be born. A simple assumption, although not necessarily the most realistic biologically, is to assume the young are normally distributed in space with the centre of the distribution at the birth site of the parent. Given that we have specified the centre, the distribution of the young is completely described by the variance of the distribution. There are several measures for this, including the diffusion coefficient D which is related to the average distance moved by an individual (technically, D is one-half the mean squared distance moved in a time unit) (Okubo 1980).

The model described above is known as a reaction–diffusion equation (Okubo 1980) and has been extensively studied in the physical sciences. It was introduced to ecology by Skellam (1951) who used it to explain the rate of spread of the muskrat *Odontra zibethecus* through Europe from an initial introduction near Prague (Fig. 3.3). The model predicts that the advancing front of the animals should be a travelling wave of constant shape, moving at a constant speed of $2\sqrt{rD}$ the units being those in which D is measured. If the front advances at constant speed, this implies that a plot of the square root of the total range against time should be linear. Note that if the initial inoculum is small, there will be a lag (the establishment phase) during which the population builds up before range expansion begins (Shigesada *et al.* 1995).

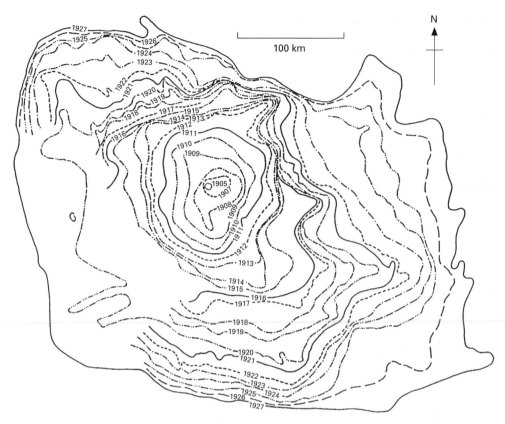

Fig. 3.3 The spread of muskrats from a focus of a few individuals released near Prague in 1905. (From Elton 1958.)

Skellam (1951) found that the rate of advance of the muskrat was roughly locally constant, although the speed of the travelling wave was influenced by local topography (the muskrat is a swamp lover and disperses more slowly in uplands). At the broadest geographical range, the relationship between the square root of area and time was remarkably linear and similar results have been found for many other biological invasions (Elton 1958; Hengeveld 1989, 1990). A second good example of the spread of a mammal is Lubina & Levin's (1988) study of the sea otter *Enhydra lutris* off the coast of California (Fig. 3.4). This was not an introduction, but a reinvasion after near elimination. The rate of advance of the animal was constant within similar habitats, but there were differences among habitats and in the average speeds of the northerly and southerly advancing fronts. Spatial heterogeneity such as this is likely to influence the course of most invasions. One interesting theoretical result is that the average rate of range expansion through a habitat composed of random patches of variable quality is given by $2\sqrt{r_a D_h}$ where r_a is the *arithmetic* mean of the intrinsic growth rate across patches, and D_h is the *harmonic* mean of the diffusion parameter across patches (Shigesada *et al.* 1986).

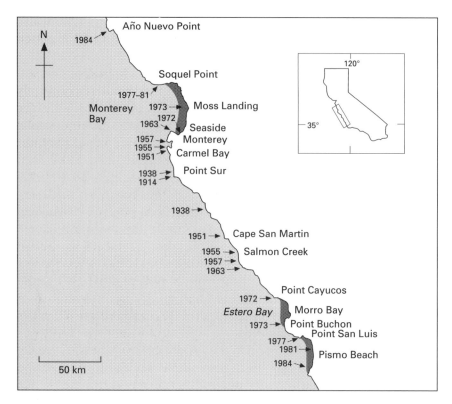

Fig. 3.4 Expansion of the range of the sea otter in California. The darker shade indicates soft-bottom habitats. (From Lubina & Levin 1988.)

While these results are encouraging, they do not prove that the underlying reaction–diffusion model is correct since several different mechanisms might give rise to travelling waves of constant speed. A more stringent test of the theory is to measure r and D experimentally and to try to predict the speed of the travelling wave. The first attempt to do this was by Andow *et al.* (1990) who studied three systems: the muskrat, the cereal leaf beetle *Oulema melanopa* (a European chrysomelid beetle introduced into the USA in 1958) and the small white butterfly *Pieris rapae* (a European pest of cabbages introduced several times to North America in the last century). In each case, they obtained estimates of r and D from a life-table and dispersal experiments and estimates of the travelling wave speed from distribution maps. For the muskrat, they studied the invasion centred at Prague (see above), but also the spread from separate foci in France and Finland. The speed of advance was quite variable, ranging from 1 to 25 km/year, while the predicted parameter estimate was 6–32 km/year—a reasonable fit. For the cereal leaf beetle, the experiments predicted a rate of spread of approximately 1.6 km/year, while the observed rate of spread was much greater, between 26 and 90 km/year. Clearly, the diffusion approach fails here. Finally, the predicted spread of the small white butterfly was 13–

127 km/year, while the observed spread was 15–170 km/year—a reasonable agreement, although with considerable uncertainty in both the expected and observed rates.

Despite the obvious simplifications of the diffusion models, they performed reasonably well in two out of the three cases, but failed decisively in the third. Diffusion models assume that range expansion occurs as the sum of a series of small steps, and that dispersal distances are normally distributed. However, most measurements of actual dispersal distances suggest that the distribution is leptokurtic, that is with longer tails than expected with the normal distribution. Van den Bosch et al. (1990, 1992) have constructed a series of models in which they retain the small step assumption, but include added biological detail in the population dynamics and dispersal (their models are phrased as integro-differential equations) (see also Hengeveld 1994). For example, they couple a normal distribution of movement with a constant probability of settling which gives rise to a distribution of dispersal distances that is more leptokurtic than in a simple diffusion model. When the dispersal distribution is estimated from movements within the range of the species, the model is impressive in its ability to predict the rate of spread of a number of different bird species.

However, the reason why Andow et al. (1990) believed that their model failed to predict cereal leaf beetle movement was that it ignored the possibility of rare, long-distance dispersal, some way in advance of the main advancing front. Models that can cope with such events have been discussed for some time by epidemiologists (Mollison 1977), but have only recently been considered by ecologists. Neubert et al. (1995) have analysed how the distribution of dispersal distances influences the rate of invasion using a model that assumes discrete generations (an integro-difference equation). They show that the rate of spread can depend critically on the relatively few individuals that disperse a long way. It is unfortunate that this is the hardest part of the distribution on which to obtain good information. The potential value of this approach is illustrated by Veit & Lewis' (1996) analysis of the spread of the house finch Carpodecus mexicanus in the New York City area. A native of the western USA, in 1940 a ban was introduced in the trade of house finches as pets and, in response, a large number were illegally liberated, so that by 1950 there was a population of around 250 in New York. During the next 10 years, the bird spread slowly into New Jersey and Connecticut, but then its range expansion accelerated rapidly and in the last 25 years it has colonized most of eastern USA (Fig. 3.5). At the same time, the western range of the bird has also increased. Veit and Lewis' model of local population dynamics included a carrying capacity set by competition for nesting spaces, and an Allee effect caused by the difficulty of finding mates at low density. They estimated movement patterns using continent-wide ringing (banding) data which suggested a markedly leptokurtic dispersal distribution. They also assumed that the probability of dispersing at all was density-dependent.

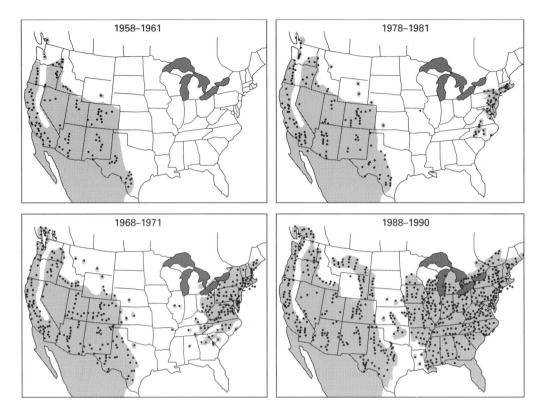

Fig. 3.5 Spread of the house finch in North America. The points are confirmed sightings and the shaded region the inferred range. (From Veit & Lewis 1996.)

The model provided a good description of the spread of the house finch, in particular predicting that the rate of spread would accelerate over time, and to a lesser degree predicting the increase in density at the centre of the range. The model also provides a good qualitative description of the texture of the range expansion which consists of a ragged expansion with some populations being established in advance of the main range, rather than the stately advance of a travelling wave. The acceleration in range expansion is largely caused by the combination of the Allee effect and density-dependent dispersal. Veit & Lewis suggest that these factors may also be responsible for other invasions with similar patterns (for example the collared dove *Streptopelia decaocto* in Europe; the Himalayan thar *Hemitragus jemlahicus* in New Zealand; and the Japanese beetle *Popillia japonica* and starling *Sturnus vulgaris* in North America).

Another way of combining dispersal at different scales is through stratified diffusion. Cheat grass *Bromus tectorum* is an introduced grass in northwestern USA and Canada that exists in scattered colonies where it is locally abundant. Over time, the number of colonies has increased, as have their individual sizes (Mack 1981). If the square root of the total range (i.e. summed across all colonies) is plotted against time, then one obtains an

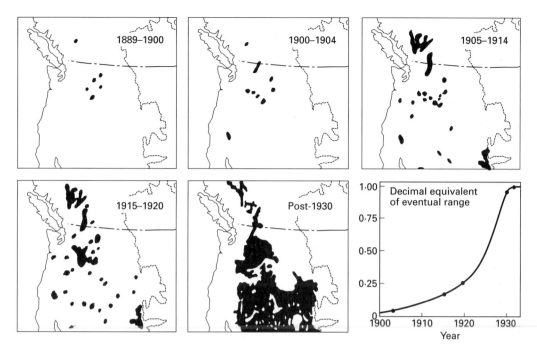

Fig. 3.6 Expansion of the range of cheat grass *Bromus tectorum* in northwestern North America. (From Mack 1981; Hengeveld 1989.)

accelerating curve rather than the linear relationship predicted by simple models (Fig. 3.6). A model that combines the diffusive spread of individual colonies (as in the Skellam model) plus rare long-distance dispersal events can explain this pattern (Shigesada *et al.* 1995). Interestingly, the model resembles the classic age-structured models of population ecology with colony foundation taking the place of birth and colony growth taking the place of ageing. If new colonies are established not too far from existing colonies, then in time neighbouring colonies will coalesce and the increase in the square root of range will change from accelerating to linear. Both types of spread (travelling wave and 'starburst') are seen as two ends of a single continuum; the key parameter is the distance between successive foci of establishment. When this distance is small, the assemblage behaves like a travelling wave, but when it is large starburst effects predominate.

3.5 Consequences

An organism has successfully colonized and spread in a new area. What does ecological science tell us about the consequences of the introduction for the existing communities? To phrase the question in a different way, can general statements be made about the impact of invasion by certain classes of organisms or must each introduction be treated as a special case? There are two ways in which this question can be approached. First from a consideration of ecological theory and principle; and second statistically, through an analysis of the observed consequences of the introduction of different types of animals (Williamson & Brown 1986; Kornberg & Williamson 1987; Drake *et al.* 1989; Williamson 1996).

It is useful to think of a sequence of types of consequence of increasingly radical importance. A novel organism that establishes itself in a new habitat may influence the population dynamics of any species on which it directly feeds, and any species for which it acts as prey. It is also plausible that native species with which it competes for food, space or other resources will be affected by the introduction. Harder to decide is how the introduction will affect other species with more tenuous trophic links to the new species. It is possible that a specific natural enemy of a competitor of the introduced species declines in abundance because its prey or host density is reduced. Similarly, the introduction of a novel species may lead to an increase in abundance of its predators and hence increased mortality of the original native prey (apparent competition) (Holt & Lawton 1994). Such chains of interactions could extend some way through the community, possibly damping, but conceivably amplifying, through feedback loops. Finally, the consequences of the introduction may extend to radical ecosystem functions that affect directly or indirectly all species in the community. This would occur if the introduced species had a major effect on such things as nutrient cycling, soil stability or the susceptibility of the community to fire.

There are many examples where introduced animals have severely affected the densities of their food or prey in their new range. Numerous introduced herbivorous insects have become major pests of crops and there are some, although fewer, examples where herbivorous insects have become important defoliators of plants in natural communities (Crawley 1997a). The introduction of fish such as the Nile perch *Lates nilotica* into African lakes has caused a dramatic drop in the density of native prey species, including the probable extinction of some 200 species of haplo-chromine cichlids (Witte *et al.* 1992; Goldschmidt *et al.* 1993). Now, it is quite true that native herbivores and predators can also defoliate their food plants or severely reduce their preys' density, but there does seem to be a strong disposition for this to occur with introductions. A common explanation is that in moving to a new region, organisms escape their natural enemies. Whereas in their native range, natural enemies regulate consumer numbers at densities some way below the carrying capacity set by resources, in their absence numbers rise and obvious resource depletion is observed. We have already discussed natural enemy release as a factor that may influence establishment. Clearly this hypothesis could be tested by the subsequent introduction of the natural enemy. While they are not rigorous experiments, natural enemy introductions as part of biological control programmes provide interesting evidence on this question. Greathead (1995) estimated that approximately 10%

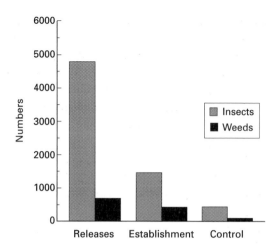

Fig. 3.7 The number of insects released as biological control agents against other insects and against weeds, with the numbers that became established and exerted good control.

of biological control introductions against insect and plant pests led to economically successful control (Fig. 3.7). These data provide strong support that natural enemy escape is often responsible for severe resource exploitation by introduced consumers, although caution must be exercised against reading too much into these data. Almost never in biological control programmes (for quite understandable reasons) are control treatments set up. A newly arrived pest causing substantial economic damage often results in the rapid implementation of biological control which may be followed by a decline in pest density. But pest densities may have fallen anyway, perhaps due to local natural enemies being recruited to the introduced species, or possibly through changes in host plant genetics or demography.

Just as interesting as the introduced species that have major effects on their resources are the numerous introductions that have barely perceptible effects. In the last 150 years, approximately 50 species of Lepidoptera have colonized the British Isles, mostly from continental Europe but some from further afield. In no case has the introduction been linked with substantially increased herbivory. Closer examination of individual examples suggests the reasons why. Two leaf mining moths (*Phyllonorycter* spp.) that invaded in the mid-1980s accumulated a parasitoid fauna (probably the most important natural enemies for this type of animal) of almost identical size and impact to that of closely related native species within a few years (Godfray *et al.* 1995; Nash *et al.* 1995). Britain has also been invaded by a number of species of cynipid gall wasp that attack the English oak *Quercus robur*. The majority of these introductions have had little effect on their host plant, probably because they are attacked by native natural enemies (again parasitoids). An exception is the species *Andricus quercuscalicis* whose asexual generation forms 'knopper galls' on acorns. These galls are quite unlike those made by native species, and perhaps in consequence were not colonized by native natural enemies, leading to high gall densities and

a substantial reduction in the acorn crop. Interestingly, in recent years, parasitoid and inquiline species have begun to attack knopper galls, and it appears that it has taken some time for native natural enemies to adapt to a potential new resource (Schönrogge *et al.* 1995, 1996). These examples suggest that an introduction will only lead to resource exploitation if the invader is sufficiently dissimilar to species already present in its new locality that it is not attacked by local natural enemies.

There are many examples where introduced organisms have had substantial impacts on native species through competition. In plants, the importance of competition from introduced species varies markedly from area to area. The direct role of invasive alien plants in causing loss of native plant species is relatively trivial in Britain (forestation with exotic conifers accounts for about 4% of reported cases of species decline, while overgrowth by aggressive alien plants like *Spartina anglica* and *Fallopia japonica* is cited in about 1% of cases) (Crawley 1997b). In contrast, the islands of Hawai'i boast more than their share of problem aliens (Vitousek *et al.* 1988; Stone *et al.* 1992):

1 Species which 'short circuit' primary succession by recruiting directly onto bare lava (e.g. the nitrogen-fixing tree *Myrica faya* (Myricaceae)).

2 Vines which smother native trees (e.g. *Passiflora mollissima* (Passifloraceae) and *Rubus argutus* (Rosaceae) which can overrun forests of *Acacia koa* (Fabaceae) or *Metrosideros polymorpha* (Myrtaceae)).

3 Thicket-forming, evergreen woody species which exclude all native plants (e.g. *Psidium cattleianum* (Myrtaceae), *Leucaena polycephala*, *Ulex europaeus* (both Fabaceae) and many others).

4 Trees which replace native canopy dominants in forest reserves (e.g. *Spathodea campanulata* (Bignonaceae) and various melastomes including *Miconia calvescens*).

5 Replacement of the native forest understorey (e.g. *Hedychium gardnerianum* (Zingiberaceae), *Lantana camera* (Verbenaceae) or the Australian tree fern *Cyathea cooperi*).

The seeds of many Hawaiian alien plants are dispersed deep into the heart of natural habitats by alien, fruit-feeding birds. Recruitment conditions favouring regeneration of the aliens are provided by feral populations of alien ungulates (principally pigs and goats but there are feral wild populations of cattle, sheep and horses as well) (Stone *et al.* 1992).

A number of factors influence the ability of an introduced plant to outcompete natives. First, the plant may simply be a superior competitor, possibly because it has evolved in a plant community or environment where there was a greater selective advantage in being a strong competitor. Second, the plant may have escaped herbivores and pathogens and hence be in better condition to compete with other plant species. Lastly, the plant may obtain an advantage not only through direct competition, but by altering the physical environment in a way that is conducive to its own success; something we return to below.

To what degree does the introduction of a novel organism influence species beyond those with which it interacts directly? Most ecologists have favourite examples of involved chains of influence, biological counterparts of the old rhyme proving that 'for want of a nail the kingdom was lost'. To give one example, the introduction of the myxoma virus into Britain in the 1970s decimated the rabbit population (itself a much older introduction). The loss of rabbits led to reduced grazing and a change in the floral composition of calcareous grasslands. A consequence of the new plant assemblage was a reduction in the abundance of certain ant species. These ants reared the larvae of a rare lycaenid butterfly, the large blue *Maculinea arion*, which is unable to survive without the ant and subsequently became extinct in the British Isles (Elmes & Thomas 1992).

Community ecologists have recognized that certain species in a community have particularly important effects on overall community structure and have termed them keystone species. The concept originates with studies of food webs of seashore animals, where it was found that removing a top predator could lead to a marked reduction in species diversity as the competitive dominant, free of predators, increased in abundance (Paine 1988). In many grassland communities, the keystone species is a dominant herbivore, for example rabbits in European grassland or (formally) buffalo (bison) in the American prairies whose grazing maintained a characteristic floral assemblage (Shelford 1963). A different kind of keystone species are those plants that can arrest the normal course of succession. Large areas of the tropics are covered by monocultures of introduced grasses which inhibit the growth of all potential competitors (Kareiva 1996). Thus, in assessing the likely risk or importance of an introduction, a critical issue is whether it will influence the abundance of a keystone species, or whether the introduction might itself become a keystone species; this, of course, is easier said than done.

The types of interaction we have discussed so far are largely mediated through population dynamics. Perhaps the most radical consequence of the introduction of a novel organism is a change in the abiotic environment in which the whole community exists. For example, alien perennial grasses can pose a serious fire risk to native trees in forest ecosystems, often representing twin hazards: (i) they produce a much greater bulk of flammable material than was produced by the native plants and therefore cause more intense fires; and (ii) the timing of their flammability is different (e.g. they stay green further into the dry season) so that fires occur at different times of year, with further damaging consequences to native vegetation (Cowling 1992). The incursion of alien grasses along roadsides allows fires to spread into previously unburned (and hence extremely vulnerable) parts of native forest ecosystems (Vitousek 1988). Invasive trees are potentially disastrous for nature conservation and watershed management. Alien invasive trees are the canopy dominants in many parts of the tropics, especially on islands. Some of the most highly valued conservation areas in the world are seriously jeopardized by invasive trees. For example, the South African fynbos has one of the richest floras on Earth and is threatened by a suite of trees, including pines from America (*Pinus radiata*) and the Mediterranean (*P. halepensis*), as well as a host of woody plants from Australia, including proteas (*Hakea, Grevillea*) and legumes (several species of *Acacia*). Watersheds dominated by alien trees transpire much more water than the native vegetation and water yields are greatly reduced. This has major economic consequences (Cowling 1992). A zoological example concerns the reintroduction of beavers (*Castor fiber*) to areas where they have been absent for centuries; beavers, the civil engineers of the animal world, can completely change the hydrology and general abiotic environments of large tracts of land (Naiman *et al.* 1988). Jones *et al.* (1994) review many examples of animals and plants that influence ecosystem structure and function in an article entitled 'Organisms as ecosystem engineers'.

It is abundantly clear, however, that the real keystone species in almost any environment is man. Where man has altered the environment, species from other regions adapted to the new conditions are likely to establish and spread. Probably the majority of introductions, plant and animal, are only successful because man has altered the environment to their benefit. Perhaps the clearest example of this comes from weeds of arable and wasteland. In a blind test, a botanist would be hard pressed to identify the provenance of a square metre of turf dug up from the UK and from the east coast of New Zealand because of the number of shared weeds (Wardle 1991). It has been argued that the long history of agriculture in Europe and Asia has led to selection for plants adapted to do well in association with man, and certainly more Eurasian temperate species have invaded Australasia and America than vice versa (Beard 1969; Fernald 1970). Grazed, mesic grasslands in Britain represent something of a paradox; they support no alien plant species at all, and yet these same grasslands were the source of many of the most pestilential pasture weeds introduced into other parts of the world (*Hypericum perforatum, Cirsium arvense, Senecio jacobaea, Pilosella officinarum, Plantago lanceolata, Rumex acetosella, Ulex europaeus, Cytisus scoparius* and *Hypochoeris radicata* in Australia, South Africa, the Pacific islands and the Americas). The probable cause of this asymmetry is the long association between Old World grasslands, people and grazing ungulates, and the importance of introduced domestic livestock as primary agents of disturbance in New World habitats, paving the way for invasion by preadapted, grazing-tolerant, Old World pasture species when these were introduced in hay and seed mixtures imported by the early European settlers. But the passage is not all one way; there are a number of American and New Zealand wasteland weeds in Europe, and the asymmetry may partly be due to differential transport. Interestingly, the exchange of weeds associated with man between the Old and New World tropics is far less biased (Holm *et al.* 1977; Crawley 1997b).

In this section, we have studied the community consequences of the introduction of a new species, beginning with simple trophic interactions and finishing with ecosystem engineering. We end by discussing a potentially more subtle consequence of some invasions: the introduction of new genetic material into the environment. One of Europe's rarest birds is the white-headed duck *Oxyura leucocephala* which is restricted to small populations in southern Spain and the Danube basin (there is also a small population in Turkey). A much commoner bird is the closely related ruddy duck *O. jamaicencis* which was introduced from North America and is found commonly in Britain and northwestern Europe. Unfortunately, if the two species meet they interbreed, and the abundance of the ruddy duck suggests that the outcome would be the eventual extinction of the native species by genetic dilution (Fig. 3.8). To prevent this, programmes have been set in place designed to eliminate the introduced species. Such programmes are not without problems: as we write a British conservation organization, the Royal Society for the Protection of Birds, has agreed to cull ruddy ducks in the UK at a meeting that was picketed by animal rights campaigners arguing against any slaughter of 'wild' birds. A second example of a conservation problem caused by genetic contamination is the red wolf *Canis rufus* in southeast North America which hybridizes with the coyote *C. latrans* (Jenks & Wayne 1992). The picture here is complicated by the fact that the red wolf may itself have arisen as a hybrid between coyotes and gray wolves *C. lupus* (Wayne & Jenks 1991).

Hybridization is even more important in plants than animals. *Spartina alternifolia* is a saltmarsh grass which was introduced to the British Isles from North America where it hybridized with a local native species, *S. maritima*. The infertile hybrid *S. × townsendii* spread by vegetative propagation, but at the end of the last century underwent chromosome doubling to produce a new, fertile species, *S. anglica*. *S. anglica* has now largely replaced *S. alternifolia* and *S. × townsendii*, and has reduced substantially the range of *S. maritima*. Moreover, it has colonized areas of muddy salt flat that were previously devoid of any higher plant (Gray *et al.* 1991). *Senecio squalidus*, Oxford ragwort, was introduced into the Oxford Botanical Garden in 1794 and has spread, reputedly along railway lines, throughout much of England and Wales. It too has hybridized with a native species, groundsel *S. vulgaris*, which, through chromosomal rearrangement, has become a new species, *S. cambrensis*. Curiously, the species has arisen on at least two occasions (once in north Wales and once in central Scotland), an example of a disjunction between phyletic and biological species concepts (Ashton & Abbott 1992).

These examples make it clear that, where two species are sufficiently similar, genetic introgression can have important consequences for native species and, at least in plants, can give rise to new, sometimes invasive species. But what are the risks of introducing new genes through genetic manipulation?

(a)

(b)

(c)

Fig. 3.8 Males of (a) the American ruddy duck, (b) the European white-headed duck, and (c) a fertile second-generation hybrid produced as a result of the release of ruddy ducks into Europe. (Photos by M. Hulme, Wildfowl and Wetlands Trust.)

3.6 Introduction of genetically engineered organisms

There is enormous current interest in the application of genetic engineering in agriculture and biotechnology which offers the prospect of designer crops expressing specific genes, and of enhanced effectiveness of biological control agents. Examples in plants include the insertion of genes for herbicide resistance, insect toxins, virus resistance and the production of proteins that enhance fruit and vegetable shelf-life. Valuable proteins such as pharmaceuticals can be produced in large quantities by insertion of appropriate genes in plants, animals or microorganisms; for example sheep can be engineered to secrete a foreign protein in their milk. Microorganisms have been engineered to produce superior viral insecticides, fungal antagonists or to improve plant frost resistance. The introduction of a 'transgene' creates a new patentable construct, and hence is an attractive commercial venture. But what are the risks of releasing genetically manipulated organisms (GMOs) into the field?; how can we be sure we are not creating Frankensteins?

The risks involved in the release of GMOs fall into two categories: (i) the general problems associated with the release of any novel organism; and

(ii) the specific problems arising from the genetic manipulation. As regards the first type of risk, exactly the same conceptual issues are involved as in the introduction of an alien organism (see above). Judging the risk of the invasion of a GMO is simplified if the unmodified organism has already been shown to be non-invasive (for example if an unmodified crop plant has been widely grown without escaping into natural habitats). Then what needs to be shown is that the genetic manipulation does not enhance invasibility. Generally, genetic manipulation for agricultural purposes is unlikely to increase Darwinian fitness, and Darwinian fitness is likely to be highly correlated with invasiveness; but there always will be exceptions and such arguments, while comforting, cannot be used as a substitute for specific studies and experiments. For example, introducing frost-resistance genes into plants may allow them to increase substantially their range into colder latitudes. Assessing the risks of GMOs that are known to be invasive (the manipulation may be designed to reduce this risk), or of organisms introduced to an area for the first time, is much more difficult and involves many of the problems discussed earlier in this chapter.

The second type of risk relates specifically to the genetic manipulation. The most straightforward problems relate to whether the introduced gene expresses a protein that is toxic to humans or non-target beneficial organisms. These are relatively easy to assess in the laboratory. More subtle are problems relating to the mobility of the transgene. To place the gene in the target organism, a vector such as a transposon or virus is often used. But if the transgene can 'jump in' then perhaps it can 'jump out' and even cross species barriers. Such risks are particularly acute in the case of engineered prokaryotes and viruses where genes can be exchanged between phylogenetically quite distant strains. More prosaic gene mobility problems arise if the GMO can mate or exchange pollen with conspecifics or related species. Rape (canola; *Brassica napus* ssp. *oleifera*) is an important target crop plant for bioengineers, and the ease with which hybrids can occur within the genus is a potential problem. Manipulated plants may act as novel selection pressures on their natural enemies or diseases which may evolve to become even worse pests or pathogens. In 1996, over 2 million hectares in the USA (13% of the national crop) were planted with an engineered form of cotton that includes the *Bacillus thuringiensis* (BT) insect toxin. The aim is to protect the plant from moth larvae, particularly *Heliothis armigera* (cotton bollworm), *H. virescens* (tobacco budworm) and *Pectinophora gossypiella* (pink bollworm). To counter the risk of the moths developing resistance (and thus rendering traditional BT sprays on other crops useless), an elaborate resistance management scheme has been implemented which relies on the manipulated crop killing nearly all pests that feed on it. However, whether this is working is unclear as recent reports suggest heavy infestations of *H. armigera* on the manipulated crops (Macilwain 1996). Not only does this undermine the economic value of the engineered crop, it also makes the evolution of resistance more likely.

As with introduced alien organisms, there are enormous practical difficulties in assessing the environmental risks of GMOs (Tiedje *et al.* 1989; Mooney & Berbardi 1990; Williamson 1992, 1994; Kareiva & Stark 1994; Godfray *et al.* 1995; Kareiva *et al.* 1996). To date, there have only been a few studies explicitly designed to compare the ecological performance of manipulated and natural organisms. In the largest study, Crawley *et al.* (1993) estimated the invasion criterion for several strains of rape (canola) containing transgenes over a wide variety of ecological and geographical conditions. Except in ploughed fields that are protected from herbivores by fences, the plant was not invasive and the presence or absence of the transgene had no effect on its intrinsic growth rate. This very large field study involved considerable resources, which led Parker & Kareiva (in press) to ask whether the estimates of invasiveness could have been obtained using fewer sites (<12) or in a shorter period of time (<3 years). They found that, while the number of sites was not critical, obtaining data on plant growth over a number of years (ideally more than three) was essential.

Bacteria in the genus *Pseudomonas* live on the surface of leaves and contain a protein that acts as a nucleus for ice formation. The ice ruptures the surface of the leaf causing damage and sometimes death, but also providing nutrients for the bacteria. Strains of *Pseudomonas* have been developed—ice-minus genotypes—that lack the nucleation protein. Can these be sprayed on the plant to replace the ice-plus genotype and thus increase the plant's ice tolerance, but without escaping into the environment? Kareiva *et al.* (1996) have carried out ecological competition experiments in which they have modelled the competition between ice-plus and ice-minus genotypes. The results are encouraging: the ice-minus strain is competitive enough to persist for some time on the plant, but is likely to be eliminated eventually and thus not persist in the environment. However, as the authors suggest, these conclusions come from only a single experiment in one environment, and extrapolation is dangerous.

Baculoviruses are insect viruses that are useful as biological control agents of pests. However, they take some time to kill their hosts and this limits their attractiveness to farmers and foresters. Insect-specific toxins have been engineered into the viruses that cause the much speedier death of the host. The changes made to the virus severely reduce its intrinsic rate of increase and thus make it unlikely to survive in the wild. This prediction is being tested by semifield studies in large cages of a virus called AcNPV which contains a gene for a toxin isolated from a scorpion. The results so far support the prediction that the manipulated virus is ecologically crippled (Cory *et al.* 1994). This is a further example of an attempt to assess ecologically a transgenic organism, but is also notable for the intense public concern it has raised—in part due to the emotive conjunction of scorpion toxins, viruses and genetic manipulation (Godfray 1995).

Like it or not, transgenic organisms are probably here to stay, and thus ecologists must contribute to the debate about their safety. It should also not be forgotten that some classes of GMOs could have major environmental benefits, for example reducing the farmer's reliance on synthetic chemical pesticides. Comparisons of GMOs with alien introductions give us both cause to worry and some comfort. Some alien introductions have caused enormous ecological damage, but many other classes of alien introduction, for example insect biological control agents, have been largely or wholly benign. It could be argued that the attention paid to GMOs is disproportionate compared with most countries' lax quarantine regulations for the importation of non-manipulated alien organisms. While the ability of genetic engineering to create problem plants is not in doubt, we are sanguine that the chances are small that well-regulated, ecologically informed and carefully overseen genetic engineering will produce unintentional problems.

CHAPTER 4

Pollutants and pesticides

Ian Newton

4.1 Introduction

The many toxic chemicals added continuously to the natural environment, either as pesticides, industrial effluents or combustion emissions, have had some devastating impacts on the natural world, as well as on human well-being. Some of these chemicals are entirely man-made, while others are natural substances whose concentration in the biotic environment has been increased by human action. Some such chemicals have now become important agents of plant and animal population declines, influencing distribution and abundance patterns on both local and widespread scales.

Three main mechanisms of population decline have emerged. First, some pollutants act directly by imposing an additional form of density-independent breeding failure or mortality. This extra form of loss reduces population levels only if it adds to natural losses, and is not offset by reduced losses from other causes. Second, other pollutants act indirectly on animals by reducing food supply and thereby causing declines in numbers. Third, other chemicals, notably atmospheric pollutants, act by altering the physical or chemical structure of habitats, making them less suitable for certain species.

4.2 Some general points about pesticides

The use of pesticides has done much to increase food production around the world and to improve the appearance of fruit and vegetables. But it has also caused contamination of soil and water, human health problems and declines in non-target species of plants and animals. If pesticides destroyed only the target pests, and then quickly broke down to harmless by-products, the problems would be minimal. But most pesticides are non-specific, killing a wide range of organisms. Second, while some break down rapidly, others last for weeks, months or even years in animal bodies or in the physical environment. Third, some fat-soluble pesticides readily pass from prey to predator, causing secondary poisoning, or even pass along several steps in a food chain, affecting animals far removed in trophic position from the target pest. Their chemical properties place all pesticides somewhere within this three-feature spectrum of variation, with respect to specificity, persistence and propensity to bioaccumulate. In addition, by contaminating soils, water and atmosphere, pesticides (like other pollutants) can reach areas, and affect biotas, far removed from their points of application (witness the presence of dichlor-diphenyl-trichlor-ethane (DDT) residues in penguins and other Antarctic birds (George & Frear 1966)). Other problems are caused by pesticide manufacture which, through accidents and discharges, has led to pollution of rivers, lakes and coastal areas, with loss of aquatic life. And these problems are accentuated by the many local accidents and abuses, excessive application, drift and careless disposal. Some of these same points apply to some other chemical pollutants, but because such chemicals were not designed as biocides, and have not been deliberately applied to large land areas, the widespread problems they cause have been slower to appear.

Pesticide usage is increasing year by year, both in terms of the area treated, the number of applications per year, and the variety of chemicals in use. For all these reasons, then, we can expect that pesticide impacts are likely to have increased progressively over the past 50 years, and will continue to increase into the future. Moreover, pesticides are now manufactured and marketed mainly by international companies, so that the same chemicals are used simultaneously in many parts of the world. This, together with the fact that some pesticides can be transported in the bodies of migratory animals or in air or water currents, means that few parts of the globe are still entirely free from their effects.

The non-specificity of certain pesticides has been examined in experimental trials. One of the earliest involved the application of a commonly used carbamate insecticide, carbaryl (trade name Sevin), to one of two adjacent 1-acre (0.4 ha) patches of millet (Barrett 1968). Applied in mid-summer, the carbaryl remained toxic for only a few days, but in this time the numbers and biomass of the arthropods declined almost to nil (Fig. 4.1). Their populations took 7 weeks to recover in biomass, and longer to recover in numbers and diversity. The chemical is also likely to have affected other organisms not studied, including microarthropods, earthworms and other soil dwellers, as evidenced by the build-up of leaf-litter. Similar results were obtained in trials with other chemicals, some of which confirmed another effect common on farmland. When arthropod populations recover from a pesticide application, certain species sometimes become more abundant than before, and even previously rare species can increase to pest proportions. This can happen when the arthropod predators take longer to recover than their prey. This can push the farmer to use even more pesticide—the so-called pesticide treadmill.

The phenomenon of secondary poisoning by pesticides first became apparent in the USA in the late 1940s when DDT was applied to elm trees to control the insect vectors of Dutch elm disease (Barker 1958). The insecticide was picked up from leaves and soil by earthworms which

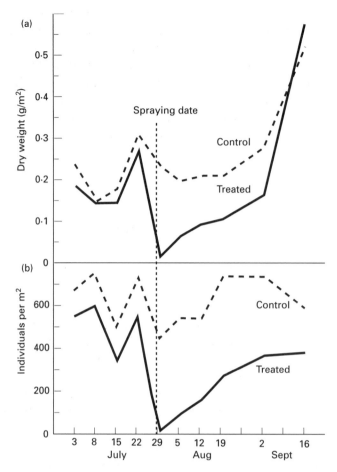

Fig. 4.1 Impact of one application of the insecticide carbaryl (applied at 4.5 kg/ha) on the invertebrates of millet. (a) Shows the effect on biomass; (b) shows the effect on arthropod numbers. (From Barrett 1968.)

were then eaten by American robins, *Turdus migratorius*, with lethal consequences. The dead birds were found to contain 50–200 ppm residues in the brain, a fatal dose obtained by eating about 100 worms. Most deaths occurred, not at the time of spraying, but in the spring or after heavy rain, when worms were most available to the birds. On the campus of Michigan State University, the robin *Turdus migratorius* population was reduced from 370 to 15 individuals, and virtually no young were raised (Wallace *et al.* 1961). Such secondary poisoning is particularly associated with DDT and other organochlorines, but has been recorded from some other pesticides.

The organochlorines are particularly noted for their tendency to pass up food chains. Increasing concentrations at successive trophic levels were evident in many (but not all) studies in which different kinds of organism from the same place were examined (Newton 1979). Within areas, concentrations were lowest in plants (first trophic level), higher in herbivorous animals (second trophic level), higher still in carnivores (third trophic level) and so on up the food chain. It is not trophic level as such that is important, but the rates of exposure (or intake) which, because of accumulation, tend to be higher in carnivores. In addition, rates of accumulation were often greater in aquatic than in terrestrial systems because many aquatic animals absorb organochlorines through their gills, as well as from their food. Fish rapidly pick up fat-soluble pollutants from water, and concentration factors of 1000 or 10000 times between water and fish are not uncommon (Stickel 1975). At the other extreme, the bioaccumulation properties of carbamate pesticides are almost zero, although some can last for years in the physical environment (Brown 1978).

4.3 Direct impacts on populations

4.3.1 Organochlorine pesticides and predatory birds

Two direct effects of pesticides on bird population levels can be illustrated by different types of organochlorine compound. This group of chemicals contains such well-known insecticides as DDT, and the more toxic cyclodiene compounds, such as aldrin, dieldrin and heptachlor. DDT was first introduced into widespread agricultural use in the late 1940s, and the cyclodienes after 1955. For a time they were widely used throughout the 'developed' world, but during the 1970s and 1980s, they were banned progressively in one country after another as their environmental effects became increasingly apparent. Now they are used mainly in tropical and subtropical areas with few restrictions.

Being fat-soluble and persistent, these chemicals readily accumulate in animal bodies, and pass from prey to predator, concentrating in meat-eating and fish-eating species. In fact, three groups of bird have been particularly affected by organochlorines, namely: (i) raptors, especially bird-eating and fish-eating species, such as the peregrine *Falco peregrinus*, sparrowhawk *Accipiter nisus*, osprey *Pandion haliaetus* and bald eagle *Haliaeetus leucocephalus*; (ii) various other fish-eating birds, such as cormorants *Phalacrocorax* and pelicans *Pelecanus*; and (iii) various seed-eating species, such as doves *Columba* and *Streptopelia*, geese *Anser* and *Branta* and cranes *Grus*, which eat newly sown seeds of cereals and other plants which have been treated directly with organochlorines (as a protection against insect attack).

DDT is not especially toxic to birds, and very high exposures (as in the early forest spraying in North America) are needed to kill birds outright. The main effects are on breeding (Fig. 4.2). At a sublethal level, DDE (the main metabolite of DDT) reduces the availability of calcium carbonate during eggshell formation, so that the eggs are thin-shelled and break, thus decreasing the reproductive rate. Other thin-shelled eggs survive incubation, but the embryo dies from dehydration caused by excess water

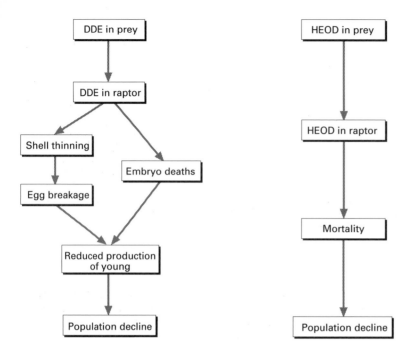

Fig. 4.2 Modes of action of DDE (from the insecticide DDT) and HEOD (from aldrin and dieldrin) on raptor populations. (From Newton 1986.)

loss through the thinned shell. If the reduction in the mean breeding rate of individuals is sufficiently marked, it leads to population decline, because recruitment can no longer match the usual mortality (Newton 1986). The effects of DDT/DDE, which were initially deduced from field studies (Ratcliffe 1970), were subsequently confirmed by experiments on captive birds (Cooke 1973; Newton 1979; Risebrough 1986). All these effects were via the female, but DDT has also been found to reduce sperm production in domestic fowl (Albert 1962).

Some other organochlorines, such as aldrin and dieldrin, are up to several hundred times more toxic to birds than DDT (Hudson *et al.* 1984). These chemicals kill birds outright, increasing the mortality rate above the natural level, and in this way can cause population decline. In many countries, these chemicals were used mainly as seed treatments, applied to protect newly sown seeds against insect attack. They affected two main groups of bird—the seed eaters which ate the treated seeds and the predators which ate contaminated seed-eating birds and mammals.

The massive declines in the numbers of some bird-eating and fish-eating raptors, which occurred in Europe and North America in the 1960s, were thus attributed to the combined action of DDE reducing breeding rate and HEOD (from aldrin and dieldrin) increasing mortality rate. The relative importance of these mechanisms of population decline seems to have differed between regions, depending on the relative quantities of

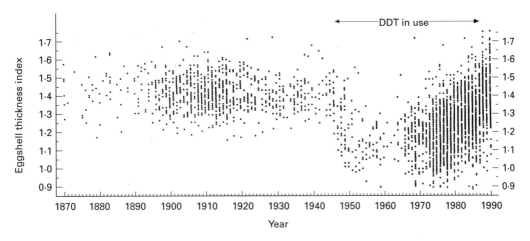

Fig. 4.3 Shell thickness index of British sparrowhawks, 1870–1990. Shells became thin abruptly from 1947, following the widespread introduction of DDT in agriculture, and recovered from the 1970s, following progressive restrictions in the use of the chemical, which was banned altogether from 1986. Each spot represents the mean shell index of a clutch (or part clutch), and more than 2000 clutches are represented from all regions of Britain. Based on shells housed in museum and private collections, and at Monks Wood Research Station. (Updated from Newton 1986.)

the different chemicals used. In much of North America, DDE-induced reproductive failure seems to have been paramount, but in Western Europe HEOD-induced mortality was probably more important (Newton 1986). As the use of these chemicals has been reduced, shell thickness, breeding success, survival and population levels of affected species have largely recovered (sparrowhawks: Fig. 4.3; peregrine *Falco peregrinus*: Cade *et al.* 1988; Crick & Ratcliffe 1995). In Britain, the recovery of sparrowhawk numbers in different regions followed the decline in geometric mean HEOD residues in liver tissue to below 1 µg/g in wet weight (Fig. 4.4). Other beneficiaries of the ban on aldrin and dieldrin in Europe included several seed eaters which fed on treated grain, such as the stock dove *Columba oenas* (O'Connor & Mead 1984).

An indication of the worldwide contamination of ecosystems with DDT can be gained from patterns of shell thickness in a single species, the peregrine falcon, *Falco peregrinus*, which breeds on all continents except Antarctica (Peakall & Kiff 1988) (Fig. 4.5). The greatest degree of shell thinning (26%) was found in the eastern USA, a region of heavy DDT use from which peregrines disappeared altogether within 20 years after the chemical came into widescale use (Cade *et al.* 1988). Marked shell thinning also occurred in peregrines nesting in Arctic North America and Eurasia (17–25% in different regions), reflecting the fact that these falcons and their prey migrated to winter in regions using DDT further south. The smallest levels of shell thinning occurred in falcon populations which were resident in areas with no DDT use, and which at the same time fed on prey

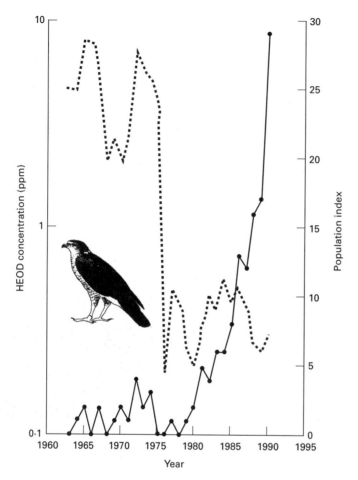

Fig. 4.4 Trend in geometric mean HEOD levels (dotted line) in the livers of sparrowhawks found dead in eastern England in relation to population levels in the same area (continuous line). HEOD is derived from the insecticides aldrin and dieldrin. (From Newton & Wyllie 1992.)

species which were themselves year-round residents in these areas. An example was the peregrine population of the Scottish Highlands, dependent mainly on the red grouse *Lagopus l. scoticus*. The falcons showed only 4% shell thinning in this region, compared with more than 19% in most of the rest of Britain (Ratcliffe 1980).

Comparison of shell thinning and population trends in different peregrine populations from around the world showed that all populations with an average of less than 17% shell thinning maintained their numbers, while all those with more than 17% declined, some to the point of extinction. In the one exception, extra eggs and young were added by biologists, so as to maintain numbers. Hence, an average of 17% shell thinning emerged as critical to population persistence; it was associated with an average of 15–20 µg/g DDE in the wet weight of egg content.

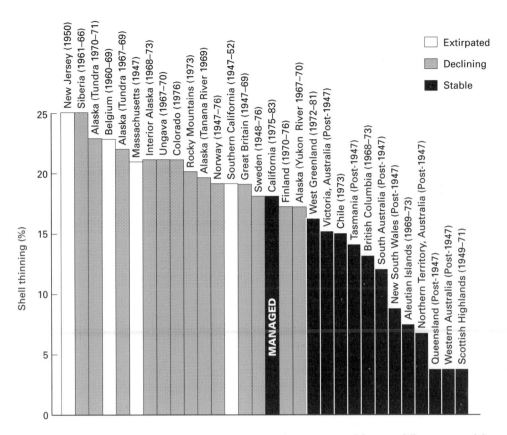

Fig. 4.5 Shell thinning and population trends in peregrine falcons in different parts of the world. All populations showing more than 17% shell thinning (associated with a mean level of 15–20ppm DDE in egg content) declined, some to the point of extinction. (Redrawn from Peakhall & Kiff 1988.)

In North America, the peregrine had been eliminated from about one-third of the continental area, and was much reduced in the remaining two-thirds. Recovery of the population was accomplished not only by a ban on organochlorine use, but also by a massive reintroduction programme. Over a period of 20 years, more than 5000 peregrines were raised in captivity and released to the wild, as chicks placed in artificial nest sites and fed until they could hunt for themselves. In this way, populations over wide areas were restored much more rapidly than could have occurred naturally (Cade & Temple 1995).

The problem caused by organochlorines results partly from their extreme persistence, a quality that adds to their effectiveness as pesticides. The longevity of chemicals in any medium is usually measured by their half-life, the period taken for the concentration to fall by half. The half-life of DDE in soils has been variously calculated at between 12 years in some cultivated soils and 57 years in some uncultivated soils (Cooke & Stringer 1982; Buck *et al.* 1983). So even after the use of

DDT is stopped, soil-dwelling organisms can remain a source of residue for raptor prey species for years to come. HEOD is much less persistent, with a half-life in soil estimated at 2.5 years (Brown 1978). Organochlorines can disappear more rapidly from animal bodies than from the physical environment, but again DDE lasts longer than HEOD. For example, in pigeons, the half-life of DDE has been estimated at 240 days, compared with 47 days for HEOD (Walker 1983). These rates vary between species, and with the condition of the individual. In laboratory conditions, some black ducks, *Anas rubripes*, which had been fed DDE with their food, were still laying thin-shelled eggs after 2 years on clean food (Longcore & Stendell 1977).

Persistence in the body means that the effects of organochlorines can become manifest weeks or months after acquisition. Mortality is most likely to occur at times of fast, as during hard weather or migration, when body fat is metabolized and organochlorines from within the fat are released into the circulation to reach lethal levels in the nervous system (Bernard 1966). This type of delayed mortality occurred among female eider ducks, *Somateria mollissima*, nesting on the Dutch Wadden Sea where nest numbers declined by 77% between 1960 and 1968, mainly through the deaths of incubating females (Swennen 1972). The birds became contaminated via their food (mussels) from organochlorines discharged into the River Rhine, but died mainly during incubation when they did not feed but depended on their body fat. Discharges were stopped in 1965, and within 3 years the eider population began to recover.

In general, birds have a less effective detoxification system for organo-chlorines than mammals have, which explains why birds are generally more contaminated than mammals from the same area. It also explains why avian predators of birds, such as the sparrowhawk *Accipiter nisus* and peregrine *Falco peregrinus*, have been more affected by organochlorines than have avian predators of mammals, such as the kestrel *Falco tinnunculus*. The bird feeders also top longer food chains than the mammal feeders do, giving more opportunities for residues to concentrate.

Organochlorine pesticides are also among the oestrogenic chemicals that are suspected to have caused the decline in sperm counts and some other fertility problems in humans. The case against such chemicals is not yet proven, and as yet the evidence is equivocal.

4.3.2 Other pesticides and pollutants with direct effects on vertebrates

Other groups of pesticides in common use include the carbamates, organophosphates and pyrethroids. Some chemicals in these groups are highly toxic to vertebrates, and have caused large-scale mortalities in fish, birds and mammals. One of the most toxic is the carbamate insecticide, carbofuron, which is thought to kill at least two million birds per year in the USA. In general, however, the carbamates and pyrethroids are much less toxic than other groups of pesticides to birds and mammals (Tucker & Crabtree 1970; Hill *et al.* 1975; Hudson *et al.* 1984). Whatever the chemical, large birds usually need bigger doses to kill them than do small birds, but birds from different families also differ greatly in their sensitivity, regardless of body size.

The persistence and bioaccumulation properties of the organochlorines are shared by some other chemicals, such as industrial polychlorinated biphenyls (PCBs) and alkyl-mercury pesticides. PCBs have been suspected to affect mammalian reproduction, but are not known to have had major effects on bird populations despite their frequent presence in bird bodies. Alkyl-mercury pesticides were widely used as seed dressings in the 1960s, when they killed many seed-eating birds and raptors, contributing to population declines, notably in Sweden (Borg *et al.* 1969).

Animals are often exposed to combinations of pesticides, either because two or more have been applied together, or because the animals themselves move from one crop to another. For some chemicals, the combined effect is much greater than expected from their individual toxicities. Such synergism happens where one chemical activates another or slows its detoxification. An example is provided by prochloraz fungicides which can accentuate the effects on birds of organophosphate insecticides (Johnson *et al.* 1994). So far this aspect has been studied only in laboratory conditions, but could clearly occur as a widespread phenomenon in nature. Similarly, water bodies near to industrial sites may receive cocktails of chemical pollutants, together with run-off from farmland. Organisms living there are exposed to so many pollutants that, when effects are observed, it is often impossible to pinpoint the causal ones.

Although numbers of most animal species can quickly recover from direct pesticide impacts, if usage continues, local populations may remain permanently depressed or die out. The wider the scale of repeated usage, the wider the scale of serious impacts. During the past 50 years, many bird species over large areas have evidently been held by direct pesticide impacts well below the level that contemporary landscapes would otherwise support. As indicated already, most examples are from raptors, fish eaters and seed eaters.

4.3.3 Lead poisoning in waterfowl

Lead poisoning of waterfowl results from the deliberate ingestion of spent gunshot. Like other birds that eat plant material, waterfowl habitually swallow small grit particles which remain in the gizzard and function in food breakdown. The gunshot is taken apparently as grit. It is then gradually eroded and absorbed to cause lead poisoning. The digestion of a single to a few lead shot is considered sufficient to kill any species from duck to swan.

In two major surveys, more than 230 000 gizzards from waterfowl collected throughout the USA and Canada were examined for lead shot (Bellrose 1959; Sanderson & Bellrose 1986). Lead was found in 7–9% of the gizzards. Allowing for under-recording (which is usual in manual examination) and the erosion rates of shot in gizzards, these findings led to speculation that up to one-third of the continental duck population could consume lead in a given year. Further, because most deaths from lead poisoning occurred in late winter, they were not necessarily offset by reduced deaths from other causes, and could thus have reduced breeding numbers (Sanderson & Bellrose 1986).

In the 1950s, Bellrose (1959) estimated that no less than 3000 tonnes of lead shot were deposited into the wetlands of North America every year; and, based on samples from several areas, the numbers of spent pellets averaged nearly 70 000/ha. These figures are likely to have increased greatly since then, as more shot has been expended. Because of species differences in feeding behaviour, however, not all species are equally vulnerable. Generally, bottom-feeders, such as the redheads *Aythya americana*, canvasbacks *Aythya valisineria* and lesser scaup *Aythya affinis* have the highest rates of ingestion (12–28%), whereas mallards *Anas platyrhynchos*, American black ducks *Anas rubripes* and northern pintails *Anas acuta* are intermediate (7–12%) and most other ducks range from 1% to 3% (Bellrose 1959; Sanderson & Bellrose 1986). Regardless of intake, lead does not become toxic until it is eroded and absorbed, so waterfowl feeding on hard foods, such as corn, experience greater exposure than those eating soft foods. In addition, protein and calcium are thought to reduce the absorption of lead by the gut. Because of the problems, the use of lead shot is now illegal in some countries, notably the USA, and steel shot is used instead. Among European countries, only Denmark has so far banned lead shot, though the incidence of lead in European waterfowl is as high, or higher, than in the USA (Pain 1991).

In Britain, a related problem arose from the use of lead sinkers on fishing lines. Over the years, such sinkers discarded by fishermen became a major cause of mortality in mute swans *Cygnus olor*, causing marked population declines on several rivers (Birkhead & Perrins 1985). The sinkers were found in the gizzards of dead swans, presumably again swallowed as grit. The problem has been solved by the use of alternative materials to weight the lines, and swan numbers have begun to recover. Ingested lead has also killed some predatory birds which ate birds and mammals which had been shot but not retrieved. Such secondary poisoning may have contributed to the decline of the California condor *Gymnogyps californianus*, but by the time the problem was recognized, the wild population was down to 10 individuals. Within 2 years, three of these birds had died from lead poisoning (Snyder & Snyder 1989).

4.3.4 Oil spills and seabirds

In recent decades increasing numbers of seabirds have died from oil pollution. Major incidents, resulting from the wrecking of laden tankers, often cause local catastrophic losses involving many thousands of birds. The number killed in any one incident depends not only on the numbers of vulnerable birds in the area at the time, but on the amount of oil spilled, how long it persists (varies with sea temperature and clean-up operations), and whether it drifts onshore or offshore. Those casualties washed onto nearby shores, where they can be counted, form an unknown proportion of the total. Other birds reach more remote shorelines or sink, so that estimates of the total numbers killed can seldom be more than informal guesses. In any incident, surface-swimming species, notably auks, seaducks and grebes, are most at risk. The biggest tanker incident of recent years involved the Exxon Valdez in Prince William Sound, Alaska, in March 1989. By August that year some 30 000 birds had been washed ashore, but from surveys the total casualties were estimated at 100 000–300 000 (Piatt *et al.* 1990).

It is surprisingly difficult to assess the impact of such large-scale local kills on seabird populations. This is partly because relevant preincident counts are not always available, and also because it is seldom known from how wide a breeding area the casualties are drawn, or what proportion consist of breeders, as opposed to non-breeders. Incidents in the breeding season are likely to reduce local nesting colonies, but within a few years the losses could be made good by immigration and by non-breeders starting to nest at an earlier age than usual. Incidents in winter could involve migrant birds drawn from wide (and usually unknown) breeding areas. The loss is therefore borne by a wider range of populations. Whether in the breeding or non-breeding season, however, major oiling incidents increase the frequency of catastrophic mortalities to which many seabird populations are exposed naturally. Whether they tip populations into long-term decline will depend on their frequency relative to the recovery powers of the population.

In addition to major incidents, thousands of smaller incidents, accidental and deliberate, occur at sea, both from shipping and from offshore platforms. These ongoing minor incidents form a continuous source of mortality, so that oiled seabirds nowadays continually appear on shorelines. In addition to the obvious seabird mortality, oil spillage affects invertebrate populations, which can take years to recover, and longer if toxic detergents and dispersants were used in the clean-up.

4.3.5 Some general comments

A major problem in ecotoxicology is how to translate laboratory findings to field situations. Knowing the dose of a chemical necessary to kill an individual animal, or a given proportion of a population, as determined in laboratory tests, is not much use unless we know what proportion of a population is killed in the wild, and to what extent those losses are additive to natural ones. Like any other form of loss, mortality from pollutants may be offset to greater or lesser extent by reductions in any natural losses which are density-dependent. Experience in Britain with the sparrowhawk and peregrine showed that production of young could be reduced to less than half their normal levels without causing a decline in breeding numbers. This was because, with reduced competition, the survival of the remaining young was higher, and with more territories than usual falling vacant, these young could begin to breed at an earlier mean age than usual. However, further reproductive failure could not be fully compensated for, and the population declined, to nil in some regions. Using field experience of this type, it should be possible to calculate for other populations the degree of additional breeding failure or mortality that could occur from pollutants without causing population decline; in other words, to calculate in the current environment the 'maximum sustainable yield', as described in Chapter 5. The next step is to relate this level of breeding failure or mortality to some threshold level of pollution that will cause it (Newton 1988), as in the DDE response in the peregrine or the HEOD response in the sparrowhawk discussed above.

The direct impact that any toxic chemical has on an animal population is greatly influenced by the life-history features of the species concerned. Large species, which tend to be long lived and slow breeding (K-selected), are likely to withstand longer periods of reduced breeding success than are small, short-lived, fast-breeding (r-selected) ones. However, once reduced in numbers by a given percentage (through mortality or breeding failure), K-selected species would be expected to take longer to recover because of their lower intrinsic rates of increase. Five years of complete reproductive failure would (on the usual 64% adult survival) reduce a sparrowhawk population to 17% of its former level, a peregrine population (90% adult survival) to 66% of its former level, and a golden eagle Aquila chrysaetos population (95% adult survival) to 81%. As the populations of larger species would normally contain a substantial proportion of non-breeders, able to replace any breeders that died, decline might be missed for several years if counts were restricted to breeders only. The resilience of long-lived species to reproductive failure may have enabled golden eagles to maintain their breeding density in western Scotland through the 1960s, despite reproductive failures resulting from the use of organochlorines in sheep dips (Newton 1979).

New pesticides are screened before acceptance in an attempt to eliminate at an early stage those of high toxicity to vertebrates, and are then further tested in field trials before they are approved for general use. However, laboratory tests are inevitably performed on a minority of captive species, while field trials are carried out in restricted areas, so new chemicals cannot be tested before release on all the species that could be exposed once they come into general use. Hence, we can expect that occasional chemicals will get into use that subsequently prove to affect certain non-target vertebrate species, in either typical or unforeseen circumstances. This happened with the organophosphate pesticide, carbophenothion, which, when first used as a seed dressing, killed thousands of wild geese in Britain in several different incidents. It was introduced for this purpose as a replacement for organochlorines, and had been extensively tested beforehand. But further laboratory testing on a greater range of species revealed great variability (by some orders of magnitude) in its acute toxicity to birds (Stanley & Bunyan 1979). In the reported incidents, pigeons and gamebirds feeding in the same fields as geese were not affected.

4.4 Indirect effects on populations through food shortage

4.4.1 Pesticides and farmland birds

The second main mechanism of population decline is indirect, resulting from the removal by pollutants of a crucial food supply. Because pesticides are to greater or lesser extent non-specific, they kill a wider range of organisms than the target pests themselves. Some insecticides, for example, are also highly toxic to earthworms or molluscs, while certain fungicides kill insects. In many countries, the use of pesticides is now so ubiquitous, often with several applications per growing season, that invertebrate populations over huge areas must be held permanently at much reduced levels—not only in the crops themselves, but in hedges and other adjoining habitats to which the chemicals are carried by wind.

   The best-documented example of population effects through reduction of food supply concerns the grey partridge *Perdix perdix* (Fig. 4.6), whose widespread decline has been attributed largely to poor production of young, in turn due mainly to lack of insect food for the chicks (Potts 1986; Potts & Aebischer 1995). This is because the broad-leaved weed species that form the food plants of the relevant insects are now absent from most modern cereal fields, as a result of herbicide use, and many other insects are killed directly by insecticide use. The survival of young partridges (up to 3 weeks old) was found to be directly correlated with the insect supply in cereal fields. This proposed mechanism was confirmed by experimental field trials in which reduced herbicide use resulted in improved weed and insect populations, better chick survival and greater partridge densities (Rands 1985). Research findings led to the recommendation that farmers keen to conserve partridges should leave an unsprayed strip of crop around the edges of fields—so-called 'conservation headlands'. This procedure enables weeds to grow, which in turn supports the insects necessary for partridges and other birds.

During the past 30 years, as shown by the monitoring programmes of the British Trust for Ornithology, most of the bird species found on British farmland have declined (Marchant *et al.* 1990; Gibbons *et al.* 1993). The same seems to have been true over much of Western Europe. In the absence of detailed studies, it is hard to say how much these declines were due to pesticide use, and how much to other agricultural changes that occurred at the same time. However, declines in many species steepened from the mid-1970s when insecticide use increased greatly in attempts to control aphids in cereal fields. In addition, herbicide use has enormously reduced the populations of various farmland weeds, on which many seed-eating birds depend. Some, once common, arable weeds have now become rare in lowland Britain. Most herbicides are not themselves cumulative, but their effects are, as they lead to progressive depletion of the seedbank in the soil. Each year seeds turned to the surface germinate to produce plants which are killed before they can seed. Several seed-eating birds, notably the linnet *Carduelis cannabina*, once fed extensively on weed seeds (especially *Polygonum* and *Chenopodium*) turned to the surface each time a field was cultivated (Newton 1972). Such seeds could once be counted at the surface of newly turned farmland soil at thousands per square metre (references in Salisbury 1961), but after more than 30 years of herbicide use, they have now almost disappeared from the soil of cereal-growing regions. It is not surprising, then, that the bird species most dependent on arable weeds have declined too, and that their declines did not become apparent until herbicides had been in use for several years.

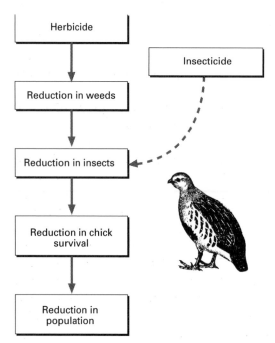

Fig. 4.6 Proposed means by which herbicides have caused a population decline in the grey partridge *Perdix perdix* in Britain. (From Potts 1986.)

Widespread monitoring of bird populations, followed by appropriate research and experiment, has played a crucial role in highlighting some of the long-term consequences of pesticide use and in confirming pesticides as a major factor in reducing, not just pest numbers, but the entire spectrum of biodiversity in farmed landscapes (and beyond in the case of organochlorines). The number of well-documented case studies is few because declines in bird numbers have mostly been gradual, and in recent years have not involved conspicuous large-scale kills. Also, any effects of pesticides have proved hard in retrospect to disentangle from those of other procedural changes.

4.5 Alteration of habitats by pollutants

4.5.1 Acidification of wetlands

Combustion of coal and other industrial activities over the last 200 years have greatly increased the amounts of sulphur and other pollutants in the atmosphere. Such pollutants can be blown long distances, affecting areas up to several hundred kilometres from their source. One major identified consequence is the acidifying effect of sulphur dioxide on rain (Fig. 4.7). The impact of 'acid rain' on well-buffered soils is probably negligible, but, on granite or other poorly buffered soils, the pH of soils, streams and lakes is lowered from 6–7 to 4–5 or less. This in turn has affected the mobility of toxic metals (notably aluminium, mercury, cadmium and lead) which move from soils to waters, and become generally more available to plants and animals, while calcium and magnesium become less available. The resulting declines in invertebrate and fish populations have led in turn to declines in the bird populations that depend on them. Other identified effects on birds result from metal toxicity and reduced dietary calcium. The problem is particularly severe in parts of eastern North America and northern Europe.

In any one region, the effects on the aquatic fauna are progressive, and involve threshold responses as pH falls. In general, when the water is around pH 6, crustaceans and molluscs—creatures with mineralized shells—find it hard to survive. Such creatures form an important source of calcium for many birds. At pH 5.5, many insects favoured by birds disappear, including mayflies, caddis and damselflies. Among fish, salmon, trout and roach start to decline as the pH falls below 6, whitefish and grayling at pH 5.5, followed by perch and pike at pH 5, and eels at pH 4.5. Different species decline at different stages because of differences in their physiology and susceptibility to toxic metals, as well as differences in their diets. The eradication of fish makes the waters unsuitable for fish-eating birds, but for a time leaves more insect food for ducks, so goldeneye *Bucephala clangula* and others may benefit (Parker *et al.* 1992). But the end result, as acidification continues, is a sterile ecosystem capable of supporting only a limited range

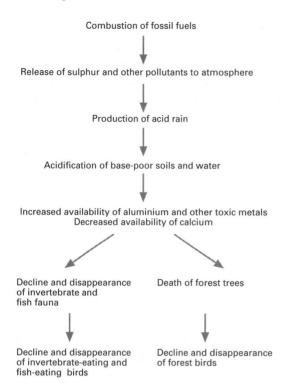

Fig. 4.7 Proposed mechanism of action for effects of sulphur dioxide on the fauna and flora.

of species. Thousands of lakes in eastern North America and northern Europe have lost their fish in the last 50 years as the pH has fallen (Freedman 1995).

4.5.2 Forest dieback

Air pollution is held responsible for forest dieback in several parts of the world, including montane areas in central Europe, eastern North America, West Africa, China and South America. These are again all areas with heavy rainfall on base-poor soils. The problem became evident in the 1970s and has increased subsequently since then. The extent to which the trees die from exposure to air pollutants as such, from exposure to aluminium and other toxic metals, or from shortage of calcium and other essential metals, probably varies with area, but the net effect is progressive loss of forest habitat, together with the acidification of streams, as discussed above. In forests, as in aquatic habitats, the effects are progressive, depending on the initial base status of the soils and the rates of acid deposition.

In Europe, evergreen trees (such as spruce) appear more sensitive than deciduous trees, and die at an earlier stage in the acidification process. The geographical scale of the problem is continually changing, but, in 1986, more than 19 million hectares, or 14% of the total forest area in Europe west of the former USSR, was considered damaged (Nilsson & Duinker

1987). This damage was at various levels, but included some areas where whole forests had died, mainly near point sources of pollution.

Pollutant impact on the trees represents a direct effect, but the animal inhabitants are affected indirectly, through destruction of their habitat. Most research has concerned birds, among which species dependent on insects from the foliage usually show the greatest declines. Species dependent on insects from dead wood (such as woodpeckers) may benefit temporarily, as may seed eaters (because seed production is a natural response of trees to adversity). As the trees collapse, open country species move in, but the net effect of forest dieback is a reduction in the numbers of species and individuals (Graveland 1990).

4.5.3 Acidification and calcium shortage

Other effects of acidification on birds result from reduced availability of calcium (through the disappearance of calcium-rich food sources) and increased exposure to toxic metals. Some passerine species, which breed in recently acidified areas, have shown shell thinning, leading to breakage and desiccation of eggs, non-laying (with birds incubating empty nests) and reduced breeding success (Carlsson *et al.* 1991; Graveland *et al.* 1994). In one area of The Netherlands, the proportion of great tit *Parus major* females that produced defective eggs rose from 10% in 1983–84 to 40% in 1987–88 (Graveland *et al.* 1994). This was associated with the leaching of calcium from acidified soils, reduction in vegetation and consequent loss of snails, whose shells form a major source of calcium for laying songbirds. In one experiment, Graveland *et al.* (1994) provided broken shells from snails and domestic chickens, which resulted in an improvement in the eggshells and breeding success of local titmice. In another experiment, the liming of an acidified area resulted in a recovery of snail populations and restoration of a calcium source for songbirds.

These various problems, which have so far surfaced only in areas of naturally low pH, can be expected to spread more widely as further acid deposition brings more areas below the critical pH threshold. Although limited measures have been taken in some countries to reduce sulphur outputs, these are not enough, and the problem continues. Meanwhile research has concentrated on finding 'critical loads', the maximum acid deposition that can be sustained by a given region without adverse effects. This naturally varies greatly from area to area depending on soil type.

4.5.4 Eutrophication

Not all pollutant effects are entirely negative. Over a certain range, some pollutants may merely result in environmental change which favours some species at the expense of others. In recent decades, the widespread eutrophication of inland and coastal waters, mainly by nitrates and phosphates (derived from fertilizers, sewage and domestic effluent), has led to changes in plant, invertebrate and fish populations, with resulting effects on birds. Impacts vary with the initial nutrient status of the water, as lakes and rivers vary naturally on a gradient from poor (oligotrophic) to rich (eutrophic). As nitrates and phosphates leach into nutrient-poor waters, macrophytic plants (water weeds) may grow, providing new substrates for invertebrates, more food and cover for fish and a greatly increased food supply for herbivorous birds, such as swans. Salmonid and coregonid fish give way to coarse fish, initially percids and then cyprinids (such as roach *Rutilus rutilus* and bream *Abramis brama* in European waters). As nutrients continue to increase, however, algae multiply and cloud the water, reducing light penetration and causing losses of water-weeds, invertebrates and fish. Certain algae and other microorganisms produce toxins that poison fish and the birds that eat them. Massive bird mortalities have been recorded in both inland and coastal waters, from which affected species can take several years to recover their numbers. In deep lakes, the dense algal populations may sink, and their decomposition may deoxygenate the water. Under extreme circumstances, the water may become almost lifeless.

The addition of organic matter to lakes and rivers, and its decomposition to support new life, is a natural process, but the addition of too much human sewage and other organic waste can cause problems. The bacteria that cause decomposition need oxygen to survive, and the more sewage is added, the more oxygen is consumed. Within limits, a river naturally re-aerates itself, but under heavy organic input the water can become totally devoid of oxygen, resulting in the deaths of fish and other aquatic life. The bacteria which cause decay are sometimes replaced by anaerobic bacteria, which can trigger other processes inimical to most organisms. The problem is increased if river flow is reduced by the extraction of water for irrigation and other purposes.

Eutrophication can thus be viewed as a process leading to changes in flora and fauna, initially with an increase in overall biomass, but above a certain level leading to decline and sterility. With the increasing use of fertilizers, more and more inland waters are approaching the highly polluted end of the spectrum, and hence are declining in their ability to support aquatic life, especially fish. The problem is also becoming increasingly apparent in shallow coastal areas.

4.5.5 Global warming

Some other pollutants, in the amounts released, have no apparent direct effects on any living organism, but can alter the physical and chemical environment in such a way as to affect population levels. The most far-reaching of these is probably carbon dioxide, whose concentration in the atmosphere is steadily rising, as a result of the burning of fossil fuels, coupled with extensive deforestation. It has increased from about 290 ppm in preindustrial times to about 350 ppm at present, and is expected to reach about 700 ppm around the year 2050. Its accumulation, together with that of other heat-trapping gases (such as methane and nitrous oxide), is expected to raise global temperatures, thus resulting in radical changes in the numbers and distributions of species worldwide. At the same time, moreover, the progressive destruction of stratospheric ozone by chlorofluorocarbons (CFCs) is steadily increasing the amount of ultraviolet radiation reaching Earth; and continuing deforestation is contributing increasingly to the carbon dioxide problem because trees contain more carbon per unit area than do small plants, and they convert carbon dioxide to oxygen more rapidly than does any other vegetation.

The rise of greenhouse gases in the atmosphere is undisputable, and it is accepted now that this will lead to a temperature rise more marked at high than low latitudes. Less certainty surrounds the rate of global temperature rise and its effects on other aspects of climate. Most current estimates suggest about 1.0–2.0°C per 100 years, but some suggest rates up to 4.0°C per 100 years. This may not seem much, but as a global average, it is close to the $4 \pm 1°C$ change which occurred from the last glacial climax to now (Webb 1992). It is equivalent to a poleward movement of temperate climatic belts of 200–300 km or an altitude change of 200 m. The main effects are likely to be on plants, with animals affected secondarily. In the change from glacial to interglacial, the main vegetation belts (together with their fauna) shifted position slowly over the Earth's surface, so as to remain in the appropriate climatic zones. However, at the rates of human-induced climate change predicted, many plant species would be left behind, unable to move fast enough (Barkham 1994), and their passage through modern landscapes would anyway be impeded by human land use. Island species may be even more vulnerable: if the latitudinal or altitudinal movement required by climate change exceeds the limits of the island, extinction may follow—although the sea around oceanic islands might moderate the change in air temperature.

As the Earth warms, rises in sea level can be expected from the expansion of water in warmer oceans and from the melting of glaciers and polar ice (Schneider et al. 1992). The rate of mean rise is estimated at about 50 cm/ 100 years (range 10–100 cm). Again, this may not sound much, but the associated increase in wave heights and tidal surges can be expected to inundate large areas of coastal habitat.

Despite the consensus that global warming will occur, it is not yet possible to confirm it with existing data. This is because, although the temperature rise recorded over the last 100 years is consistent with predicted rates, it is still within the range of natural variation. Uncertainty over the reality or otherwise of the phenomenon can be expected to decline greatly in the coming years. Similarly, while many plant and animal species are likely to change greatly in abundance and distribution, it is impossible to do more than guess about the precise impacts or about which species will be most affected. Almost certainly, however, the major changes will result from the human response to climate change, rather than from climate change itself or the natural vegetation changes that result from it.

4.6 Genetic aspects

In nature, individuals of any species are exposed to widely differing doses of particular chemicals. If one measures DDE (say) in a population of birds or eggs, the distribution of residues is usually skewed, with most individuals containing small amounts (perhaps reflecting low exposure) and a few containing large amounts (reflecting high exposure). Differential exposure is one reason why some individuals are more affected than others. In addition, however, in any population, individuals usually vary in their response to a given dose of the same chemical. For this reason, the toxicities of particular chemicals to a given species are often expressed as LD_{50} values—the lethal dose expected to kill 50% (or any other stated percentage) of a population. If any of this variation in response is under genetic control, an exposed population could in time change genetically, through the selection of resistant genotypes. However, known examples of the development of resistance (such as many insects to DDT, some plants to toxic metals or some rodents to warfarin) have so far come from only a small proportion of exposed species and from only a small proportion of toxic chemicals.

The first known vertebrate to develop resistance to organochlorines was the mosquito fish *Gambusia affinus* in the lower Rio Grande valley in Texas, which became resistant to DDE and toxaphene. These fish could then accumulate large body residues, and presented a greater hazard than before to the predatory fish and birds which ate them. Andreason (1985) suggested that this change in the mosquito fish may have contributed to the shell thinning and population declines in fish-eating birds observed at the time in southern Texas. This is an instance where resistance in one organism may protect that organism, but at the same time harm its predators.

Pollutants can also change populations genetically by indirect means. The best-known example concerns industrial melanism in moths. Lichens are very sensitive to air pollutants, and have disappeared from trees in many parts of Britain, changing the appearance of bark from light to dark. Many moth species, notably the peppered moth *Beston betularia*, which rests on bark, exist as light and dark morphs. Following the industrial revolution in Britain, dark morphs replaced light ones over much of the country. Experiments showed that predation by birds was the main selective factor, as light morphs, which were well camouflaged on lichens, became extremely conspicuous on dark sooty backgrounds (Kettlewell 1973).

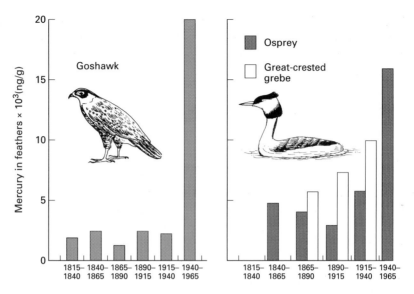

Fig. 4.8 Mercury levels in the feathers of some Swedish birds, showing temporal pattern of contamination of terrestrial and freshwater habitats. (Redrawn from Jensen *et al.* 1972.)

4.7 Use of historical material

In examining long-term trends in contamination, it has often proved helpful to examine dated material present in museum collections. This procedure has proved useful, not only in dating the start of eggshell thinning in birds (see Fig. 4.3), but in tracing the increase, since the nineteenth century, in the contamination of birds with mercury (Fig. 4.8). Whereas the feathers of the goshawk *Accipiter gentilis* showed a sudden rise in mercury content following the introduction of alkyl-mercury pesticides in the 1950s, those of the fish-eating osprey *Pandion haliaetus* and great-crested grebe *Podiceps cristatus* showed a more gradual rise, starting near the end of the nineteenth century, reflecting the progressive contamination of aquatic habitats from increasing industrial activity (Jensen *et al.* 1972). Use of historical material has thus helped to distinguish between different possible causal factors in bird population declines.

4.8 Conclusion

In this chapter, I have touched on only a minority of the chemical impacts on plants and animals that have been described, mainly those that have led to obvious population declines. They illustrate some of the key features of pollutant impacts, including the indirect, delayed, unexpected and widespread nature of many. Other pollutant impacts may remain undetected, either because they are indirect (through effects on competing or food organisms), because they have no effects obvious to the casual observer, because they occur in remote areas, or because they are yet to become apparent. Almost all the impacts discussed above are products of the last 150 years, and those of pesticides mainly of the last 50 years. They are therefore new forces in the population ecology of all organisms. One might reasonably ask, then, why pollution has reached the serious stage it has.

One reason is that, traditionally, the polluter has not been required to pay the cost of his actions. The factories that pollute the atmosphere have not paid for the deaths of remote forests, for the loss of fish from rivers or for the adverse effects on human health. Nor has the farmer paid for the side effects of his pesticides, the loss of wildlife, the pollution of water and the effects on human health. This is partly because such costs have been considered acceptable in the development of the modern industrial state, and partly because they are often intangible, difficult to pin down and to value in cash terms. Moreover, many problems, such as acidification, involved threshold effects that did not become apparent until the polluting activity had continued for many years. The layman might reasonably wonder how many other ecological time bombs lie in the offing.

Another reason for the problem is the application of false economics. The value of pesticide use, for example, is judged primarily on whether the extra crop yield that year exceeds the cost of the chemical used. It takes no account of any longer term effects on soil productivity, future crops, water quality, loss of wildlife and human medical expenses. The cost of all these, where they can be expressed in cash terms, is borne by society at large, and at some later date, even by future generations. Few other modern-day problems offer politicians more excuse for inaction.

In other cases, it is extremely difficult to apportion responsibility. As I write this chapter, another oil tanker, the Sea Empress, has just gone aground off the Welsh coast, spilling 80 000 tonnes of oil. A newspaper headline sums up the situation: 'Built in Spain; owned by a Norwegian; registered in Cyprus; managed from Glasgow; flying a Liberian flag; carrying an American cargo; and pouring oil onto the Welsh coast. But who takes the blame?' Add to this rough seas, poor controls, aged tankers and captains with alcohol problems, and you have all the common ingredients of this type of disaster.

4.8.1 Summary

Pesticides and pollutants have reduced plant and animal populations directly through adverse effects on reproduction or survival (e.g. organochlorine pesticides and some predatory birds), or indirectly through reduction of food supply (e.g. insects for partridge chicks, weed seeds for finches) or through the physical or chemical alteration of habitats (e.g. atmospheric pollutants and acidification). Some pollutants cause ongoing breeding failure or mortality, while others cause periodic catastrophic mortality. They thus act in ways that mimic various natural forms of mortality, except that they usually operate in a density-independent manner. Their impact depends on the frequency and scale of losses, and on the extent to which these losses can be offset by reduced natural mortality or enhanced re-production. The destruction of habitats and food supplies by pollution reduces the abundance and distribution of populations that depend on them. Because some pollutants can disperse long distances in air and water currents, they can affect populations hundreds or thousands of kilometres from the areas of production or usage. Moreover, because of threshold effects, some major problems created by certain atmospheric pollutants did not become widely apparent until many decades after their emission began. Major problems in the future are likely to result from climatic changes expected to result from the human-induced rise in atmospheric carbon dioxide and other heat-trapping gases.

CHAPTER 5

Sustainable and unsustainable exploitation

William J. Sutherland and John D. Reynolds

5.1 Introduction

Easter Island, in the Pacific Ocean, is famous for its spectacular carved statues each weighing up to 95 tonnes. Pollen analysis shows that the island was forested when first colonized by Polynesians about 400 AD. Historical evidence shows that the forests were cleared to provide areas to grow crops, for dug-out canoes for fishing and to roll the statues from the quarries and lever them into position. This fertile island enabled an extraordinary civilization to develop, with the population reaching about 7000 people around 1500 AD. Yet by 1722, when visited by a Dutch explorer, all of the trees had been cut down, resulting in soil erosion, poor crop yields and no boats for fishing. The food shortage had led to warfare and cannibalism, with the population having declined to about a third of what it once was (Kirch 1984).

Evidently, unsustainable exploitation by humans has a long history. How far have we now progressed towards exploiting our natural resources sustainably? A contemporary case in another overpopulated island is far from reassuring. In Britain, 94% of raised peat bogs have been destroyed over the last few centuries, largely due to drainage followed by conversion to agriculture and forestry. This loss is continuing, but is now largely due to demand by gardeners and horticulturists for peat moss to improve their soil. This occurs despite the availability of alternatives to peat, as well as the fact that people managed perfectly well without peat before marketing began over the last few decades. If developed countries are unable to protect such an important habitat from such a trivial threat, then one wonders how we can solve more pressing problems of overexploitation elsewhere. Note, too, that in this example both the state of the resource and the reason for its decline are absolutely clear. Evidently, unsustainable exploitation cannot always be ascribed to ecological ignorance.

The main aim of this chapter is to consider why unsustainable exploitation is so ubiquitous, and what can be done about it. By 'exploitation', we are concerned with the removal of organisms or their products from the natural environment, without managing the habitat specifically for that organism. Thus, we are not concerned with agriculture or aquaculture.

Indeed, most of the world's temperate forestry operations are also not examples of pure 'exploitation', since they are usually accompanied by some sort of site preparation or re-planting (Freedman 1995). By 'sustainable' we mean that resources are used at rates within their capacity for renewal.

We begin by considering how to manage resources sustainably, including a review of the theory of how populations respond to mortality. We then examine various causes of overexploitation, followed by a review of solutions to prevent this from happening. In the final section we consider briefly the subject of intentional eradication. Although our concerns about the negative consequences of overexploitation will be apparent throughout, we have intentionally avoided the thorny issues of animal welfare and ethical considerations of which species should or should not be exploited.

The extent of exploitation that actually takes place usually depends upon far more complex issues than simply calculating the levels of exploitation that produce the maximum sustainable yield (MSY). For example, the level of exploitation of whales has been determined by the interplay between the financial short-term gains from overexploiting (for reasons explained in Sections 5.3.1 & 5.3.2) and the ethical concern of conservationists in certain countries, particularly in Western Europe and North America. In this case, it is likely that few of the politicians and fishers on one side or conservationists on the other side would wish to exploit to produce the MSY. Scientific figures such as those developed in the following section are typically used as a debating point to justify stances rather than as rigid targets.

Much of the debate about sustainable exploitation refers to a few species of fish, whales, seals or gamebirds for which there is often a considerable amount of scientific data. It must be remembered that there are also an enormous number of species exploited throughout the world, often collected in a low-key way, for which there may be little or no information. As one trivial example, of the 173 million condoms used in Germany each year, 80% are dusted in *Lycopodium* spores so they can unroll easily. Germany alone imports 4 tonnes of *Lycopodium* each year for this purpose from China, Russia and India (Balick & Beitel 1989; Berkefeld 1993). The principles described in this chapter apply equally to such poorly documented exploitation.

5.2 How to manage sustainably

As individuals of a population are shot, trapped, netted, felled or otherwise removed, the number remaining will usually be less than in an unexploited population. There is good evidence from a wide range of species that, at reduced densities, either the birth rate will increase or the natural mortality (i.e. not due to exploitation) will decline (Sinclair 1989). Indeed, if there were no such density-dependent compensation in fecundity or survival, even the smallest elevated mortality inflicted by humans would drive the population to extinction. The persistence of sensibly exploited populations thus shows that there must be some compensation. Sustainable exploitation therefore relies on removing individuals within the limits that can be accommodated by density-dependent responses. In order to manage sustainably, we must understand the biology of the plants or animals being targeted, and incorporate this information into models of population dynamics to find the level of exploitation that produces the highest long-term yield.

5.2.1 Surplus yield models

The simplest models, which date back to Schaefer (1954), combine processes that increase population biomass, namely growth of individuals and recruitment, with processes that have the opposite effect, through natural mortality and exploitation. Schaefer was concerned with fisheries, and assumed logistic population growth, but the principle is general. If exploitation is too high, populations may be driven so low that their ability to grow is impaired, resulting in a low long-term yield. Conversely, if exploitation is very low, populations that could have compensated by increased growth of individuals or recruitment will provide a lower yield than they could have, and instead they will be limited by resources such as food and space (Fig. 5.1). The MSY will therefore be at an intermediate level between these extremes. It will be determined by the population's ability to replace lost individuals.

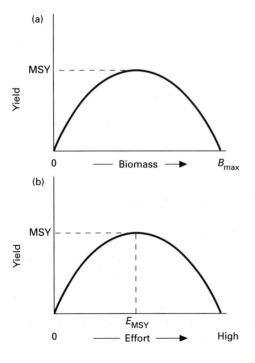

Fig. 5.1 The surplus production model for finding maximum sustainable yield (MSY). (a) Theoretical relationship between yield and stock biomass, ranging between 0 and the maximum expected in an unexploited population (B_{max}). (b) Relationship between yield and exploitation effort, showing level of effort that maximizes yield (E_{MSY}). For derivation, see Schaefer (1954), Pitcher & Hart (1982) or King (1995).

The data needed for this approach, also called 'surplus production models', are disconcertingly easy to collect. Indeed, no substantial information on the biology of the exploited species is required. One needs only a time series of some measure of effort devoted to exploitation (e.g. number of hunting permits issued per season or hours spent fishing per year), and the resulting yield (e.g. number of deer shot or kilograms of fish caught per year). These data can then be used to derive parameters that give the theoretical level of exploitation that maximizes yield, as in the example shown in Fig. 5.2.

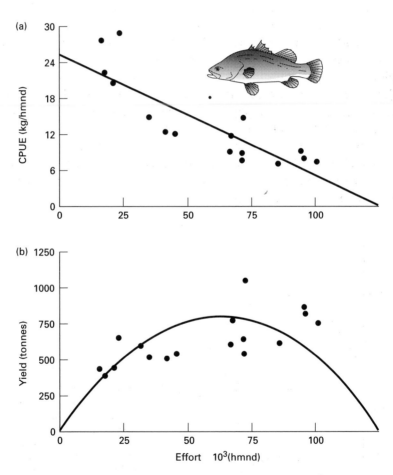

Fig. 5.2 An example of the surplus yield model applied to a fishery for barrumudi *Lates calcarifer* in the Northern Territory of Australia. Each data point represents 1 year. (a) Data for catch (kg) per unit effort (CPUE) plotted against effort, measured in units of 100 m of net/day. The regression equation can be represented as CPUE = $a + bE$, where b is the slope of the regression against effort, E, and a is the intercept. These parameters are used to fit a Schaefer curve through the data in (b), showing the level of fishing effort that maximizes yield, based on the equation $Y = aE + bE^2$. (After King 1995.)

Note that this method relies on data from a population that has already been exploited for a number of years. Even when such data are available, in practice it may be difficult to calculate precisely where the level of effort giving MSY actually lies (Hilborn & Walters 1992). Indeed, this is most reliably found once the population has already been overexploited! By the time this becomes apparent, social, political and economic considerations always make it very difficult to bring in regulations that reduce exploitation, due to short-term losses to individuals who rely on the resource.

This brings us to an important caveat that also applies to other sections in this chapter. MSY should be used as an upper limit to exploitation, rather than an actual target to be achieved. This is because even with accurate data, clever models and deft management leading to an accurate hit on the MSY, there is an in-built lack of stability caused by any environmental fluctuations that cause an overshoot, and hence overexploitation. In other words, while exploitation above the MSY is clearly bad, failsafe management requires exploitation under the MSY, not on it.

The simplicity of surplus yield models masks the complex biological processes underlying changes in yield with effort and the contributions made to the total yield by individuals of different ages and sizes. It also ignores annual variation in recruitment, as well as other population processes that will inevitably be prone to environmentally induced fluctuations.

5.2.2 Production models from population parameters

Another approach to estimating potential yields is to consider rates of population increase from life-history parameters. By working from first principles of population dynamics, one avoids the need for data on yield and exploitation effort from an already exploited system. Robinson & Redford (1991) suggested such an approach for sustainable exploitation of tropical mammals, based only on maximizing numbers of individuals hunted. They calculated maximum production over a specified period, P_{max}, as $(gD \times l_{max}) - gD$, where g (our term) is the fraction of the carrying capacity at which maximum production is likely to occur. D is the population density and l_{max} is the maximum finite rate of population increase over the specified time period. The latter depends primarily on the number of adult females reproducing and the average number of offspring produced per female per year. The population density can be estimated through surveys of the species of interest, or by extrapolation from general regressions between animal body size and density (e.g. Cotgreave 1993). We have seen earlier that maximum production occurs at population densities below the carrying capacity, but to determine the exact value at which this occurs, g, requires an educated guess. In species with logistic population growth, maximum production occurs at one-half of the carrying capacity ($g = 0.5$). Robinson & Redford suggest a value of $g = 0.6$ for very short-lived species (age of last reproduction <5 years), 0.4 for short-lived species (age of last reproduction between 5 and 10 years) and 0.2 for long-lived species (age of last reproduction > 10 years).

An example of this method was provided by FitzGibbon *et al.* (1995) for subsistence hunting in the Arabuko-Sokoke Forest, Kenya. The authors used line transects through the 372 km² forest to determine the relative abundance of the key hunted mammals in three main habitat types. Several methods were used to convert relative abundance to size of prey populations, including trapping, mark-recapture and estimates of rates of dung production and decay (for dung piles of antelopes). Data used to calculate l_{max} for six species are given in Table 5.1, and data for calculating production and maximum sustainable exploitation rates are given in Table 5.2. The results show that two species—yellow baboons *Papio cynocephalus* and Syke's monkeys *Cercopithecus mitis*—are overexploited, but that the exploitation rates of the other species appear to be sustainable.

This method is appealing because it is derived from population parameters which theory and intuition predict to be important for sustainable exploitation. However, the predictions are only as good as the data, and because estimates of potential production are directly proportional to estimates of population density near carrying capacity, there is considerable scope for error due to difficulties of estimating abundance (Peres, in press). Nevertheless, an ambitious series of surveys comparing the abundances of hunted mammals and birds over a 10-year period in Amazonian forest sites has confirmed predictions from the Robinson–Redford method that hunting in most areas is not sustainable (Peres, in press). Note that, as

Table 5.1 Reproductive data used in estimating the intrinsic rate of population increase, r_{max} and hence l_{max}, as well as exploitation rates of mammal species in Arabuko-Sokoke Forest, Kenya (FitzGibbon *et al.* 1995).

Species	Body weight (g)	Age at first reproduction (years)	Female young/year	Age at last reproduction (years)	l_{max}	Exploitation rate (%)
Four-toed elephant shrew (*Petrodomus tetradactylus*)	200	1	2	4–5	3.0	60
Golden-rumped elephant shrew (*Rhynchocyon chrysopygus*)	535	1	2	>5	4.06	40
Syke's monkey (*Cercopithecus mitis*)	6000	5	0.25	25	1.15	20
Yellow baboon (*Papio cynocephalus*)	17500	5	0.25	>20	1.15	20
Duikers (four species)	10000	1	0.5	10	1.48	20
Squirrels (two species)	30	?	?	<5	?	60

r_{max} is calculated from Cole's equation, which is solved iteratively:

$$1 = e^{-r_{max}} + b\,e^{-r_{max}(a)} - b\,e^{-r_{max}(w+1)}$$

where a = age at first reproduction, b = annual birth rate of female offspring and w = age at last reproduction. l_{max}, the maximum finite rate of population increase, is the proportional increase in population size from time t to time $t + 1$.

Table 5.2 Estimates of production and maximum sustainable exploitation levels for six mammal species in Arabuko-Sokoke Forest, Kenya (FitzGibbon *et al.* 1995). See Table 5.1 and text for calculations. The rate of maximum sustainable exploitation assumes that populations are at carrying capacity and that maximum production occurs when population density is at 60% of carrying capacity (see text). If exploitation has already reduced populations to their levels of maximum production, these figures can be increased.

Species	Density (per km^2)	Production (no/km^2)	Maximum sustainable exploitation	Current exploitation
Four-toed elephant shrew (*Petrodomus tetradactylus*)	391.2	782.4	102 000	5687
Golden-rumped elephant shrew (*Rhynchocyon chrysopygus*)	58.9	180.2	7455	3146
Syke's monkey (*Cercopithecus mitis*)	57.7	8.7	387	1202
Yellow baboon (*Papio cynocephalus*)	16.1	2.4	112	683
Duiker (four species)	62.7	30.1	1 339	181
Squirrel (two species)	11.0	?	2 530	484

with surplus yield models, this approach does not allow one to keep track of effects of exploitation on age structure. This is dealt with by the next technique.

5.2.3 Dynamic pool models

Older plants or animals are often more valuable: older stags yield better antler trophies to deer hunters, older trees provide more timber, and older tuna yield more sushi. Dynamic pool models, often used to derive yields per recruit, have been developed to account for this (Beverton & Holt 1957). As with surplus yield models, the MSY maximizes both the number of individuals taken and some age-dependent feature of the organism, usually weight. As each cohort matures, individuals increase in size, but are reduced in number due to natural mortality. Dynamic pool models consider explicitly the number of individuals of each age and the size of each and derive the exploitation strategy that yields the highest long-term biomass. If exploitation mortalities are too high, many individuals will be taken before they have had a chance to grow, and if mortalities are too low, although the individuals will be large, the total yield will be too low because too few are caught (Fig. 5.3).

Calculations for dynamic pool models are done easily with a computer spreadsheet, in which estimates of natural mortality are combined with a range of potential mortalities affecting separate age classes of individuals (reviewed in Pitcher & Hart 1982; Hilborn & Walters 1992). The resulting yields, such as biomass of fish caught, are based on the number of fish caught and their sizes at each age. These are summed to give a total yield for each potential fishing mortality. When yields are plotted against fishing

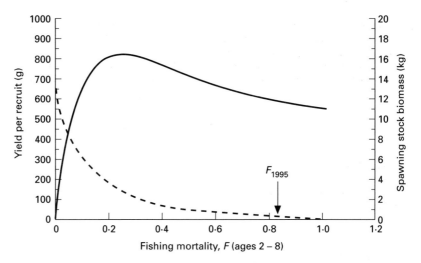

Fig. 5.3 The dynamic pool (yield per recruit) model for North Sea cod *Gadus morhua* showing predicted long-term yields per recruit (solid line) at various levels of instantaneous fishing mortality, *F*. The predicted spawning stock biomass is also shown as a function of *F* (broken line). The actual level of fishing mortality in 1995 (F_{1995}) was estimated to be 0.85. This provides a lower yield than MSY, because too many individuals are taken before they have had a chance to grow to a large size (ICES 1994).

mortality, the MSY can be determined (Fig. 5.3). Dynamic pool models assume that the population is in a steady state, and thus the numbers of individuals alive in each age cohort in any particular year are equivalent to the decline in numbers of any cohort over time.

This technique, first developed explicitly by fisheries biologists, remains the mainstay of fisheries predictions in developed countries which can afford the considerable time and expense of collecting data on the numbers and sizes of fish at each age. Surveys at sea may use gears that sample a wider range of sizes and ages of fishes than found in commercial landings, thereby providing information on the numbers of young fish that will recruit to the fishery in subsequent years. For example, in Europe, at least eight countries participate in annual surveys of the North Sea, under the coordination of the International Council for the Exploration of the Sea (ICES). Data from such surveys, combined with information from commercial landings sampled by government biologists at ports, provide a good snapshot of population structure. This information is used to assess the current state of the stocks and for modelling the effects of various levels of exploitation on population structure and yields to the fishery.

Sophisticated techniques have been developed to refine estimates of population sizes and mortalities. In developed countries, virtual population analysis (VPA) and an approximation of VPA, cohort analysis (Pope 1972), are used routinely to calculate numbers of fish at each age, and rates of fishing mortality. The details of these methods differ (for a review see

King 1995), but the principle is the same. One uses initial estimates of mortality in a given age cohort (due to natural and fishing mortality) to work backwards year by year in calculating the numbers of survivors and previous mortality rates due to the fishery. Thus, for a given age cohort of fish, one calculates the number of fish that must have been alive the previous year by adding the number caught by the fishery in the current year, t, to the number estimated to have died of natural causes over the same time period. The fishing mortality that must have occurred during the year $t - 1$ is then calculated as the proportional reduction in numbers between year t and year $t - 1$, minus losses due to natural mortality. This procedure is then repeated for year $t - 2$, and so on back to the year in which the cohort was born. When estimates from all age cohorts in a given year are added together, one can derive historical fishing mortalities and population sizes, which are important for deducing how fishes of various ages are affected by fishing pressure.

Although the resulting estimates from VPA of annual fishing mortality and population sizes depend on the initial estimates for the most recent year, their proportional dependency on these becomes smaller because the number of survivors will be higher as one works back through time. This gives greater confidence in estimates from earlier years, which can be fine-tuned as new population estimates are generated each year (Hilborn & Walters 1992).

Although this method, like many other techniques for deriving effective limits of exploitation, was developed by fisheries biologists, in principle it can be applied to any population in which a time series of data is available for abundance of individuals whose age can be determined. It should be borne in mind, however, that data required for sophisticated techniques such as VPA which inform dynamic pool models are a luxury that few developing countries can afford. Indeed, even the catch and effort data needed for less sophisticated surplus yield models are rarely available in developing countries in the tropics.

In principle, this concept could be applied to exploitation of resources in which the adult organism is not killed, but produces useful resources such as eggs, leaves, seeds or resins. For example, in the Maya Biosphere Reserve in northern Petén, Guatemala, leaves of palm trees (*Chaemaedorea* spp.) are clipped from the plants on a regular basis for export to the USA and Europe where they are used as decorations in cut-flower arrangements (Reining & Heinzman 1992). The plants are not killed, and people can return to the same plants every 3–4 months for an exploitation cycle that can be indefinite. One could think of an MSY in leaf production in the same manner as the trade-off between growth and natural mortality of individuals in a dynamic pool model, where removal of leaves is equivalent to removal of the entire individual in more conventional models.

5.2.4 Recruitment

Although dynamic pool models are very good at accounting for the growth of individuals, they do not deal explicitly with absolute recruitment (indeed this is why calculations are 'per recruit'). Thus, the ability of the population to replace itself over a large range of adult population sizes may be taken for granted. This ignorance has contributed to some spectacular disasters in fisheries, and it is a serious risk in many other forms of exploitation. The crucial question is: how far down can you push a population without impairing its ability to replace itself? In fisheries, the typical answer has been that no one knows until it is too late. Recruitment is often highly variable among years as a result of the combination of biological events, such as predation and food limitation, and annual fluctuations in abiotic factors, such as salinity, changes in water currents and temperature. A great deal of effort has been devoted to understanding stock–recruitment relationships, because if future recruitment could be predicted, advice could be given to safeguard it. A recent example comes from an analysis of North Sea cod, *Gadus morhua* (Fig. 5.4).

Note that there have been strong annual fluctuations in recruitment, with little relationship to the size of the adult population in the early years. Since the mid-1980s, however, as spawning stocks have declined under fishing pressure, there is evidence of a decline in recruitment, and a potential for commercial collapse (Cook *et al.* 1997).

Some of the most convincing patterns between stock size and recruitment

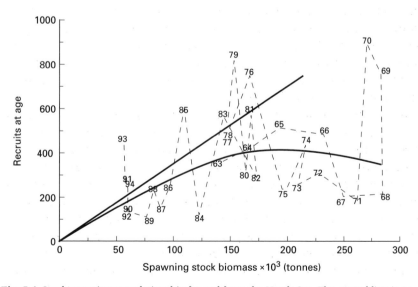

Fig. 5.4 Stock–recruitment relationship for cod from the North Sea. The curved line is a fitted stock–recruitment function, and the straight line shows the estimated levels of recruitment required to sustain the population. Note that whether or not one believes in the accuracy of the fit of the data to the stock–recruitment function, in 23 years out of 31, recruitment has been too low to replace the population. (After Cook *et al.* 1997.)

have been described for species such as salmon and trout (Hilborn & Walters 1992; Elliott 1994), where high population sizes may cause overcompensating density-dependence, due, for example, to females digging up each others' nests on the spawning grounds, or cannibalism. In these species, as well as long-lived, late-maturing species, which cannot replace themselves as quickly as species with high potential intrinsic rates of increase, an understanding of how recruitment relates to adult populations can be important for managing long-term sustainable exploitation.

For species with relatively low fecundity, in which the number recruiting into the population is likely to depend upon the number of eggs or young produced by the adults, Leslie matrices can be used for calculating the consequences of exploitation (Getz & Haight 1989). As an example, Milner-Gulland (1994) explored the consequences of various management strategies for hunting saiga antelope *Saiga tatarica* in Central Asia. The available information on age-specific and sex-specific survival and fecundity was incorporated within a Leslie matrix along with assumptions about the mortality caused by occasional severe winters and droughts. Males are hunted selectively as they have horns, which are used for traditional Chinese medicine, and are larger, thereby producing more meat and larger hides. Milner-Gulland showed that exploiting a larger proportion of males increased the yield and reduced the population variation, although the population is likely to collapse if too high a proportion of the males is killed. This model also showed that a few years of heavy exploitation combined with droughts and bad winters could severely reduce the population. High exploitation levels are only likely to be maintained if the exploitation levels are related to the climate.

5.2.5 Simulation models

An alternative approach is to create simulation models. These have the advantage that it is possible explicitly to include all potentially relevant components of the population biology. One can then use these models to explore the consequences of changing parameters.

Potts (1986) used a simulation model to calculate the maximum sustainable yield for the grey partridge *Perdix perdix* on agricultural land in the UK. Ecological studies showed that egg loss increased with nesting density. The reason for this seems to be that encounters with nests resulted in some carrion crows *Corvus corone* becoming nest-finding specialists and thus, at high nest densities, more crows specialized on partridge eggs. There was much less of an increase in egg loss with nest density in areas where gamekeepers killed crows. Fewer hens also died during incubation in areas with gamekeepers as a result of a reduction in densities of red foxes, *Vulpes vulpes*. The survival of the chicks was known to depend upon the abundance of food for the chicks, which in turn largely depended upon the use of insecticides and herbicides (which kill the weeds upon which the preferred insects feed).

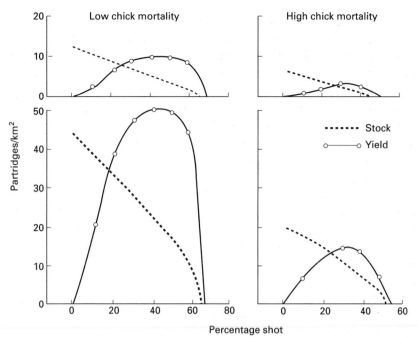

Fig. 5.5 Predictions from a model of grey partridges showing how the predicted stock and yield vary with the percentage shot. The simulations were run for all combinations with predator control (lower graphs), without predator control (upper graphs), with low chick mortality (left graphs) and with high chick mortality (right graphs) (Potts 1986).

A model incorporating all this information, including the best estimates for the survival of partridges at each age as well as clutch size, was run with different shooting levels to predict the consequences for the stock and yield (Fig. 5.5). This shows how the control of crows results in a fivefold increase in exploitable yield, while leaving strips of agricultural field unsprayed with pesticides results in a doubling of the yield. The advantage of this kind of approach is that the model can be tailored precisely to a specific situation, providing a sound scientific basis to management advice. The underlying calculations in simulations are inevitably less clear than in analytical approaches, however, and one must be cautious about generalizing such results to new situations.

5.2.6 Short cuts to measuring maximum sustainable yield

We have seen that one hindrance to holding exploitation within sustainable limits is that it is often impractical to collect the data and employ the sophisticated techniques necessary to find the MSY. Faced with such difficulties, researchers are often forced to fall back on rules of thumb, derived from a variety of theoretical and empirical studies.

It seems likely that short-lived species can sustain higher annual levels of exploitation than long-lived species (which tend to be slow growing, late in maturing and produce few offspring). As a rough approximation, derived from models using a range of parameter estimates, the annual sustainable exploitation mortality is thought to be about 0.1–0.15 times the annual natural mortality rate (Kirkwood *et al.* 1994).

Another rule of thumb used in fisheries is the '$F_{0.1}$' concept, in which the population is exploited at a level of instantaneous mortality, F, where the slope of the yield curve is one-tenth of the slope at the origin (i.e. where there is little fishing pressure). This point is called $F_{0.1}$, and is more conservative than F_{MSY} (Gulland 1983).

In species in which males do not contribute parental care, a greater yield can usually be taken from males than females. For example, for pheasants *Phasianus colchicus*, the MSY results from killing 90–95% of males and 20% of females. In the USA the law prevents females from being hunted, which initially suggests that females are underexploited. However, in practice, as a result of ignorance, mistakes and disobedience, 17% of the females are killed and thus this law produces about the right solution (Hill & Robertson 1988).

One approach frequently taken in studies of exploitation is to consider whether the population has declined; this is viewed as evidence for over-exploitation. However, as we saw with surplus yield models, a sustainably exploited population may have fewer individuals than an unexploited one. For example, the MSY for elephants *Loxodonta africana* occurs at 37% of the unexploited population size, and the equivalent figures are 56% for red deer *Cervus elaphus* and 40% for reindeer *Rangifer tarandus* (Beddington & Basson 1994). Thus, a reduced population size cannot by itself be taken as evidence of overexploitation. The opposite may be true of species with indeterminate growth such as trees; an unexploited population may have many large individuals, but little recruitment. As exploitation increases, the number of large individuals declines, but recruitment increases, resulting in a similar or higher biomass comprising more individuals.

5.3 Causes of overexploitation

Enhanced growth of individuals and increased recruitment are two main potential mechanisms through which populations compensate for in-creased mortality. Exploitation beyond these capabilities has been given the corresponding names of 'growth overexploitation' and 'recruitment overexploitation'. Whereas the former implies inefficient use of resources by exploiting before the species' growth potential has been realized, it is the latter that keeps managers and conservationists awake at night, because of the possibility of population collapse. We now review some of the key causes of these forms of unsustainable exploitation.

There is a widespread and popular belief that non-technological cultures exploit wild populations in a sustainable manner. There are undoubtedly cases in which taboos and rules do prevent overexploitation. In the Warlpiri and Arenda tribes, in the central Australian desert, individual men and women are responsible for particular plant and animal species

(P. Jarman, personal communication). The responsibilities include teaching stories relating to their species, carrying out ceremonies at sites relating to that species, passing on the instructions for maintaining the species and undertaking management. For example, spinifex grass is burnt so that the rufous hair wallaby *Lagorchestes hirsutus* can feed on the new shoots. There are areas where hunting is illegal and it turns out that these are areas where the mammals retreat in times of drought and would thus be very susceptible to overexploitation (P. Jarman, personal communication). These social rules allow some flexibility, for example, an individual facing starvation can hunt in a forbidden area. These rules thus seem a sensible compromise to reduce starvation of the aborigines. For some of the species, the government is running recovery plans and the relevant aborigines have been given key responsibilities in implementing them.

Despite such attractive examples, deliberate restraint to conserve populations is probably not the norm. The evidence is that overexploitation has resulted in many extinctions both historically and recently and in both technologically advanced and traditional cultures (Chapter 2). Indeed, there is a problem with evidence used to infer that traditional cultures exploit species in a sustainable manner. It is likely to be generally true that, with the continuation of traditional levels of exploitation, the targeted populations will not decline. However, cynics can argue that this does not mean that all past exploitation has been sustainable, since species that had been overexploited will have already collapsed. Therefore the remaining community will be in equilibrium with current rates of exploitation. An increase in either the human population or exploitation efficiency due to improved weapons, access or transport, as has happened in many developed countries during the last century, may then result in a new lower equilibrium. The difference between traditional cultures and more technological cultures may therefore be a matter of timescales: we can currently witness and record massive shifts in ecosystems due to overexploitation by developed countries today, whereas any such shifts in traditional cultures may have happened long ago.

5.3.1 Tragedy of the commons

Open access to a resource by more than one person or collective often leads to overexploitation because of competition among individuals (Gordon 1954). This, the well-known 'tragedy of the commons' (Hardin 1968), is illustrated in Fig. 5.6.

Here the cost of exploitation is added to the surplus yield model (Section 5.2.1) of exploitation. As well as including the obvious costs of equipment and fuel, this also includes the opportunity cost, for example of being unable to claim compensation for unemployment or carry out alternative work. As shown in Fig. 5.6, the expectation is that more and more effort, for example logging trucks or fishing boats, will be added until the economic

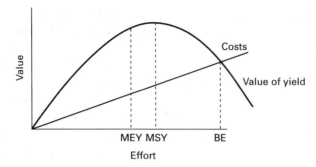

Fig. 5.6 The tragedy of the commons. The value of the yield increases with the exploitation effort up to the maximum sustainable yield (MSY), after which they decline again. In this model, the costs are assumed to be directly proportional to exploitation effort. The greatest reward, the maximum economic yield (MEY), is at the point at which there is the greatest discrepancy between these lines. With open access, individuals will continue to invest more effort until the bioeconomic equilibrium (BE) point is reached, where the costs equal the value of the yield.

benefits equal the costs. At this break-even point, the profits have gone, the population is usually overexploited and excessive and wasteful effort is going into the industry.

Although the model presented here obviously includes far too little ecological or economic detail to give precise predictions, the general result, that populations are overexploited with too much effort while generating low profits, seems to be a depressingly accurate description of most of the world's fisheries.

The control of common access exploitation is difficult. The central idea is that, with many individuals behaving rationally and seeking the most profitable solution for themselves, any possible profits are always dissipated eventually through increased expenditure. For example, some governments subsidize fishers. The consequence is to lower the cost line in Fig. 5.6 so that fishing becomes more profitable. It therefore pays more individuals to enter the fishery until the profits have again disappeared, but the stock is now even more overexploited. Another technique is to restrict the total effort. If successful and the population recovers and is exploited at the maximum sustainable effort, controls can be relaxed. Controls on effort are widespread for bird and mammal hunting in the USA and Canada, where people readily accept the need for hunting seasons and licences. In commercial operations such as fisheries, however, recovery of the population will encourage others to participate, and as the total effort is limited, each can fish for only a restricted period. Such inefficiency will raise the cost until the profits have gone. It will also encourage each fisher to invest more in the technology that will increase the number of fish they can catch per day out fishing. Whether restrictions are imposed on either total catch or effort, the result will be economic problems for fishers, who have invested in expensive equipment, often at the original prompting of governments.

One logical solution is to tax people because this raises the cost line of Fig. 5.6, so yielding more fish, for example, with less wasted effort and providing a profit to the government (Clarke 1989). After short-term losses a new equilibrium will be reached at which the profits to the fishers would be the same. It would, however, take a very brave politician to suggest taxation as the solution to an impoverished industry seeking subsidies!

An obvious way to overcome the tragedy of the commons is to provide unambiguous single ownership rights to resources. This ought to give people incentives to manage the resources sustainably for the ongoing benefit of themselves, their families or their community. However, in reality, the opposite may sometimes occur. For example, in Vanuatu, a country consisting of numerous islands in the South Pacific, many of the areas of surviving forest occur where there are land ownership disputes. If these disputes are resolved, a frequent consequence is that the timber rights are sold to a logging company and the area is deforested. This pattern has been repeated in many other parts of the world.

5.3.2 Time discounting

People usually prefer to have money now rather than in the future, even when corrected for inflation (Schmid 1989). As a result, we expect to be paid extra if we wait for our money, which is why banks pay interest if we leave money with them and charge interest if we wish to spend money before earning it. The discount rate measures the decline in the future value of money. Thus, politicians who gain (politically or financially) from selling a resource now are unlikely to be those who would gain from conserving it for the future. These people therefore place a high discount rate on the resource.

The consequence of discounting is that, as resources are worth more now than in the future, it can make economic sense to overexploit them. For slow-growing species such as whales or rainforest trees, the interest gained from exploiting them all and investing the money may exceed the annual profits from sustainable management (Clarke 1973; Lande *et al.* 1994, 1995). Even if not calculated directly, it is common for government management authorities to use the argument of 'current economic needs' for increased quotas regardless of the consequences for future yields (Clarke 1989). Time discounting can thus serve as a serious impediment to sustainable exploitation.

Economists argue sensibly that, with a constant annual discounting rate, the value of a payment or cost must decline exponentially with time. Thus if the benefits of restraining the extent of forest logging or the high costs of decommissioning a nuclear power station are recouped far in the future, they have a negligible impact on considering whether it is currently cost effective. However, there is evidence that members of the public do not have such a high discount rate for long-term projects and the manner in

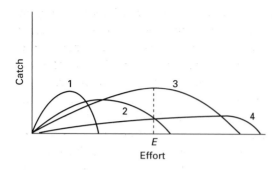

Fig. 5.7 The yield for a number of coexisting species in relation to exploitation effort. It is not possible for them all to be exploited in the most effective manner. At exploitation level E, species 1 is made extinct, species 2 is overexploited, species 3 is exploited at an appropriate level and species 4 is underexploited.

which they discount future values fits a hyperbolic decline (Henderson & Sutherland 1996). Curiously, although economists always use exponential rates, they may choose lower rates for long-term projects, such as forestry schemes or water development projects, in accordance with the hyperbolic relationship (Henderson & Sutherland 1996).

5.3.3 Killing non-target species

Multispecies exploitation and by-catches

Many species are not exploited in isolation. Those that will suffer most are those that are disproportionately likely to be caught and whose theoretical MSY is at a low level of exploitation (Fig. 5.7).

As an example, the saola *Pseudoryx nghetinhensis* is a newly discovered bovid (cow) first seen by scientists in 1992 in Vietnamese forests (Dung *et al.* 1993). Local hunters gain considerable esteem in their villages for returning with such a valuable species (Kemp *et al.* 1997). Saolas are so rare that they are not worth targeting and are only captured opportunistically while hunting other game. As a result of this opportunistic hunting the saola is now restricted to remote inaccessible forest and is probably highly vulnerable to global extinction. Furthermore, the considerable interest from the world's scientific community may also increase the pressure to obtain dead or live specimens such that it may become worthwhile targeting this species and further increase the likelihood of extinction.

The unfortunately named common skate *Raja batis* was virtually extirpated from the Irish Sea by commercial fisheries, in part because it was a by-catch of commercial fisheries targeting other bottom-living species such as flatfishes (Brander 1981). They were vulnerable because of a combination of advanced age of maturity (11 years), low fecundity (40–70 eggs/year) and high mortality, due to their broad shape and large size which made them susceptible to fishing nets from hatching onwards. This species thus has the dubious honour of being the first, and still one of the few, fishes to become virtually extirpated locally due to commercial fisheries.

The overall impact of fishing on non-target fish species is still unclear. A recent analysis of population trends in 10 non-target species of fish in the

North Sea revealed that, from 1970 to 1993, although fishing mortality on some non-target species in the region was quite high, eight of the species had actually increased in number (Heessen & Daan 1996). Data for the other two were either inconclusive or indicated no overall trend. The size distributions of the species also remained fairly constant. Much remains to be done to understand ecosystem-level effects of commercial fisheries and ways to mitigate their effects (Norse 1993; Hall 1996; Laevastu 1996).

By-catches that include birds, mammals and turtles have caused particular concern and this is particularly a problem with drift netting to catch tuna and squid. In the 1970s, the total length of nets was estimated at an astonishing 50000 km per night. A public outcry over by-catches led to a United Nations ban of large-scale deep-sea drift netting in 1992. Since then high-seas drift nets have largely disappeared from most waters, a major problem being some countries in the European Union.

An alternative means of catching tuna is by long lining, which also has a by-catch problem. Long lining consists of 2400–3000 baited barbed hooks on lines 100–125 km long. In the 1980s, over 100 million hooks were set annually. The lines sink to 60–300 m but, when the bait is first cast overboard and is close to the surface, pelagic seabirds are attracted to the bait and large numbers are hooked and drown. One conservative estimate is that 44000 albatrosses were killed each year in the southern oceans by Japanese tuna longlines (Brothers 1991). The population of wandering albatross *Diomedea exulans* at Bird Island, South Georgia has declined by 1% per annum for almost 30 years (Croxall *et al.* 1990), apparently due to deaths from longline fishing for southern blue-fin tuna *Thunnus maccoyii* (Brothers 1991). The annual losses of albatrosses from this fishery are estimated to be the equivalent of 2–3% of adults and 14–26% of the juveniles (Croxall *et al.* 1990) and this is sufficient to account for the long-term decline. Long lining is likely to increase and become an even more important conservation issue; for example, China is currently building hundreds of new long-lining boats.

There are a number of measures that reduce the by-catch of seabirds to longlines, including using a line with streamers. The best solution seems to be to shoot the line out from below the waterline and Norwegian designers have produced a boat that does this. It is in the fishers' interest to prevent a by-catch of seabirds as the loss of bait taken by seabirds cost the southern blue-fin tuna fishery seven million Australian dollars annually in the 1980s (Brothers 1991).

A process analogous to by-catches explains the link between rainforest exploitation and declines in primates. In one study (Wilkie *et al.* 1992) in the northern Congo only the most valuable timber was removed. This destroyed 6.8% of the canopy. Such a small level of habitat degradation might be expected to have little effect on the primate populations, yet line transects showed they were very scarce in logged forest. The real reason for the decline was the construction of roads and logging camps which

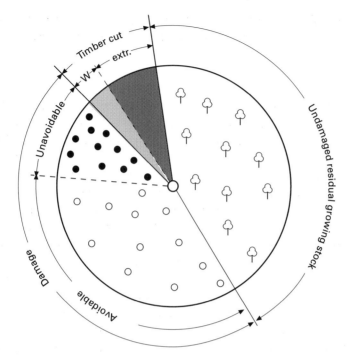

Fig. 5.8 Damage to timber by selective logging in a lowland tropical forest, typical of mixed dipterocarps. The circle depicts the biomass of commercial and potentially commercial species, divided into the following sections: timber cut, damaged stock and undamaged residual growing stock. Unavoidable damage is due to trails and felling activities, and avoidable damage is due to poor management, poor enforcement and lack of skill by loggers. extr., extracted timber; W, excessive waste due to breakage, abandoned logs and other errors. (From Bruenig 1996.)

increased the opportunities for exploitation of primates. Furthermore, logging vehicles, boats and planes transported hunters, guns, ammunition and the game they had shot.

Forestry can also cause important direct damage to other species of commercially important trees. For example, in typical lowland mixed dipterocarp forests of the tropics, a low rate of selective felling by tractors of 10–20% of timber biomass often leads to damage to another 20–30% of the remaining stand (Fig. 5.8).

Discarding target species

In managed fisheries there are often strict regulations concerning the size and species of fish that can be landed. Fish that are too small are thrown overboard, and rarely survive. Of the 85.7 million tonnes of fish caught in 1983, the most conservative estimate of the total waste and by-catch thrown back is 17–39 million tonnes (Dunn 1995). This estimate excludes the by-catch of turtles, seabirds and marine mammals. In addition to the obvious waste of fish which could have been landed more profitably at a larger size, the disquiet that this causes among the fishing community leads to dishonest reporting and hampers stock assessments based on commercial landings.

5.3.4 Variability and uncertainty

In Section 5.2.1 we noted that exploitation near the theoretical MSY is risky because of environmental fluctuations. Beddington & May (1977) showed that, for a given degree of environmental randomness, as the level of exploitation increases, the fluctuations in population size and yield also increase. An example of the importance of variability was the crash of the Peruvian anchovetta *Engraulis ringens* (Clarke 1981). The fishery where they were caught was the largest fishery in the world during the period 1960–72 and made a considerable contribution to the economy of Peru. Following scientific advice, the authorities restricted the annual total catch to the estimated MSY of about 10 million tonnes. However, they did not restrict the numbers of fishers employed, so that there was massive overcapacity, with the annual quota being caught in just 3 months and the industry remaining inactive for the rest of the year. The anchovetta population crashed in 1972–73, probably due to a combination of the MSY figure being set too high and an El Niño event—a change in the water current with an influx of warm water blocking nutrient upwellings from below—which had considerable deleterious ecological consequences. Whatever the reason for the decline, the presence of 20000 people dependent upon the fisheries made it politically difficult to reduce effort even after the problem was recognized. Fishing continued and pushed the yield down to below 1 million tonnes. The fishery has not yet recovered.

The second risk of using an MSY is that the effort necessary to achieve it may be overestimated due to the failure of scientists to make the correct predictions. The recent collapse of the Canadian Grand Banks cod fishery is a sobering example (Hutchings & Myers 1994; MacKenzie 1995; Walters & Maguire 1996). This was blamed in part on gross overestimations of stock size: too much emphasis was placed on commercial data, which yielded an overly optimistic estimate because the fishers were congregating around the few remaining shoals. There was also a failure to appreciate that recruitment failure due to small population sizes of adults, in combination with changes in water currents and salinity, was a real possibility. As noted in Section 5.2.4, few commercially important species show clear relationships between spawning stock size and recruitment; though again with hindsight this is now clear for cod stocks off eastern Canada, and there are indications that this may be true for the earlier-maturing stocks in the North Sea (see Fig. 5.4).

The lesson from the Peruvian anchovetta and Canadian cod is that there are risks in operating near the theoretical point of MSY, due to environmental fluctuations and uncertainty in population parameters. It is better to underestimate rather than overestimate the yield that can be taken.

5.3.5 Corruption, dishonesty and politics

As a result of the tragedy of the commons and discounting, there are often
conflicts between the interests of the individual and the wider community.
Corruption and dishonesty can be common, with tree-felling licences being
sold illegally and fish catches being landed in secret.

Yablokov (1994) and Zemsky *et al.* (1995) exposed the fact that the former
Soviet Union invented much of the whale catch data they submitted to
the International Whaling Commission. Blue whales *Balaenoptera musculus,*
humpback whales *Megaptera novaeangliae* and right whales *Balaena glacialis*
were all exploited even though all three were fully protected. Furthermore,
about 90 000 more whales were killed than reported. High officials and
the KGB were involved in the deceit and Soviet whaling ships were even
designed so that illegal carcasses could be hidden by a burst of steam if a
plane or boat came near.

Yablokov (1994) suggested that whaling data from countries other than
the former Soviet Union may also be inaccurate. An illegal consignment of
3.5 tonnes of minke whale *Balaenoptera acutorostrata* meat labelled as prawn
was discovered at Oslo airport due for export in 1993; in 1994 a Norwegian
whaling vessel with a government inspector on board was observed catching
an extra whale. The inspector claimed he had been asleep while the whale
was chased, harpooned and flensed! (Papastavrou 1996). Molecular genetic
studies of meat for sale in Japan in 1993 showed that it included protected
species such as humpback whales (Baker & Palumbi 1994).

As dishonesty increases, this will affect the quality of the data. Thus it is
accepted that the quality of the fisheries data obtained by the European
Union through commercial landings has decreased as dissatisfaction with
regulations has grown and quotas have become more restrictive. This
hampers accurate stock assessments and the management regulations that
depend on them, thus further increasing the likelihood of mismanagement.

Bruenig (1996) suggests that there is a long tradition in forestry of
sustainable management of tropical forests, but that worldwide current
management practices are usually very damaging both to the forest and
long-term exploitation. The problems include overexploitation and especially removing trees of 40–80 cm in diameter which are in the fastest
period of growth, failure to leave some mature trees, damage in extraction
to other trees and the soil, returning to extract further trees from logged
areas after a few more years' growth, and use of the forest after extraction
by farmers or land speculators. Much of the damage is a result of the demand
for short-term profits, bad planning and untrained workers. Bruenig
suggests that the motivation for sustainable use will only occur with strong
incentives and shared benefits with the responsibilities shared by the
government, civil servants, private forestry, forestry industry sectors,
labourers and the local population. This will only occur with the security
of long tenure, conditional on compliance and good behaviour, which can

result in the creation of a skilled and motivated workforce. By contrast, in most areas, the concessionaire acts as a multiple fee-collecting licence holder who contracts out the exploitation to the contractors. The contractors then hire logging subcontractors who then hire subsubcontractors, and it is these often unskilled casual labourers who actually do the road construction, felling and extraction. The last two stages tend to be squeezed financially and there is little incentive for those carrying out the cutting to obey laws or exploit in the responsible and caring way that is essential for sustainability.

Politicians regularly use predictions from sustainable exploitation models as a starting point for negotiations rather than as a fixed upper limit. This represents a simple conflict between the interests of long-term sustainability and the short-term needs of constituents and election cycles. There is also good evidence of political corruption causing overexploitation for profit. An example at the time of writing concerns logging operations in Cambodia, where senior politicians have been accused of circumventing the ministries of environment and finance in handing out secret concessions to foreign timber companies in return for personal payments. In total 6.5 million ha (one-third of the country's land area) is available for logging concessions.

5.3.6 Genetic changes in exploited populations

A reasonably straightforward and common solution to prevent overexploitation is to exploit the large individuals and conserve the smaller ones. However, recent work has shown that concentrating on larger individuals can act as an evolutionary selection pressure that can result in changes in growth rate and age of maturity which will, in turn, result in a reduction in yield (Stokes *et al.* 1991).

A classic example concerns five species of Pacific salmon (Ricker 1981). For each species the mean size has shown a steady decline over the past 50 years. The pink salmon *Oncorhynchus gorbuscha* returns from the sea to breed when 2 years old and all individuals then die. As a result of this strict biannual cycle, the odd-year fish can be considered as a separate population from the even-year fish. The data shown in Fig. 5.9 were derived from seine nets which are unselective for body size, so that the changes must be due to the changes in the size structure of the population.

Of the 97 populations for which such data are available, there was a significant decline in 57 between 1951 and 1974. For the even-year populations, mean body weight dropped from 2.1 to 1.4kg. This decline in salmon body size could not be related to changes in salinity or temperature. Calculations show that the decline in size can be explained entirely by the exploitation being strongly biased towards large fish if it is assumed that 22–30% of the variance in size is inherited, which is within the range for other fish.

Law & Grey (1989) have analysed a similar situation for the Arcto-Norwegian population of cod *Gadus morhua* which spawn near the Lofoten

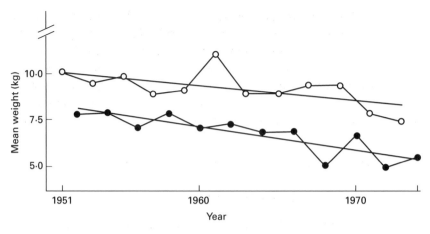

Fig. 5.9 The decline in mean weight of 2-year-old pink salmon caught using seine nets between 1951 and 1974. The upper line shows the data for odd years while the lower line shows the data for even years. (From Ricker 1981.)

Islands of Norway, where they have been fished for at least 1000 years. A fishery used to be concentrated on the spawning grounds until the 1930s, when fish on the Barents Sea feeding grounds started to be caught as well. As a result of this new fishery, only 2% of fish survive the period between 3 and 8 years old (8 years was the age at maturity), whereas 40% survived this period prior to fishing on the feeding grounds. It is inevitable that this mortality must favour early reproduction and the age of sexual maturity has advanced by about 2 years over two decades. Law & Grey (1989) predicted that, with the current policy, in which the instantaneous fishing mortality is about 0.2 (18%) on the spawning grounds and 0.6 (45%) in the Barents Sea, individuals will mature at 4 years old and the annual MSY will be 0.21 million tonnes. The strategy of only exploiting them on the spawning grounds, with a mortality set at 0.6, should result in fish maturing when 9 years old thus providing an annual yield of 1.44 million tonnes. These models show that evolutionary changes could have considerable effects on yield. However, with the current short-term perspective of fisheries management, it seems very unlikely that many will be persuaded to take into consideration the impact of genetic changes.

One slightly more hopeful point may be worth considering. The evolutionary changes discussed above have consequences analogous to growth overexploitation: individuals at the time of capture are smaller than optimal for maximum yield. On the other hand, this evolutionary response to selection may help preserve recruitment if losses in fecundity due to smaller size of females are offset by more females surviving to breed. Thus, an evolutionary response to (artificial) selection may ultimately help the fishery.

There may be other contexts in which evolutionary responses could aid in long-term maintenance of populations. In eastern Zambia, the proportion of adult female elephants that were tuskless increased from 10.5% in 1969 to 38.2% in 1989. This was apparently a result of illegal ivory poaching that selectively killed the tusked individuals. This was primarily a within-generation change in frequency. However, there is evidence from studies of families that tusklessness is sex-linked and inherited (Jackmann *et al.* 1995). Thus, given sufficient time, there is the potential for an evolutionary response to selection against tusked individuals.

5.4 Solutions to prevent overexploitation

5.4.1 Difficulties of reducing effort

Attempts to reduce exploitation are often resisted even though the yield will be greater in the long term. One reason relates to the tragedy of the commons, as it is in each individual's interest to exploit at a high rate while discouraging others from doing so. The second, and probably more important, reason is that overexploited populations take years, and usually decades, to recover, which is little use to the people who have lost their livelihoods in the short term. Many of the current exploiters will thus not gain from enhanced yield in the future.

5.4.2 Legislating to reduce mortality

The two main ways of reducing mortality are to reduce the total amount of effort (for example by limiting the number of hunting licences or the length of fishing seasons) or efficiency (for example by limiting the total daily bag limit or net size).

Reducing effort in many ways seems the most sensible, because it can be easier to enforce and it discourages the incidental killing of non-target species and undersized or too-young individuals. One method that is sometimes used is to limit the number of days that individuals can hunt or fish, but this tends to be unpopular. Another method is to limit the total allowable catch within a given time period and stop exploitation when that catch is reached. However, both of these measures miss the essential point that people like to go hunting or fishing and resent constraints which are often perceived as unnecessary interference.

5.4.3 Exclusion areas

Another method of reducing exploitation that is currently popular is the use of reserves in which all exploitation is banned. This solution is usually easier to enforce than restrictions on landings. In theory, reserves such as marine protected areas are meant to protect part of the population, while 'leaking' individuals to hunting or fishing areas for exploitation. Marine reserves have often been suggested for areas that may then export larval fish. Reserves also have the benefit of providing control areas so that the consequences of exploitation can be seen. Two review groups have suggested that 20% of the area marine regions should be set aside as reserves and it has been suggested that even higher percentages may also be needed for heavily overexploited fisheries (Roberts 1997).

5.5 Eradication

In this chapter, we have assumed that management of exploitation is aimed towards sustainability. In the case of pest species, the objective may be to overexploit or even, as in the case of smallpox, to eradicate the species.

One example is the coypu *Myocastor coypus*, a large South American rodent which was farmed for its fur. Feral populations appeared in Japan, North America and Europe as they escaped or were released. The population in south east England increased to 70 000 by 1960 (Boorman & Fuller 1981). They caused problems by burrowing through river banks, and eating crops and certain marshland plants. They were therefore controlled in Britain from 1961. After successful trial eradication in one area, it was calculated that eradication from Britain was possible within 10 years with 24 trappers. This programme was started in 1981. One key element in this project was that the minimum time in which it was realistic for the trappers to eradicate the coypus was calculated and a bonus scheme was then devised so that it was in the hunters' interest to eradicate the coypu as soon as possible (Gosling & Baker 1987). This was essential as making the coypus extinct would make the hunters unemployed. The reward was a lump sum equal to 3 years' salary if the exploitation took place within the minimum term that the simulation predicted was possible. Eradication was successfully achieved in 1989.

The natural history of a species may often be crucially important in determining the success of a control scheme. For example, there have been attempts to control the feral pigeon *Columba livia* in towns by means of toxic baits. However, Murton *et al.* (1972) showed that the dominant pigeons are the ones that usually breed and use the predictable food resources. It is thus only subdominant non-breeders that use new food resources, such as those provided in control programmes. As a result, large numbers of birds can be killed without a great effect on the population. For example, in one study, although 9000 birds (mainly non-breeding immigrants) were killed, the total resident population of 2600 birds was only halved.

The results of previous eradications may not always be obvious. In Europe and North America, spectacular predators, such as the gray wolf *Canis lupus*, brown bear *Ursus arctos* and white-tailed eagle *Haliaeetus albicilla*, are largely restricted to mountains and high latitudes. It is not widely appreciated that many of these species once occurred well to the south, but were eradicated by farmers or gamekeepers. This persecution still persists. The hen harrier *Circus cyaneus* is heavily persecuted in the UK on moors managed for the shooting of a gamebird, the red grouse *Lagopus lagopus*, with an estimated 11% of adults killed each summer in Scotland (Bibby & Etheridge 1993). In the absence of persecution, hen harriers breed very successfully on grouse moors (Etheridge *et al.* 1997) and it was calculated that, in the absence of interference, they would initially increase by an estimated 13% per year until a new unknown equilibrium was reached.

5.6 The future

The future for many exploited populations might look bleak with many being severely overexploited and many facing additional threats such as habitat loss and degradation. It does not follow that these processes will continue. A major factor influencing exploitation is the attitude of the public, which has already affected the exploitation of whales (Fig. 5.10), seals and the fur-trapping, drift-netting and egret-hunting industries. Many currently consider that shooting an adult seal is unethical, but are happy to eat cod that has been gutted alive on a trawler deck. It may be that changes in public attitudes and fashions will have the greatest influence on future patterns of exploitation.

Acknowledgements

We thank Nicola Crockford, Hanna Kokko, Simon Jennings, Mike Pawson, Carlos Peres and Carl Smith for discussions and comments.

Fig. 5.10 Nagaoka Tomohisa was once a professional whale harpooner working in the Antarctic who won awards for some of the largest catches on record and killed a total of 4000 whales during his career. He now runs his own very successful whale watching company in Japan. Forty thousand people went whale watching in Japan in 1996 and interest is increasing. As a result of ventures such as this, exploitation is likely to be increasingly unacceptable. (Photo by M. Carwardine.)

CHAPTER 6
Small and declining populations
Daniel Simberloff

6.1 Introduction

Small populations and declining ones have been the focus of conservation scientists because they are assumed to be under the highest threat of imminent extinction. However, it may not be automatically true that such populations are in great danger. Some small populations have persisted for very long times. And some populations currently in decline are of species typified by drastic vicissitudes of population size, so that one would expect them to be declining often even if there were no new threat causing the decline. Even if we can establish that small populations and declining populations are in the greatest danger, just how small does a population have to be in order to be immediately threatened with extinction? Exactly what threats are so greatly exacerbated in small populations? And how rapid and sustained a population decline is needed before we should assume there is a conservation problem? These are the questions addressed in this chapter.

6.2 Persistence times and the minimum viable population size

Much recent conservation biology has focused on the concept of a minimum viable population size or MVP (references in Simberloff 1988; Caughley 1994). The term implies that as populations get smaller and smaller their existence becomes increasingly tenuous until, at some critical threshold (the MVP), swift extinction is assured. Several forces may weigh disproportionately on small populations, and which one comes into play first as a population declines, and thus sets the MVP and becomes the immediate cause of the extinction, depends on the particular population. The key forces are as follows.

6.2.1 Demographic stochasticity

Demographic traits are characteristics such as sex ratios and birth and death rates that govern the population growth rate. There are mean birth

116

and death rates, and sex ratios, for the entire population, but these means are based on birth and death rates and the sexes of individuals. In a large population, the variation among individuals in these traits is unlikely to matter. For example, if the mean sex ratio is one female to one male, the population is unlikely to run out of males or females. However, the smaller the population is, the more its fate is likely to rest on one or a few individuals. If these few individuals happen to have unusual values for some demographic trait, the entire population may be threatened. Consider the sex ratio example: if the number of individuals, in any generation, in a population with non-overlapping generations is N, the probability that they will all be of one sex (thus dooming the population) is 2^{1-N}. With a population of, say, 100, this number is infinitesimal (1.6×10^{-30}), but in a population of, say, six individuals, this fate is a real possibility—its probability is 0.03125.

A convenient way to characterize the relationship between population size and the likelihood of extinction owing to some force, like demographic stochasticity, is to plot the expected persistence time of the population (that is, the expected number of generations to extinction) as a function of population size. For demographic stochasticity, the plot curves steeply upwards (Fig. 6.1). This is equivalent to saying that, if demographic stochasticity were the only threat to a population, extinction would be expected to be fairly quick in a very small population, but even a modest increase in population size defers this expected demise until far into the future—so far, in fact, that in nature, where many other forces are also operating, demographic stochasticity is unlikely to cause extinction.

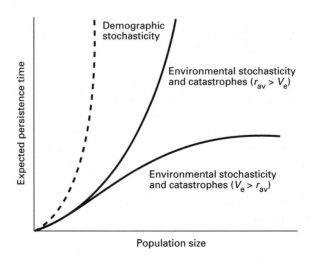

Fig. 6.1 Expected persistence time as a function of population size for populations subjected to demographic and environmental stochasticity and catastrophes. V_e is the variance in population growth rate caused by temporal environmental variation and r_{av} is the average population growth rate. (After Lande 1993; Caughley 1994.)

Another way of looking at this relationship is to say that demographic stochasticity is unlikely to set the MVP, because other forces will come into play with decreasing population size well before demographic stochasticity would cause extinction.

6.2.2 Environmental stochasticity

One such force may be fluctuations in the environment—the physical environment, like weather factors, and the biotic environment, like predators, parasites and competitors of the population in question. Of course, such environmental variation affects the demographic parameters of the population—for instance, an unusually cold winter may lower the average birth rate or increase the average death rate. But environmental stochasticity affects the entire population—in a cold winter the probability of death is increased for all individuals. Demographic stochasticity, on the other hand, refers to the variation among individuals within the population—some will die and others will live, no matter how severe the winter.

One may imagine that a force that affects all individuals in a population simultaneously, like bad weather or a population explosion of a predator, would be more likely than demographic stochasticity to eliminate the whole population, especially given the above result that demographic stochasticity is unlikely to cause extinction unless a population is minuscule (see Fig. 6.1). Indeed, in the 1980s, a consensus emerged among conservation biologists that environmental stochasticity, on average, is a more important threat and that conservation planning should aim to counteract it.

Subsequent modelling (Lande 1993) suggests that the importance of environmental stochasticity, and the way in which it varies with population size, depends crucially on the extent of the variance in population growth rate caused by temporal environmental variation (V_e) and on the average rate of increase of the population (r_{av}). If r_{av} exceeds V_e, the expected persistence time is a monotonically increasing function of N, and one that curves sharply upwards (see Fig. 6.1), although probably not as sharply as the curve for demographic stochasticity. This is to say the MVP would more likely be set by environmental than by demographic stochasticity, because, for any N, the expected persistence time is shorter. However, as with demographic stochasticity, the upward curve suggests that, as long as the population is not very small, environmental stochasticity is not very likely to cause extinctions in nature, because other forces will come into play at even lower population sizes.

If, on the other hand, V_e is greater than r_{av}, the situation is rather different. The relationship between N and the expected persistence time is still monotonically increasing, but now it is asymptotic at a rather low value of persistence time, rather than curving sharply upwards (see Fig. 6.1). In short, environmental stochasticity is still more likely than demographic stochasticity to cause extinction, and expected time to extinction is now so short, even for large population sizes, that, in nature, such extinction might actually occur. The implication of the shape of this curve for the design of nature preserves is also important. It had generally been assumed, through the 1980s, that the way to insure against short-term extinction was to have a large enough population size so that the point on the environmental stochasticity curve of persistence time as a function of population size was on the steeply rising section. However, if the curve is asymptotic and has no steeply rising section, some other strategy will have to be adopted, as even a very large preserve (and correspondingly large population) does not greatly increase the expected time to extinction over that in a substantially smaller preserve. More appropriate strategies will be discussed below in the discussion of metapopulations.

6.2.3 The meaning of stochasticity

Whether any process is really 'stochastic' is not really very important for conservation. One could argue that the variation described above as 'stochastic' is really deterministic in the sense that, with sufficiently good tools, we could predict it completely. Consider demographic stochasticity. There are certainly physical reasons why one individual lives and another dies, why one female gives birth to x offspring and another to y offspring, why one offspring is a male and another a female. Genetics and physiological condition might be expected to play key roles. Similarly for environmental stochasticity—there are physical reasons why one winter is more severe than another, why a predator population is extraordinarily high in some years, etc. However, to determine all these things would take resources and time that are normally unavailable. Even a quintessentially 'random' event, like the exact spot at which a raindrop falls, is ultimately deterministic in the sense that, with enough instrumentation, we could perhaps explain and predict its exact trajectory.

However, in all these matters, we do *not* have the practical ability to make precise predictions, so we characterize our prediction as a random draw from a specified distribution—this is what we mean by 'stochastic', not that there are no physical determinants of the process or phenomenon we are predicting. Ultimately, whether anything in nature is stochastic, and at what level it is stochastic, is a matter for philosophers to discuss, and rests on the meaning of such terms as 'cause' and 'random'.

6.2.4 Catastrophes

Many small populations in nature have been eliminated or nearly eliminated by single catastrophes such as fires, volcanic eruptions or hurricanes. For example, Hurricane Iniki, striking the Hawaiian island of Kauai in 1992, extinguished five species and subspecies of bird, while an earlier hurricane on St Kitts in the West Indies eliminated an endemic bullfinch. In the light of such observations, the consensus of the 1980s saw catastrophes as more likely to cause extinction than environmental stochasticity, while environmental stochasticity would be more likely to cause extinction than demographic stochasticity. There was less mathematical modelling of catastrophes than of demographic and environmental stochasticity, but common sense seemed to dictate that population size would be less of a factor in the effect of catastrophes than of demographic and environmental stochasticity. After all, if a population is in the path of a meteorite, lava flow or hurricane, it probably does not matter what size it is, because all individuals will die. Rather, what matters is spatial dispersion, as will be discussed below.

More recently (Lande 1993), the distinction between environmental stochasticity and catastrophes has been blurred. The distinction was artificial, as catastrophes *do* constitute variation in the environment; they are simply extremes of variation, rather than the 'normal' range of variation that had been considered as 'environmental stochasticity'. Given this similarity of the two forces, it is not surprising that a single model subsuming both forces (Lande 1993) shows them to scale similarly in plots of persistence time as a function of population size (see Fig. 6.1).

6.2.5 Means, variance and skewness

The analytical mathematical models leading to the plots of persistence time vs population size in Fig. 6.1 are simple ones with few parameters, and the prediction is in terms of mean persistence time. However, with such high stakes as extinction of a population or even an entire species, mean persistence time might be an inadequate characterization of risk. Further, the availability of fast computers with vast storage capabilities has led to an alternative sort of model: a simulation with many parameters and equations tailored to the biology of the species of interest and with complicated relationships of the parameters and variables.

These simulation models sometimes yield surprising results. For example, a simulation model used to propose conservation strategies for the grizzly bear (*Ursus arctos*) predicted a mean persistence time less than a tenth, as long as previously used analytical models had. Moreover, variance among simulation runs was large, and the persistence times were skewed. A population of 50 individuals persisted on average for 114 years, but in 56% of the runs, the population persisted for less than this mean, while the population lasted for more than three centuries in 6% of the runs. A manager might decide in such a case that the mean persistence time is of less importance than the fact that the probability of surviving for less than that time is greater than 50%.

Although such simulation models are more widely used in conservation management than the analytical ones, they have two serious drawbacks.

First, it is harder to understand the assumptions and logic of such a model than it is to understand a simple analytical model, so that it is difficult to know how much faith to put in the results. If no such insight is possible, managers may tend to accept predictions even if the model would appear very unrealistic to an expert. This problem is of still greater concern because of the second drawback—these models are generally not directly testable in the field, at least in the time frame available to conservationists. A model that predicts events centuries or millennia in the future could be adopted as a management guide with irrevocable results, yet there may be no prior way of verifying that its predictions are likely to be accurate.

6.2.6 Ethological and physiological problems

Some animal species have characteristic behaviours that place small populations at particular risk. For example, the effectiveness of schooling behaviour as an antipredator mechanism in fishes rests on the presence of a large number of morphologically identical individuals. This mass prevents a potential predator from focusing for long enough on any one of the fish to be able to home in on it. As the number of schooling individuals declines, the predator is increasingly less likely to be befuddled. Perhaps the most fundamental problem for some species as population size dwindles is simply the decreased likelihood of finding reproductively competent mates. And in some animal species, even if sexually mature adults find one another, group displays are required to induce hormones that stimulate ovarian development and/or mating. An insufficient number of individuals may lead to reproductive failure in the entire population.

6.2.7 Genetic threats

In any real population, not all individuals are equally likely to contribute their genes to future generations. For example, if a population has more males than females, each female will on average contribute more genes to subsequent generations. This is because every individual in the next generation has one mother and one father, so the females as a group and the males as a group each contribute the same total number of genes. Of course, evolved rules of mating behaviour may also determine the relative likelihoods of each individual contributing genes. Thus, with a harem mating system, a few males will contribute fully half the genes to the next generation (again, because every individual has one mother and one father), while the other males will contribute nothing. As another example, if every so often the population size in some generation is much smaller than in other generations (perhaps because of extreme weather), each individual in that generation will contribute more, on average, to subsequent generations than will individuals in a 'normal' generation. This is because all the genes in subsequent generations descend from the few individuals in the small generation.

Because of these reproductive idiosyncrasies of real populations, in order to make comparisons among populations in rates of genetic change, biologists refer to the 'effective population size'. This is the size of an 'ideal' population with the same rate of genetic change as the population of interest, but in which all individuals would be equally likely to contribute genes to later generations. For example, there would be equal numbers of males and females in the ideal population, and the population size would not change from generation to generation.

Using effective population sizes is a way to standardize real populations for comparison, but it is easy to underestimate a threat to a population if one fails to remember that, for genetic forces, 'population size' means 'effective population size'. This is because the effective population size of populations in nature is often much lower than the censused population size. How much lower depends, of course, on the details of the breeding system and the asymmetries among individuals of contribution to future generations, but for natural populations where these details are known, effective population sizes are usually no more than half the census population size, and sometimes as little as a tenth of the census population size.

Four main genetic processes threaten small populations disproportionately.

Inbreeding depression

When close relatives mate, there is a higher probability that any gene will be homozygous—that is, both alleles will be the same. Because some recessive alleles are deleterious, this increased homozygosity can lead to lower fitness—the deleterious recessives, usually masked by a different dominant allele, are exposed. In such a case, the inbreeding is said to have led to 'inbreeding depression'. There may be another component to inbreeding depression—a lower likelihood that development proceeds normally when fewer genes are heterozygous.

The rate of inbreeding, and therefore the degree of inbreeding depression if it occurs, is higher the smaller the effective population size. This is intuitively clear; the smaller the population, the more likely it is that a randomly chosen pair of individuals are closely related. In fact, in a very small population, after a few generations, it is unlikely that two individuals will *not* be closely related. For randomly mating populations, or for particular sorts of mating systems used, for example, by animal breeders, it is possible to calculate the rate at which homozygosity increases through time. But the effect of the increased homozygosity—that is, the extent, if any, to which inbreeding depression occurs—is an empirical matter, and must be determined by experiment and observation of each species.

In zoos and animal-breeding facilities, inbreeding depression is a well-known problem and much effort is expended to avoid it. For example, pedigrees are often kept and individuals are moved great distances to prevent close inbreeding. The phenomenon has also been detected in many plant species. However, some animal species famous in conservation circles have little genetic variation—thus high homozygosity—and appear to suffer no inbreeding depression: Pere David's deer *Elaphurus davidianus*, the northern elephant seal *Mirounga angustirostris* and the wisent or European bison *Bison bonasus*. In these species, more inbred individuals are no less fit than less inbred ones. One hypothesis for this latter state of affairs is that the population sizes of these species were gradually reduced, so that natural selection was able to eliminate the deleterious recessive alleles thought to be the main cause of inbreeding depression. Thus the high rates of homozygosity now encountered do not result in the unmasking of deleterious alleles. If these population sizes had been reduced too quickly, according to this hypothesis, the populations would not have had time to purge themselves of these deleterious alleles by natural selection before they went extinct.

This hypothesis is only an hypothesis, however, and the populations of some species with no detectable inbreeding depression have been reduced quite quickly. In fact, Templeton & Read (1983) have suggested a diametrically different way to avoid inbreeding depression in captive propagation programmes—deliberately choose those highly inbred individuals that are healthiest, and mate them, generation after generation. The underlying hypothesis is that such individuals, by chance, have few or no deleterious alleles causing inbreeding depression. Thus, continually mating such individuals effectively purges the population of these alleles as quickly as possible. In fact, this method was used successfully for the entire Speke's gazelle population in US zoos; more than 50 individuals are all descended from one male and three females, and the population today exhibits no inbreeding depression.

This approach has at least two potential disadvantages. First, it can be used only on captively propagated species, which constitute a small minority of the species threatened with extinction. The second disadvantage is more speculative. One of the great evolutionary discoveries of the last 30 years has been vast amounts of genetic variation (the exact amount depending on the species and population, of course), but the significance of this variation is not well understood. Some of it consists of the deleterious genes that cause inbreeding depression. It is clear that these genes are disadvantageous now, but it may be that some of them would be adaptive in a future changed environment, as could many genes that are currently neutral from a selective standpoint. It is possible that mutation could restore purged genetic variation, which could then be favoured by natural selection when the environment changes. But mutations happen very rarely, especially in small populations. So, depending on the rate of environmental change, genes that had been disadvantageous but had become favoured could be crucial to the survival of a population. The Templeton–Read method purges the population of some of these genes.

Genetic drift

Apart from inbreeding, genetic variation also declines in populations because alleles are lost by chance; which alleles are incorporated into gametes and, to some extent, which gametes get to form zygotes are random processes. This is the phenomenon of genetic drift. In a closed population, this loss of variation is opposed by mutation, which introduces new alleles. However, mutation is a rare event, and, the smaller the population, the fewer mutations occur, so the rate of loss of variation to drift is larger. How significant this loss is for conservation is uncertain. The problem is, the less genetic variation, the less quickly natural selection can work, and the fear is that, if there is too little variation and the environment changes quickly enough, extinction might occur before natural selection has time to adapt the population to the new environment. Changes in both the physical and biotic environments could severely challenge a small population. A chemical pollutant, for example, may kill or prevent reproduction by all individuals but a few characterized by tolerant geno-types. Similarly, a new parasite or pathogen may favour the survival and reproduction of just a few resistant genotypes in a large population and select against all others. In a small population, there may be no resistant genotypes.

Genes that are favourable in the current environment are unlikely to be lost to drift unless the population is very small, because natural selection will dominate their dynamics. Genes that are selectively neutral or even deleterious are not maintained by selection, and are lost faster in smaller populations, but whether they are likely to be useful in a future changed environment is a widely debated topic.

Mutational meltdown

The third random genetic process that can threaten small populations is, in a sense, the opposite of the loss of alleles to drift. It is the fixation of harmful alleles, by drift, before natural selection can weed them out. In sufficiently small populations, just as drift can cause the loss of a favourable allele, it can cause the fixation of a deleterious allele. Lynch *et al.* (1995) have theorized that, in a small population, fixation by drift causes deleterious mutations to accumulate, and this accumulation causes the population size to decrease, which accelerates the rate of fixation of still more deleterious mutations. This accelerating build-up of harmful genes—'mutational meltdown'—eventually extinguishes the population. The process is gradual at first, but once the average viability of population members falls below some threshold value, the acceleration is quite rapid.

In fact, according to this theory, for populations with effective sizes of less than around 100 (census sizes may therefore be a few hundred or even 1000 individuals), this threshold may be reached in just a few dozen generations, and extinction will ensue in a few tens to hundreds of generations. The threat is seen as particularly great for low-fecundity species like birds and mammals. Further, this genetic process will be exacerbated by demographic and environmental stochasticity. The conservation prognosis is fairly grim, as protective measures are often not taken for a species until its total population is quite low. For example, in the USA, the Endangered Species Act has not been invoked to stem the decline of animal species until they have already fallen to *c.* 1000 individuals (census, not effective, population size) (Wilcove *et al.* 1993). According to this theory, it is then too late, and they are already doomed.

Hybridization

Many small populations of both plants and animals have suffered a sort of genetic extinction after hybridization with more numerous non-indigenous species, or sometimes with other native species when a habitat barrier to interbreeding has been obliterated by human activity (Rhymer & Simberloff 1996; see also Chapter 3). For example, the expensive current effort to reintroduce the endangered red wolf *Canis rufus* to the eastern USA is almost certainly doomed because they are interfertile and breed with the coyote *Canis latrans*, a species that is rapidly expanding its range and increasing its population in this region. Similarly, the rare Simien jackal *Canis simensis* of Ethiopia may disappear as a distinct species because it is greatly outnumbered by feral dogs *Canis familiaris*, whose males mate with jackal females.

In these cases, the great disparity between the population sizes threatens the genetic integrity and distinctness of the smaller population, even though both parental populations undergo introgression—gene flow back into the parental populations as the hybrid individuals mate with individuals of the latter. However, even without introgression, hybridization may threaten a very small population that mates with a larger one. In parts of the former Soviet Union, introduced American mink *Mustela vison* greatly out-number the highly threatened native European mink *Mustela lutreola*. Further, the American mink becomes reproductively mature earlier, and American mink males mate with European mink females as the latter mature. The resulting foetus is always spontaneously aborted, so that no gene flow occurs between the two species. But the loss of that breeding season for the female European mink is a major factor in the decline of that species.

6.3 Observations on small populations

With this battery of observed and/or hypothetical threats to small populations, and the realization that numerous forces—especially habitat destruction and fragmentation—have already reduced the populations of many species to within the ranges discussed above, the prognosis for conservation seems exceedingly grim. However, numerous observations in nature suggest that, in at least some cases, things need not be as bad as they seem, and the theory should be taken with a grain of salt.

For one thing, many very small populations have persisted for long periods of time and do not manifest symptoms of imminent extinction, unless humans introduce new threats. For example, the Socorro Island hawk *Buteo jamaicensis socorroensis* population, restricted to the tiny Socorro Island (*c.* 600 km from the Mexican mainland), has about 20 pairs. We can assume that it has been completely isolated for many millennia, both because the geological history of Socorro Island shows it has remained isolated and because the hawk has evolved to differ subspecifically from its nearest mainland relative. Further, because of the position of the hawk at the top of its food web, and the requirement of each individual for a substantial area from which to gather prey, we can deduce that the population has been this small throughout its entire history. Yet there is no reason to think it is endangered unless human activity on the island, which houses a small Mexican army barracks, poses a new problem (Walter 1990).

The Devil's Hole pupfish *Cyprinodon diabolis* (Fig. 6.2) is endemic and restricted to a pool of some 200 m² surface area near Las Vegas, Nevada, and numbers between 200 and 600 individuals. The small population size is not caused by humans, but rather because it inhabits a tiny 'habitat island' that prevents spread and also isolates it from other populations. It has been isolated in this pool for so long that it has speciated, yet there is no reason to think its situation would be precarious were it not an aquatic species living near a huge human metropolis in the desert.

Fig. 6.2 The world population of the Devil's Hole pupfish *Cyprinodon diabolis* comprises 200–600 individuals in one small pool.

Brown (1995) has tabulated a number of other plant, vertebrate and invertebrate species whose total population sizes have apparently been very low for a long time, not because of human activity but because of their geography and biology. Typically, such species live on isolated real islands or habitat islands, a location that prevents their spread or population growth. It is quite possible that, if one had data for species extinction and population size for all species, extinct and extant, that have ever inhabited the Earth, one would find that, on average, those species with minuscule population sizes were, in fact, likely to persist for shorter periods than other species. Unfortunately, such data do not exist. However, it is evident from the above examples, and others, that these tiny populations are in no way doomed to extinction in a few decades or centuries. In fact, there is no reason to believe they cannot persist for many millennia, if humans do not intervene.

Another class of evidence that a very small population need not be doomed to rapid extinction—evidence that particularly casts into doubt the automatically crucial role of genetic forces described above, that are thought to imperil very small populations—is the fate of introduced species. In many regions, some of the most common species are not native, but have been introduced deliberately or inadvertently by humans in the recent past. It appears to be true that the more individuals that are introduced, the more likely the introduction is to survive and to spread. Nevertheless, many introduced species that have large population sizes and geographical ranges originated from very few founders. For example, the initial introduction of the North American muskrat *Ondatra zibethicus* to Europe numbered four individuals, yet they quickly spread over much of Europe (see p. 51). Eurasian house sparrows *Passer domesticus* spread over much of North America from an initial group of perhaps 50 individuals; they are now one of the most common birds in the USA. Among insects, several species deliberately introduced to control pests spread to populate large parts of continents despite having had fewer than 20 founders (Simberloff 1989). To be sure, as noted above, the probability of such growth and spread is higher for larger numbers of founders, but it is apparent that some minuscule founding populations have not gone extinct, and instead have become strikingly successful.

6.4 Extinction and the numbers and spatial arrangement of small populations

The number of individuals in populations is obviously important to the persistence of both populations and species, but it is not the only determinant. The number of populations and their configurations with respect to one another can also be crucial. If a species consists of a single, isolated population, it is likely to be more endangered than if there are several, well-distributed populations. But what aspects of the spatial arrangement of populations are important for persistence?

6.4.1 Metapopulations

A metapopulation is a group of small populations connected by occasional movement of individuals between them. Although the concept is an old one, its relationship to conservation and extinction of species has recently been much discussed (Hanski & Simberloff 1997; see also Chapter 7). If the individuals of a species were arranged according to some metapopulation models, persistence time might be much higher than would have been predicted for one larger, freely interbreeding population with the same number of individuals. Consider a situation where a species consists of numerous sites housing small populations, all of them subject in the short term to some substantial probability of extinction from the various forces discussed above that might set MVP size. If there were enough of these sites, and if movement between them was frequent enough, the entire ensemble, or metapopulation, might be very resistant to swift extinction even though each component population is ephemeral.

This is because movement between the small populations may serve two functions. First, it may prevent some small population that had dwindled close to extinction from actually disappearing—this phenomenon, the forestalling of extinction, is known as the 'rescue effect'. Second, even if some of these populations go extinct, immigration into their sites from still extant populations may re-establish these extinct populations before the extant ones also go extinct (Fig. 6.3). In other words, which sites are occupied by populations and which ones are vacant continually changes, but the sites are never all vacated at once.

In this conception, in which no population is long lived, but the species itself resists extinction, the fate of an individual population and the exact reasons why each one disappears assume far less importance than in the MVP literature discussed above. Instead, the chief matter of concern is the number of populations (or possible sites they might occupy) and the rates

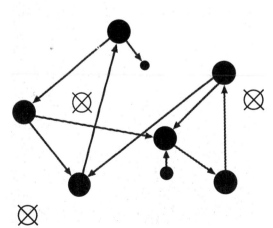

Fig. 6.3 Metapopulation structure, in which movement between ephemeral small populations maintains the entire ensemble even if each population has a substantial probability of extinction. Empty circles represent temporarily extinct populations.

of movement of individuals among them. In short, the arrangement of the species in space is the key to its survival. Individual populations may still be important. For example, if too many of them are lost, immigration into the remaining sites may fall below some crucial rate and the whole metapopulation may collapse as each successive population extinction is no longer redressed. However, no one population plays a key role.

The evidence on metapopulations in nature is mixed. The spatial dispersion of some species, particularly those occupying habitats that are successional or naturally patchily distributed, seems to accord with the notion of a population of small populations. But there is rarely enough information on rates of movement of individuals between aggregations to know whether the aggregations are really acting as quite independent populations (as is required by the metapopulation model) or just as parts of one, large, quite continuous population. For several species, such as a European butterfly, the Glanville fritillary *Melitaea cinxia*, and a North American plant, the Furbish lousewort *Pedicularis furbishae* (Fig. 6.4), information on such movement supports the idea that they are metapopulations and that their metapopulation structure helps prevent regional or global extinction. These cases are exceptional, however, so it is still uncertain how major a conservation problem metapopulation collapse is.

Fig. 6.4 Furbish lousewort (*Pedicularis furbishae*), and endangered plant endemic to Maine with high local extinction rates but a persistent metapopulation.

6.4.2 Do not put all the eggs in one basket

Even if the rates of movement of individuals between aggregations do not produce an extinction-resistant metapopulation, the threats of catastrophes and environmental stochasticity suggest that a prudent conservation strategy would be to have several populations rather than a single one whenever possible. This would be so even if each population had to be smaller than the original one and thus would have a somewhat higher extinction probability from the various forces setting MVPs. This is because of the spatially restricted nature of various phenomena.

An excellent example is the destruction wrought by Hurricane Hugo in 1989 on the one remaining population of the Puerto Rican parrot (*Amazona vittata*), one of the world's rarest birds. Some 50 individuals occupied the Luquillo Forest in Puerto Rico when the hurricane whipped through this part of the island, destroying a substantial fraction of the types of trees required for nesting. Ironically, plans were afoot by this time to establish a second population, specifically to avoid the risk of extinction that a hurricane or other localized disaster might cause. Fortunately, the species survived, but the lesson is clear: it is not wise to put all one's eggs in one basket. The establishment of a network of dispersed populations might confer resistance to several other possible threats to single populations: contagious diseases, fires or volcanic eruptions. In the several examples of extinctions or near-extinctions cited in Section 6.2.4, a network of sites would have avoided the crisis.

6.5 Proximate and ultimate causes of extinction

In a way, the focus on the exact causes of extinction in very small populations often misses the conservation point (Simberloff 1986). Soulé (1983) suggests this metaphor: 'The extinction problem has little to do with the death rattle of the final actor. The curtain in the last act is but a punctuation mark—it is not interesting in itself. What biologists want to know about is the process of decline in range and numbers? If a population has declined from a status of abundance over a widespread range to a few tens or hundreds of individuals in a geographically restricted region, it is highly likely that one of the many forces discussed above that set MVP sizes will deliver the *coup de grâce*, but it may not be very useful to chalk the extinction up to that force.

For example, the heath hen *Tympanuchus cupido cupido* (Fig. 6.5) was a common, widespread bird in northeastern USA whose range was ultimately restricted to the small Massachusetts island of Martha's Vineyard by rampant hunting and habitat destruction. A refuge was established on this island in 1908 for the last population, and there was even a temporary population build-up there. However, a series of problems combined to eliminate the population by 1932: environmental stochasticity and catastrophes (in the

Fig. 6.5 The proximate reason why the heath hen *Tympanuchus cupido cupido* became extinct was demograhpic stochasticity as the last few individuals were all males. However, the ultimate causes for extinction were hunting and habitat loss. Supplied by the Academy of Natural Sciences, Philadelphia.

form of a disease, a fire, a remarkably cold winter and unusually heavy predation), demographic stochasticity and perhaps inbreeding depression. The last few individuals were males, so that technically demographic stochasticity could be listed as the cause of the extinction, but it is quite likely that inbreeding depression helped to cause this sexual imbalance, and, in any event, the inbreeding depression and demographic stochasticity would not have occurred but for the various catastrophes and the environmental stochasticity. And the catastrophes and environmental stochasticity would have been unimportant if the population size and the geographical range had not already been greatly reduced. So the ultimate causes of this extinction, and the ones that would have to have been understood and counteracted if the species were to have been saved, were hunting and habitat destruction. The proximate cause was demographic stochasticity, but it could have been any of several factors once the ultimate causes set up the conditions that made swift extinction likely.

Caughley (1994) sees a dichotomy rending conservation biology, with some practitioners working on proximate causes of small population extinction under a 'small population paradigm' (SPP) and others seeing the causes of decline as the ultimate causes of extinction, governed by a 'declining population paradigm' (DPP). Of course, as noted in Section

6.3, some populations apparently have always been small and are not small because their range and/or numbers have been reduced by humans. However, there is little doubt that, among species currently represented by few individuals, very many are in this predicament because of human activity. Further, the threats to 'naturally' small populations may not be a good guide to the threats to populations that have suffered a swift, anthropogenic decline.

The SPP is essentially the body of theory described above for MVP size. The DPP, by contrast, has no formal theoretical basis—this is perhaps the reason that the SPP is more dominant in academic conservation biology. The problem is that the reasons for population decline are often idiosyncratic and do not easily lend themselves to simple, general modelling. The most common causes for the decline of a species are habitat destruction (including pollution) and fragmentation, harvest of various sorts (including hunting) and the effects of non-indigenous species (Simberloff 1986; Diamond 1989). But these forces and many others act and interact in various combinations, so every case has its own story, and, even if one cause is dominant, the story is usually complicated.

Further, in contrast to the SPP, the DPP has not generated many computer models or direct experiments in the laboratory or field. Rather, the usual *modus operandi* under the DPP is intensive study of the natural history of the species of concern, massive observation, and attempts to correlate either densities or rates of decline of the population to various factors that might be causal. Aside from the fact that correlation cannot prove causation, the entire approach seems rather unscientific and not very technological, compared to experiment and simulation modelling. These characteristics plus the absence of general theory combine to lower the popularity of the paradigm.

Nevertheless, it is quite likely that the real contributions to conservation under the DPP are more substantial than those under the SPP. Because the SPP is generally invoked only after a population has declined, the management procedures it suggests will usually be a form of damage control, and, if the population has declined enough, the prognosis is poor. The analogy is to treating an illness when it has already become full-blown. By contrast, the DPP is analogous to preventative medicine—or at least it is invoked at an earlier stage than the SPP, optimally at the beginning of the decline and usually before the population is already minuscule. Furthermore, if the ultimate cause of a population's decline is not recognized and counteracted, the population will never be safe, no matter how cognizant we are of the threats encompassed by the SPP—genetic problems, demographic stochasticity, etc. To return to the medical analogy, the patient is consigned to perpetual treatment, at best. Management under the DPP, by contrast, aims to counteract the ultimate threat, and thus to obviate what become the proximate threats under the SPP. The goal is cure, rather than perpetual treatment.

There is another potential, indirect conservation benefit of management under the DPP, one whose contribution may be even greater than the direct benefit. The SPP, by targeting the proximate factors affecting individual species, is unlikely to lead to management that will aid other species. Under the DPP, on the other hand, such benefits are quite likely. The single most frequent ultimate cause of extinction is habitat destruction and fragmentation. One of the next two most common causes of extinction is the introduction of non-indigenous species. Action against either of these phenomena is likely to help species in addition to the target species.

Consider the northern spotted owl *Strix occidentalis caurina* in the northwestern USA. The ultimate reason for its endangerment is the destruction and fragmentation of the old-growth forest it inhabits; for effective reproduction and population maintenance, it requires immense tracts of large, old trees. Because the Endangered Species Act of the USA prevents destruction of the critical habitat of an endangered species, numerous injunctions against planned logging of old-growth forest were based on the needs and causes of decline of the northern spotted owl. However, this forest contains many species in addition to the northern spotted owl, and some of these are imperilled for the same reason the owl is—strict requirement for this sort of habitat. For many species, such as the 6000-odd insects that inhabit this sort of forest, the exact habitat requirements are unknown, but it is highly likely that some of them are old-growth specialists as well. Thus, management of the spotted owl under the declining species paradigm has made the owl a classic 'umbrella' species—a species whose habitat requirements are so large and specific that saving it is likely to lead to maintenance of numerous other species as a by-product.

In particular instances, such as the rescue of the Lord Howe Island woodhen *Tricholimnas sylvestris* (Caughley 1994) and the rehabilitation plan for the northern spotted owl (Hedrick *et al.* 1996), the two paradigms—the declining population paradigm and the small population paradigm—have been successfully melded. A complete attack on the problem of population extinction will entail greater use of both approaches.

6.6 Conclusion

The causes of extinction of small populations are various, and may not generally be the same for populations that have always been small as for those that have recently been reduced by human activity. When a population is very small, there is clearly an elevated probability that it will go extinct quickly, although the long persistence of some very small populations in nature suggests that tiny populations are not automatically doomed by any of the forces that are theoretically believed to set minimum viable population sizes. The geographical restriction of many small populations is probably as much a threat to their persistence as is the small size itself. Thus, a good conservation strategy is to establish multiple populations of any species.

Some species may exist in nature as metapopulations, in which all populations are ephemeral. In such species, the exact reasons for the disappearance of a particular population are not nearly as important as the rates of movement between sites and the relative rates of population disappearance and establishment. In such a system, the key significance of population extinction is that, if too many local populations disappear, the entire metapopulation may collapse because movement will not suffice to replace extinguished local populations. However, it is currently uncertain how many and what sorts of species are structured as metapopulations in nature.

Finally, a more effective approach to small populations from a conservation standpoint may be to focus not on the likely threats once a species has been reduced to a very small population, but on the forces that acted to make it small—in short, the ultimate rather than the proximate causes of its distress. Such a focus is more likely to provide long-lasting conservation of the species and may well benefit other species incidentally. However, the problems facing each declining species are likely to be at least somewhat idiosyncratic and their solution will almost certainly entail detailed natural historical research. Theoretical principles may play a role in this approach.

CHAPTER 7

Metapopulation, source–sink and disturbance dynamics

Martha F. Hoopes and Susan Harrison

7.1 Introduction

Many or most remaining natural communities are now significantly more fragmented than they once were, and current rates of human population growth suggest that this trend will continue unabated (Reid & Miller 1989). The effects of fragmentation upon the dynamics of populations and communities have thus become a major focus of conservation biology. Considerable theoretical and empirical work has shown that fragmentation has impacts above and beyond the simple loss of habitat. Within the smaller areas of habitat that remain, centres of population are more isolated, leading to increased rates of local extinction and decreased rates of gene flow and re-colonization (e.g. Fahrig & Merriam 1985; Laan & Verboom 1990; Opdam 1990; Thomas 1991; Thomas et al. 1992; Sjogren-Gulve 1994; Harrison & Taylor 1997). In turn, the local extinctions of species that function as pollinators, herbivores, predators, seed dispersers or decomposers may lead to secondary extinctions, community simplification and even changes at the ecosystem level (e.g. Jennersten 1988; Klein 1989; Leigh et al. 1993; Rathcke & Jules 1993; Roland 1993; Aizen & Feinsinger 1994; Kreuss & Tscharntke 1994). The human-altered matrix around remnant habitat patches may threaten the species within them, through harbouring large populations of invasive species or native predators (Andren & Angelstam 1988; Small & Hunter 1988; Knick & Rotenberry 1995; Maehr & Cox 1995). The successful conservation of remnant natural areas must therefore take account of multiple issues arising from habitat geometry. In recognition of this, conservation biologists have begun to embrace the three spatial ecological theories which this chapter discusses: metapopulation, source–sink and disturbance dynamics.

Habitat geometry and its effect on species diversity is far from a new issue; since the eighteenth century, scientists have recognized that habitat area is strongly correlated with species richness, and there have been many attempts to quantify and explain this relationship (Browne 1983). While Lack (1976) and others observed that larger areas contained more types of habitat, Preston (1960, 1962) and others emphasized the effects of habitat area on population sizes and probabilities of extinction. MacArthur & Wilson

(1963, 1967) proposed that isolation, in addition to area, determined the diversity of habitat isolates. Their theory of island biogeography, succinctly relating habitat geometry to diversity through rates of colonization and extinction, became perhaps the most widely known and applied idea in ecology. The relevance of this theory to human-fragmented habitats was soon recognized, and the principle that reserves should be large, rounded and minimally isolated became a cornerstone of the then new field of conservation biology (Diamond 1975a, 1976; Diamond & May 1976).

The subsequent history of island biogeography theory and its application to reserve design have been well reviewed by Shaffer (1990). Tremendous debate ensued over whether natural island biotas are truly characterized by dynamic equilibria in diversity, what the theory and related evidence say about the merits of single large vs several small reserves, and whether island biogeography theory has any predictive power for conservation. Satisfying tests of the theory are few, since this requires data sets for many islands, species and generations. Most ecologists would now agree that diversity on islands or nature reserves cannot be reduced to the two simple factors of area and isolation, and the once acrimonious debate has subsided more than it has been resolved. The controversy highlighted the tremendous attractiveness and practical shortcomings of ecological theory as applied to conservation problems. As will be seen below, we suspect that the rise and fall of island biogeography theory in conservation biology carries some larger lessons that have perhaps not fully been learned.

The theories of metapopulation, source–sink and disturbance dynamics propose, in verbal and mathematical form, that species diversity and abundance depend on spatial and temporal aspects of habitat configuration. All three of these theories have been incorporated into conservation textbooks, current research and management plans for various species. We describe these theories as they are used in conservation biology, explaining their theoretical basis and empirical support and the ways in which they are being applied. All three, we will argue, share features in common with island biogeography theory: great heuristic value, considerably narrower empirical support than is generally presumed, and the potential both for practical usefulness and for damaging misapplication.

7.2 Metapopulation theory

Metapopulation theory examines the dynamics of sets of semi-independent populations connected by dispersal (Hanski & Gilpin 1991). The term metapopulation, or 'population of populations', was coined by Levins (1970), who examined evolution in a subdivided species. In Levins' model, a metapopulation is a network of extinction-prone subpopulations occupying a mosaic of habitat patches. Subpopulations inhabit identical patches, and are subject to equal and independent probabilities of extinction and re-colonization. In this model, the metapopulation has a single global equilibrium for the fraction of available patches that are occupied, an equilibrium that depends only on extinction and colonization rates. As long as new populations are founded at a rate that equals or exceeds that of local extinctions, the metapopulation can remain extant. The key to persistence is sufficient dispersal.

This model makes some important predictions that extend to many more complex metapopulation models as well. For a given average extinction rate of subpopulations on patches, colonization must be above a critical threshold for the metapopulation to persist. Similarly, for a given colonization rate, extinction must be below a critical threshold for meta-population persistence. Thus, as a given habitat becomes increasingly fragmented, represented by increasing extinction, decreasing colonization, or both, a species is expected at some point to collapse abruptly from a stable positive equilibrium to global extinction. Conversely, providing corridors or other means of increased dispersal among patches should make the entire regional metapopulation persistent.

This simple model is designed more to structure thought than to predict actual species dynamics. Some of its more unrealistic assumptions are that patches are infinite in number and identical in size, habitat quality, likelihood of extinction and likelihood of re-colonization; that from any given patch, dispersal to all other patches is equally likely (no spatial structure); and that patches are either empty or occupied to carrying capacity. More complex analytical models of metapopulations relax one or more of these assumptions; in most cases, except for 'mainland–island' models with some local population(s) immune to extinction, such modifications cause the simple global equilibrium to disappear along with analytical tractability (Ray *et al.* 1991; Hastings 1994). The most common ways to make models more realistic are through stochastic simulations and spatially explicit population models (SEPMs). Stochastic simulations allow local dynamics of occupied patches to be considered explicitly and allow temporal and spatial variation in extinction and colonization rates. SEPMs incorporate distances between patches and the dispersal distances of the organism. Patches exist on a spatial grid, and growth, dispersal, colonization and extinction are governed by patch-specific rules (DeAngelis & Gross 1992; Dunning *et al.* 1995).

In conservation applications, the usual question of interest is how much the viability of a species is affected by the number and size of patches within a region and by the rates of movement of individuals among patches. One of the first such applications was a stochastic, spatially non-explicit matrix model by Menges (1990) of the rare lousewort *Pedicularis furbishae* (see Fig. 6.4). This model demonstrated the importance of colonization of newly available sites in this species' early successional habitats, and was used to compare the effect of different management regimes. In contrast, Kindvall & Ahlen (1992) used a metapopulation model without stochasticity, but with explicit habitat geometry, to examine the persistence of the bush cricket, *Metrioptera bicolour*, in recently fragmented habitat in Sweden. This model was used to evaluate the relative importance of different patches for the cricket's survival. Finally, some of the most complex spatial models of an endangered species were the series of SEPMs for the northern spotted owl in the northwestern US Pacific. These models examined in ever-increasing detail how the size and spacing of forest fragments would affect the population viability of the owl (Fig. 7.1), and their results formed the basis for an agency management proposal calling for a network of old-growth reserves (Thomas *et al.* 1990; Lamberson *et al.* 1992; McKelvey *et al.* 1993).

As the popularity of metapopulation models as management tools has grown, however, increasing concerns have been voiced about their limitations. Several authors have noted that spatial population models are extremely data demanding, that the required data—especially on dispersal—are very difficult to obtain, and that model results tend to be highly sensitive to poorly estimated parameters (Doak & Mills 1994; Conroy *et al.* 1995; Turner *et al.* 1995; Ruckelshaus *et al.*, in press). For example, the SEPMs used to devise a proposed management strategy for spotted owls made simplifying assumptions about juvenile dispersal which proved to result in seriously overoptimistic conclusions (Harrison *et al.* 1993). Similarly, Kindvall & Ahlen (1992) noted that their conclusions regarding

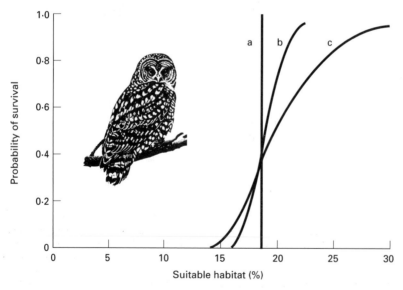

Fig. 7.1 Model of metapopulation dynamics created by habitat fragmentation: the estimated 250-year probability of survival of the northern spotted owl (*Strix occidentalis caurina*) vs the percentage of the landscape that remains as old-growth forest. The three curves represent different levels of environmental variance in population growth rates: (a) no variance, (b) low variance and (c) high variance. The probability of survival drops sharply at the point where re-colonization rates fall below those of local extinction, because forest is so scarce. (After Lamberson *et al.* 1992.)

the bush cricket were highly dependent on dispersal parameters that were difficult to measure in the field.

A more basic question is how broadly the metapopulation concept applies to species in fragmented habitats. Although most populations and habitats are patchy at some spatial scale, this does not always imply the type of dynamics portrayed in metapopulation models (Doak & Mills 1994; Harrison 1994; Hastings & Harrison 1994). Important metapopulation attributes, such as the sensitive dependence of global persistence on the rates of dispersal and re-colonization, and the spreading of the risk of global extinction, may be expected to arise when habitats are fragmented at a spatial scale comparable to a species' long-distance dispersal capabilities. But, alternatively, habitats may be fragmented so finely that populations are not really subdivided, or so coarsely that populations are completely isolated from one another. Also, very uneven fragmentation may lead to 'mainland–island' dynamics in which only the largest populations and habitat patches really matter for persistence (Figs 7.2 & 7.3). While

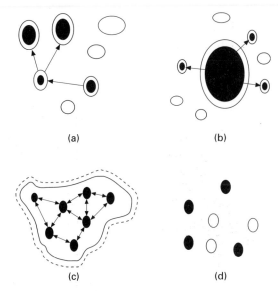

(a)   (b)

(c)   (d)

Fig. 7.2 Different population structures that can arise from habitat fragmentation. Circles represent patches of habitat; filled circles are occupied and empty ones are uninhabited. Individuals move freely within patches and typical dispersal between patches is shown by arrows. (a) Classic (Levins) metapopulation: with populations that fluctuate and go extinct independently, but are interconnected by re-colonization. Patches (populations) are roughly equal in size (persistence), and dispersal distances are comparable to the distances among patches. (b) Mainland–island metapopulation: where patches are so unequal in size that most or all turnover (extinction and colonization) occurs in the very small populations and is inconsequential. (c) Patchy population: where patches are close together relative to dispersal distances with individuals moving freely within the dashed line, so that the patches collectively support a single population. (d) Non-equilibrium metapopulation: where patches are very far apart relative to dispersal distances, so that the populations are not interconnected.

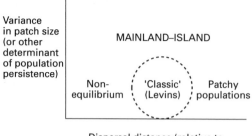

Fig. 7.3 Relationships among the population structures shown in Fig. 7.2.

some good examples of metapopulations have been documented, for example the pool frog, *Rana lessonae* (Sjogren-Gulve 1994), and the Glanville fritillary butterfly, *Melitaea cinxia* (Hanski & Thomas 1994; Hanski *et al.* 1995), many other systems that have been described as metapopulations actually behave in qualitatively different ways (Harrison & Taylor, 1997), and many more careful tests of alternative hypotheses about metapopulation dynamics in natural systems are needed.

Limitations of the metapopulation concept and metapopulation models are perhaps not important, as long as these are being used only for purposes of thought. However, their potential danger arises when they are used to justify actions entailing further reductions in habitat. For example, management plans for the US national forests must ensure viable populations of vertebrate species, typically by designating a few areas to be set aside from logging; mitigation plans under the US Endangered Species Act may allow developers to destroy some habitat of a threatened species in return for protecting or creating other areas. In both these settings, spatial models or verbal references to metapopulation theory are sometimes used to support plans that call for a great reduction in total habitat, but with the remainder to be arranged as a network of supposedly interconnected patches (for example Fig. 7.4). It is extremely unlikely that metapopulation models can ever be the basis for defining safe strategies for habitat loss, given their requirement for large amounts of usually unobtainable data, and the inherently high degree of stochasticity in their outcomes.

We suggest that metapopulation theory offers a useful framework for thought, but that it risks suffering the fate of the equilibrium theory of island biogeography—enthusiasm followed by widespread disillusionment—unless it is used only where it is biologically appropriate and supported by reliable data. Before applying it to real-world situations, researchers must question whether habitat is fragmented at a scale that produces sets of populations that are largely independent of one another, yet importantly interconnected. Even when this is the case, modelling is only reliable if considerable information exists on habitat quality and distribution, population sizes and dynamics, and dispersal behaviour. Careful attention

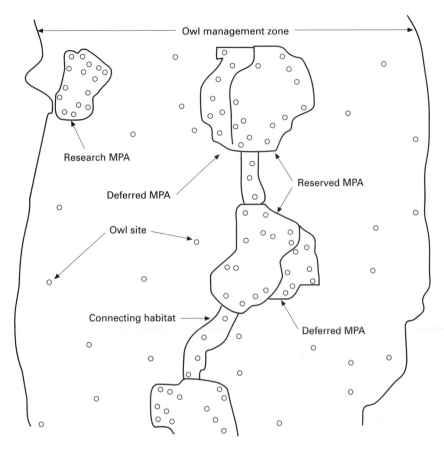

Fig. 7.4 Example of the use of metapopulation concepts: reserve design proposed for the northern spotted owl *Strix occidentalis caurina* by the US National Forest Products Association in 1991. The forest is reserved in blocks called multiple-pair areas (MPAs), which are connected by corridors twice the width of an owl's home range, and which may be reserved (no logging), deferred (logging temporarily halted) or research (used to devise logging practices compatible with owls). This proposal stated that '… a mix of suitable environments is likely to support a metapopulation with higher probability of long-term persistence than an environment comprised only of the most-preferred situations'. After peer review by conservation biologists, this proposal was rejected by the US Forest Service. (Redrawn from Spotted Owl Subgroup of the Wildlife Committee of the National Forest Products Association and American Forest Council 1991.)

must be paid to the biases and uncertainties arising from incomplete information. Metapopulation models may be extremely useful for weighing alternative conservation strategies and highlighting critical areas for further research; however, it is risky to use them to make specific predictions.

7.3 Source–sink dynamics

Source–sink models examine the dynamics of populations in habitat patches of different qualities; they are an elaboration of a metapopulation approach, in which habitat patches are no longer identical but instead differ in their ability to support populations of organisms. As first formulated by Pulliam (1988), they include subpopulations in two types of habitat: 'sources' with net positive population growth ($\lambda > 1$ in the absence of emigration) and 'sinks' with net negative population growth ($\lambda < 1$ in the absence of immigration). Note that the difference between these two habitat types is not in size but in quality. Source populations grow to some carrying capacity beyond which individuals disperse to sink habitat, and without this dispersal from sources, the sink populations would go extinct. However, sink habitats provide reservoirs for surplus individuals from sources, from which the sources can be replenished in the event of a catastrophe. Thus, the existence of sinks increases both the absolute numbers and the persistence of the combined source and sink populations (Pulliam 1988; Pulliam & Danielson 1991). Given certain assumptions about the timing of dispersal and mortality, sink habitats can harbour equal or even higher population densities than source habitats (Howe *et al.* 1991; Pulliam & Dunning 1994).

The most important modification of source–sink models consists of relaxing the assumption that dispersing individuals always select habitat optimally on the basis of its intrinsic quality and its degree of crowding. When this assumption is relaxed, so that individuals may move inappropriately from better to worse habitats, the sink habitat loses its positive role and may cause populations to decline (Buechner 1987; Pulliam & Danielson 1991; McKelvey *et al.* 1993; Doak 1995). Donovan *et al.* (1995a) illustrate this point with a model in which natal fidelity acts as a limited form of habitat selectivity. Birds always choose to return to their natal territory and can distinguish it from other areas. Natal fidelity in this model is beneficial to population survival, since sources produce more offspring than sinks. If natal fidelity is low, only high fecundity in the source habitat can keep the entire population from declining.

Models using the source–sink framework have frequently been used to examine the effects of habitat loss or degradation on species persistence. Thompson (1993) compared the possible effects of clearcutting and other timber management schemes on the predicted population sizes of a hypothetical neotropical migrant bird, distinguishing between breeding birds in a good habitat vs 'floaters' in a non-breeding habitat. Doak (1995) used demographic data and a source–sink model to examine how an increase in road density would affect viability of the grizzly bear *Ursus arctos horribilis*, as well as how soon a critical decline towards extinction might be detectable by the population census techniques (Fig. 7.5). Mladenoff *et al.* (1995) used a source–sink model to show that conservation efforts for gray wolves *Canus lupus* in the US state of Minnesota may be crucial to the development of viable populations in the states of Michigan and Wisconsin. Wootton & Bell (1992) similarly found that peregrine falcon *Falco peregrinus* restoration in southern California depends on the northern California population; using data on relative fecundity in northern and southern populations, supplemented by data from European peregrines

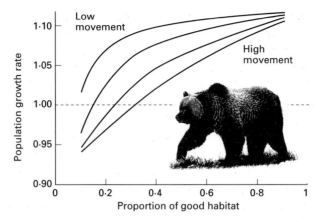

Fig. 7.5 Model of source–sink dynamics created by habitat degradation: changes in the population growth rate of grizzly bears *Ursus arctos horribilis* with changing proportions of 'source' habitats (far from roads) and 'sink' habitats (near roads). Bears move between source and sink habitats in this model randomly, because they are unable to tell them apart. Note that the decline in population growth rate is greatest with high rates of movement, but is the most sudden (non-linear) with low movement rates. (After Doak 1995.)

where necessary, they found the northern population was self-sustaining while the southern population required immigration to survive. Liu *et al.* (1995) considered the problem of multiple species in a heterogeneous, managed landscape; modelling the endangered red-cockaded woodpecker *Picoides borealia* and the Bachman's sparrow *Aimophila botterii* in south-eastern US pine forests, they found that a source habitat for one species might be a sink habitat for the other, so that management practices might easily have opposite effects on the sparrows and the woodpeckers.

Distinguishing between source and sink habitats in the field, though, has posed a problem for experimental tests of the source–sink theory. Most studies to date have simply inferred source–sink dynamics from variation in population densities or vital rates, but such assumptions are far from sufficient, as shown by a number of studies (Kadmon & Shmida 1990; Howe *et al.* 1991; Davis & Howe 1992; Beshkarev *et al.* 1994; Watkinson & Sutherland 1995). Pulliam & Danielson (1991) suggested simply measuring birth and death rates in each habitat, while Kadmon & Shmida (1990) argued that, since sinks must be supported by source populations, it was necessary to measure net immigration as well. Watkinson & Sutherland (1995) showed that, even when both demographic and immigration data are available, it is possible to draw incorrect conclusions. Using a model of two populations—each persisting without immigration—Watkinson & Sutherland (1995) showed that certain intermediate levels of dispersal between these habitats could elevate the population size in the lower quality habitat above its carrying capacity so that it would appear to be a sink by both Pulliam & Danielson's (1991) and Kadmon & Shmida's (1990) definitions. True sinks can only be distinguished from these 'pseudosinks' (in which viable populations appear to have $\lambda < 1$ because immigration leads to population densities above the carrying capacity) either by isolating the populations from one another or by measuring the relationships of demographic rates to population density (Watkinson & Sutherland 1995). For long-term data sets, Howe *et al.* (1991) suggested that sinks might be distinguishable from sources by their higher population variability. Beshkarev *et al.* (1994), however, found that variability between source-dominated and sink-dominated landscapes did not differ significantly; they concluded that a more powerful method of distinguishing between sources and sinks was to look for temporal autocorrelation in population numbers in sources and lack of autocorrelation in sink habitats.

Because of these difficulties with empirical tests of the theory, the importance of the source–sink phenomenon in nature is still far from certain. Although many researchers have examined the theoretical implications of source–sink dynamics (Buechner 1987; Pulliam 1988; Howe *et al.* 1991; Pulliam & Danielson 1991; Davis & Howe 1992; Doak 1995; Donovan *et al.* 1995a, 1995b), there have been very few experimental tests of the theory. To date, the only strong empirical evidence for source–sink dynamics comes from Keddy's (1981, 1982) experimental work with *Cakile edentula*, an annual plant growing on beaches in Nova Scotia. In this species, seeds were washed from the shore, where they germinated well and had high fecundity, to the dunes where germination and fecundity were very low. Censuses of all life stages, including seeds, showed that populations were largest in the dune (sink) habitat (Fig. 7.6). Beshkarev *et al.* (1994) offer further empirical evidence of source–sink dynamics by testing the predictions of source–sink theory with a 30-year data set of hazel grouse *Bonasa bonasia* populations in Russia; their work is purely descriptive, however, and their criteria for distinguishing between sources and sinks are not useful for researchers without long-term data sets.

Just as in the case of metapopulations, models of source–sink dynamics require information that is very difficult to obtain. This includes information on habitat-specific vital rates and movement rates, and the behavioural responses of dispersing organisms to habitat of different types; habitat selection behaviour during dispersal is an especially poorly known area for most organisms (Haas 1995). It is unlikely that, for any species, enough can be known about the complex interactions between dispersal, demography and habitat heterogeneity for source–sink models to make reliable predictions about the outcomes of management alternatives (Doak & Mills 1994; Ruckelshaus *et al.*, in press).

General rather than specific conclusions are perhaps the most important contribution of source–sink models to conservation biology. For example, Lamberson *et al.* (1992) and Doak (1995) demonstrated that, when habitat degradation leads to a sink habitat interspersed with source habitats, significant declines in population density may not be detectable until years

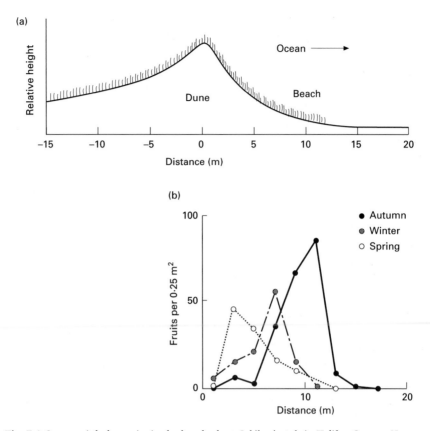

Fig. 7.6 Source–sink dynamics in the beach plant *Cakile edentula* in Halifax County, Nova Scotia, Canada. (a) The dune profile measured in metres with the dune peak at 0 m; seaward distances are positive numbers and landward distances are negative. (b) The density of seeds changes throughout the year. Because reproductive output is highest on the beach, seed densities are initially highest there; wind and waves then shift the seeds towards the dunes, where they germinate and survive but have very low reproductive output. Thus the dune (sink) population is continually subsidized by the flow of seeds from the shore (source) population. (After Keddy 1982.)

later. Pulliam *et al.* (1992) and Doak (1995) examined the form of the relationship between the decline in percentage of source habitat in the landscape and the resulting decline in population size; both studies found non-linear relationships in which small changes in habitat could lead to sudden, drastic changes in population growth rates. In some cases, these curves could be extremely steep, thereby acting as thresholds to extinction. However, while Doak (1995) found these non-linearities were greatest at the lowest rates of dispersal, Pulliam *et al.* (1992) found the reverse, illustrating that even the qualitative results of source–sink models are highly sensitive to model details.

The theory of source–sink dynamics has helped to focus attention on habitat quality, and to show that certain habitats may have a dispro-portionate importance that is not always evident from the species' abundance. It shows that poor habitat can affect the viability of populations in adjacent good habitats, and small changes in the distribution of good and poor habitat may precipitate relatively large changes in the species' viability. Conservationists must be wary of misapplications here, too, however. Given the range of data necessary for the most general conclusions, source–sink models are certainly not adequate to justify habitat loss even when the area in question is concluded to be a sink habitat. Confusion arises because the 'source–sink' label is too readily placed on any case in which population densities or habitat quality are spatially variable (Watkinson & Sutherland 1995), just as the term metapopulation is too quickly applied to any case where populations or habitats are discontinuous (Harrison & Taylor 1997).

7.4 Disturbance dynamics

Disturbance dynamics refers to the interplay between species diversity or abundance and the scale, rate and intensity of disturbance. Ecological disturbances are often defined as events causing a rapid loss of a large fraction of the standing biomass of an area. They may be abiotic, such as hurricanes, fires, avalanches or floods, or biotic, such as grazing, trampling or gopher digging. The important attribute of disturbance is its ability to change the structure of biotic communities and ecosystems through altering the abundances of individual species. One of the best-known consequences of disturbance is reduction in the proportionate abundance of competitively dominant species, as in Connell's (1978) 'intermediate disturbance hypothesis'. Connell proposed that too little disturbance leads to low diversity through competitive exclusion, and too much disturbance eliminates species incapable of rapid re-colonization, while intermediate levels of disturbance promote coexistence of species in a spatial mosaic of patches at different stages of succession. Intermediate disturbance has become perhaps one of the best-accepted principles in ecology (Connell & Slatyer 1977; Connell 1978; Sousa 1984; Ricklefs 1990; Begon et al. 1996; Meffe & Carroll 1994). The general proposition that disturbance makes all ecological communities dynamic in time and space has been termed the 'non-equilibrium paradigm' (Pickett et al. 1992).

Conservation biologists have since taken enthusiastic hold of intermediate disturbance and non-equilibrium dynamics (Meffe & Carroll 1994). Today there is broad acceptance among conservation managers that natural disturbance is desirable, and that management practices should attempt either to maintain or to mimic characteristic disturbance regimes in terms of size, frequency and intensity (e.g. Marzolf 1988; Howell & Jordan 1991; Thomas 1991; Della Sala et al. 1995; National Fish & Wildlife Foundation 1995). Applications of these ideas are inherently spatial, since the appropriate regime of disturbance depends on how large, and how widely spaced, patches at each successional stage should be. Pickett &

Thompson (1978) proposed that reserve size should be determined by the 'minimum dynamic area', i.e. the minimum area large enough to incorporate a 'shifting mosaic steady state' (Bormann & Likens 1979) of patches at different stages of recovery from disturbance.

Some of the best examples of the need for managed disturbance come from attempts to conserve species dependent on the early or middle stages of succession. The endangered Karner blue butterfly *Lycaeides melissa samuelis* in northeastern USA feeds solely on an herbaceous lupine that flourishes for only a few years following fire. None of the butterfly's life stages can survive fires, and the butterfly is a poor disperser. The recommended management strategy is to burn its habitat on a 10-year rotation, at a fine enough spatial grain to maintain a mosaic of habitat types on a scale suitable for the butterfly to disperse from inhabited to newly suitable patches. Similarly, Thomas (1991) concluded that the conservation of many declining British butterflies requires actively managing their habitats to mimic sheep grazing, coppicing or other disturbances once caused by traditional land-use practices (Fig. 7.7).

Most management regimes are aimed at multiple species, however. The basic problem with attempting to mimic natural disturbances, whose key characteristics (spatial scale, frequency, intensity and timing) are usually unknown, is that different species may respond very differently to the regime chosen. In the midwestern USA, managers have been trying to restore native tall-grass prairie with a combination of native introductions,

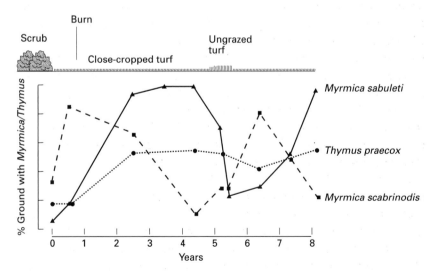

Fig. 7.7 Effects of the intensity of grazing in patches of thyme *Thymus praecox* on the abundances of two ant species. Note that the two ant species respond oppositely to disturbance. These effects are of interest because the habitat is being managed for the endangered large blue butterfly, *Maculinea arion*, which has a mutualistic relationship with one of the ants *Myrmica sabuleti*. (After Thomas 1991.)

removal of non-natives and a fire regime that favours fire-adapted native prairie species (Kline & Howell 1987; Howell & Jordan 1991). Floral diversity is increasing and is approaching that in remnant stands of prairie vegetation (Kline & Howell 1987), but sedentary native insects are possibly decreasing in response to fire regimes that are more frequent, hotter and applied in seasons when natural fires would not commonly occur (Schwartz 1994). Grazing regimes applied to promote plant diversity have another undesirable side effect: they may introduce non-native plants through animal faeces (Hobbs & Huenneke 1992). Since invasions by alien species tend to be strongly promoted by disturbance, and to have potentially devastating effects on native species (D'Antonio & Vitousek 1992), they represent a major hazard in any attempt to conduct managed disturbance.

Nowhere is the substitution of managed for natural disturbances more controversial than in forestry. Current US forest management policies claim that logging can mimic natural disturbances such as fire, and is therefore beneficial to the health of forest ecosystems (US Forest Service 1994; US Forest Service & US Bureau of Land Management 1994). Part of the basis of this argument, namely that 80 years of fire suppression have altered the structure of existing forests towards dense stands of small trees, is supported by ecological evidence (Minnich et al. 1995). However, there is little evidence that clearcuts are good substitutes for fires; they exacerbate the already severe fragmentation of old-growth stands to the detriment of forest wildlife, and they tend to decrease the plant diversity of both canopy and understorey (Della Sala et al. 1995). Overall spatial heterogeneity is lower in logged stands, and this has been shown to correlate with extinctions and declines of native species and increased invasion by non-natives (Barbour et al. 1993; Minnich et al. 1995).

Enthusiasm for disturbance mosaics may sometimes obscure the simple fact that it may be only the late successional elements within the disturbance mosaic that are rare, threatened and in need of protection. For example, preserving all of the few remaining stands of old-growth pine forest in southeastern USA is essential for the conservation of the red-cockaded woodpecker (Ligon et al. 1986), and the few remaining undammed rivers in the western USA (as elsewhere in the world) are the last strongholds for many native fish and mollusc species that are disappearing from dammed rivers around the world (Moyle & Leidy 1992). In primeval conditions, late successional species survived disturbances by virtue of there being a large amount of habitat available; but in many areas of the world today, late successional habitats are reduced to a few precious remnants, and even natural disturbances may threaten the survival of the species dependent on them. In the Sierra Nevada of California, the vertebrates of greatest concern are the California spotted owl, great gray owl *strix nebulosa*, pine marten *Martes americana* and fisher *Martes pennanri*, all old-growth inhabitants. Wildlife biologists crafted a management plan that called for protecting all old-growth trees throughout the spotted owl's range (Verner *et al.* 1992), but the US Forest Service unfortunately replaced this with an 'ecosystem management' strategy that called for clearcutting to maintain a diverse successional mosaic. Although edge-loving species may flourish under this plan, it is unlikely to benefit any of the species whose survival is presently most precarious.

Disturbance dynamics has perhaps become too popular with resource extraction industries and resource management agencies, yet it remains a valid ecological principle with an important place in the management of natural areas. It is crucial to recognize that the concept of intermediate disturbance functions only within certain limits, however. It does not explain regional diversity, nor does it suggest that all species thrive at intermediate levels of disturbance. The overall diversity that is maximized by intermediate disturbance may include few of the late successional species that are globally rarer in the human-dominated landscape, while it may promote many weedy, non-native or otherwise undesirable species (Hobbs 1991; Moyle & Light 1996). In the modern setting, managed or even wholly natural disturbances may create damaging positive feedbacks, in which invasive species alter community- or ecosystem-level attributes, in turn further disfavouring native species (D'Antonio & Vitousek 1992; Hobbs & Huenneke 1992).

Clearly, therefore, ecologists must be extremely careful in promoting this concept. Managing for disturbance is necessary in some cases, but seldom do clear guidelines exist as to the appropriate scale, intensity, frequency or timing of disturbances. Much experimentation is required to devise a successful regime, especially when one type of disturbance (e.g. logging or cattle grazing) is being substituted for another one (e.g. fire or native ungulates). Finally, it must be kept in mind that, although disturbances of some kind do occur in all ecosystems, imposing disturbances is not always an important aspect of conservation management. In some cases, only the species characteristic of late successional stages may be in need of protection, and these may not benefit from managed disturbance at all.

7.5 Conclusion

In the last two decades, a proliferation of theories has helped ecologists to understand how the spatial configuration of habitats affects the dynam-ics of populations and communities, and ultimately the distribution of biological diversity. Conservation biologists have eagerly taken up these theories and attempted to convert them into guidelines for managing species and communities in a heterogeneous world. Three of the most popular spatial ecological theories are metapopulation dynamics, source–sink dynamics and disturbance dynamics. Each one of these theories, we argue, has some important insights to offer; yet all three are being applied more broadly and less carefully than they should. Key assumptions that limit the applicability of the theories are not always fully appreciated. Empirical evidence to even support their relevance to real systems, let alone to convert them into predictive management tools, is sorely lacking. Most significantly, each one has the potential to be misapplied in damaging ways.

Metapopulation theory proposes that species exist in a dynamic balance between extinction and re-colonization of local populations in patches of habitat. It suggests that we cannot conserve individual populations separately from their regional context; that promoting corridors and other forms of landscape connectivity may be a good idea; and that the potential exists for a widespread species to collapse suddenly to regional extinction in response to a gradual degradation in the regional habitat network. However, we still need much more empirical evidence to substantiate this idea. It may be that the metapopulation phenomenon is highly scale-dependent, requiring a close match between the spatial scale of habitat fragmentation and the dispersal distance of the organism in question. If so, it follows that a given network of reserves and corridors is likely to support a functional metapopulation for at most only a few species in the community. Metapopulation models can provide some general insights about habitat fragmentation, but are seldom reliable sources of specific predictions. Misuses of the theory include attempts to use spatial models to justify further habitat loss.

Source–sink theory proposes that populations exist in heterogeneous habitats that include areas in which population surpluses are produced (sources) and areas in which the population cannot replace itself without immigration (sinks). It draws conservationists' attention to variable habitat quality, and, in particular, to the idea that population survival may be dependent on a few key habitats, not always those in which the species is most abundant. Like metapopulation theory, it highlights the possibility of sharp non-linearities in the responses of populations to the loss or degradation of habitat. Yet, to date, it is validated by even less evidence than metapopulation theory. Like metapopulation models, source–sink models are more suitable for general insights than specific prescriptions; model outcomes in both cases are highly dependent on assumptions that are rarely based on adequate data. One possible misuse of this theory is to justify habitat reduction by claiming that peripheral or low-density areas are useless or even harmful to the viability of a population.

The theory of disturbance dynamics proposes that communities exist in spatial and temporal mosaics, and that diversity depends on maintaining the appropriate level of disturbance. This theory has had tremendous value in causing ecologists to question the equilibrium approach once prevalent in ecological models, as well as in causing conservation managers to reject their former abhorrence of fire and other natural processes. Conservation plans incorporating fire, grazing and other forms of managed disturbance have enjoyed some success in maintaining or restoring natural communities. At the same time, overenthusiasm for the intermediate disturbance hypothesis has sometimes obscured the need to focus conservation efforts on late successional vegetation and species that are not promoted by such disturbances. Resource extraction interests have perhaps disingenuously promoted the fashionable idea that 'hands off' management is unacceptable and that their activities can benevolently mimic natural disturbance processes.

Our general caution and conclusion is that ecological theories can never be turned into general rules. Appropriately understood, ecological theories provide the basis for hypotheses, which sometimes have applications to specific real-world situations. One of the major problems for conservation decisions is the difficulty in adequately testing these hypotheses. In some cases, alternative management practices can be tested experimentally; when this approach is not possible, models can be used with existing data to explore the consistency of hypotheses and the consequences of alternative assumptions. In this way, these models can expand our general ecological knowledge, suggest new areas for investigation in a conservation situation, and eliminate options that seem too risky. Conservationists must remember that these theories are better for formulating questions than for answering them. When decisions must be made in the near absence of information, it is well to remember the most important general message from spatial population theory: we must be extremely cautious about allowing further reduction, fragmentation or degradation of natural habitats, since small changes may have large consequences.

CHAPTER 8

Implications of historical ecology for conservation

Oliver Rackham

'Just as no [national] park is purely a natural area without any historical
relevance, so all historical parks have some natural values.'

(Alston Chase, *Playing God in Yellowstone*, 1987)

8.1 What is historical ecology?

All natural areas have a history, usually involving hundreds or thousands
of years of interaction with people who have lived and worked in them.
Understanding that history is an essential part of conservation. How did
an area function in its 'normal' state, before recent threats to it developed?
What are the historical limits of variation with regard to, let us say, grazing,
and does the present level of grazing fall within those limits? If woodcutting
is thought to be a threat, have there been previous episodes of woodcutting,
and how did plants and animals thought to be sensitive to it get through
those episodes?

Historical ecology is the history of vegetation and landscape. It is not the
history of countryfolk, nor the history of people's attitudes to landscape,
nor the history of the things that people have said about landscape. It is
the history of plant and animal communities, not of environment, although
aspects of environmental history, such as the Little Ice Age, often come
into it. The timescale is usually tens, hundreds or thousands of years,
short enough for evolutionary changes to be a relatively minor (though
not negligible) factor. In many cases the end of the last glaciation is a
convenient starting point.

Although it would be perfectly possible to write the historical ecology of
an island that had never had human inhabitants, most work is on cultural
landscapes. A cultural landscape is neither wholly natural nor wholly
artificial; it results from centuries of usually complex interactions between
plants and animals, human activities and the environment.

8.1.1 Responses of plants and animals to human activities

Plants and animals are not, of course, environment. They are not part of
the scenery of the theatre of ecology—the passive recipients of whatever
destiny mankind chooses to foist on them—they are actors in the play,
each of which has its own agenda in life which the investigator must
understand. I shall assume (although in practice this is not always so) that
any serious article on ecological history begins by summarizing the main

Fig. 8.1 An inland live oak *Quercus fusiformis* a few months after a fierce fire. The tree was killed to the ground and has sprouted from the roots (near Temple, Texas, Angust 1996). (Photo by W. Dossett.)

plant species and their properties in relation to the relevant activities such as browsing, burning and woodcutting.

Some of the world's trees die when cut down. Others sprout in various ways, either from the stump (termed *coppicing*) or from the roots (*suckering*) (Fig. 8.1). Coppicing has been the normal method of managing woodland in many European countries for hundreds or thousands of years: it assures a permanent and ever-renewable supply of wood (Figs 8.2 & 8.3). It normally produces *wood*, poles and rods used for fencing and various more or less specialized crafts, but especially for fuel. *Timber*, logs big enough for beams and planks, comes from trees which are allowed to grow bigger and often are not expected to sprout after felling.

Coppicing and suckering are widespread and widely utilized among the trees of other continents. Often they are a response to burning rather than to woodcutting, although they are not confined to flammable kinds of tree. Coppice shoots are very palatable to wild and domestic herbivores; where browsing cannot be excluded it is the practice instead to treat trees as *pollards*, cut at 3–4 m above ground so that the animals cannot reach the young shoots (Fig. 8.4). A variant of pollarding is *shredding*, cropping the side branches of a tree leaving a tuft at the top. Pollarding and shredding are very widespread in Europe and Asia; often the main function of the trees is to produce not wood but leaves on which to feed cattle and sheep when pasture is short (Austad 1988). They are typical of savanna—grassland or heath with scattered trees—rather than of forest (Fig. 8.5).

Practices of this kind create habitats which are of importance in their own right. Pollarding and coppicing prolong the life of trees. Much of the specialized wildlife of trees—hole-nesting birds, insects associated with

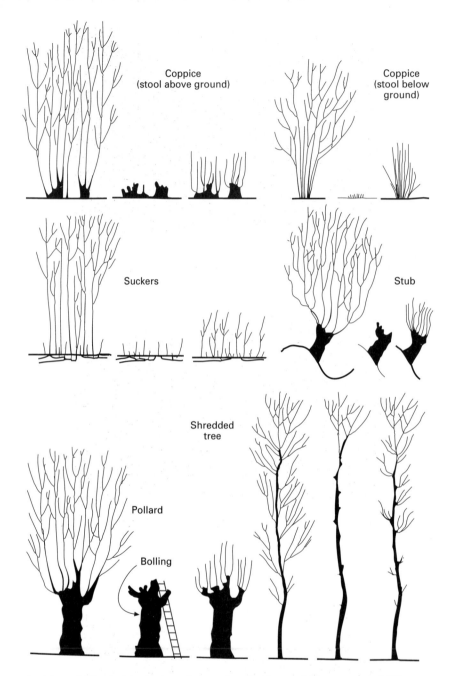

Fig. 8.2 Methods of managing trees for wood production, and the responses of different kinds of trees to them. For each method the tree is shown before cutting, immediately after cutting and 1 year after cutting. (From Rackham 1990.)

rotten wood, lichens of old dry bark—is dependent on old trees, and especially on the specialized microhabitats present on ancient pollards (Read 1992) (Fig. 8.6).

Fig. 8.3 Coppicing in a wood of chestnut *Castanea sativa*. This area shows 1 year's growth since last felling (Ligurian Apennines Italy, September 1984). (Photo by O. Rackham.)

Fig. 8.4 Pollarded oaks in a field (Risby, Suffolk, England, April 1980). (Photo by O. Rackham.)

Fig. 8.5 Savanna-like grassland with shredded deciduous oaks (Pindos Mountains, Macedonia, Greece, August 1987). (Photo by O. Rackham.)

Fig. 8.6 Pollarded hornbeam *Carpinus betulus*, some 300 years old. A single old pollard provides many microhabitats not present at all in younger or unpollarded trees (Hatfield Forest, Essex, England, June 1990). (Photo by O. Rackham.)

Coppice woods and wood-pasture savannas have diminished in many countries over the last 250 years through the rising influence of modern forestry as a profession with its own ideology. Foresters, on the whole, think forests are only for producing timber, and disapprove of browsing, multiple land uses and old trees.

8.1.2 History and pseudohistory

Conservationists and anti-conservationists demonstrate the close connection between history and conservation by eagerly deploying historical arguments regardless of whether they are true. For some reason historical ecology is particularly productive of canards and factoids* (Rackham 1990).

The public believes, in defiance of everyday experience, that trees are necessarily killed by cutting them down. In Britain, as recently as the 1970s, a minister of agriculture argued against the conservation of hedges on the grounds that hedges were no more than 200 years old; this can be shown

---

* A factoid is a statement which looks like a fact, is believed as a fact, and has all the properties of a fact except that it is not true.

to be false by anyone who spends an hour studying early maps in a county record office. In Scotland, the 'restoration' of the Great Caledonian Wood has become a political issue, regardless of the protests of historical ecologists who point out that the Great Caledonian Wood is mythical and never existed in historical times (Breeze 1992; Dickson 1993).

8.2 Historical ecology in choosing and designating sites

Biological conservationists are, rightly, much concerned with the protection of wilderness areas, such as wildwood or 'primaeval forest', which so far have had little or no human activity in them. To declare that a site is wilderness is an historical judgement, which seems often to be arrived at hastily and on no stronger basis than the absence of obtrusive cultural remains. Hitherto developers and anti-conservationists have seldom had the intelligence or skill to make a scientific case to the contrary, but it cannot be assumed that this will always be so.

As archaeological knowledge advances, human activities are shown to be more pervasive and to have started earlier than had previously been thought. People may have had profound effects on the ecology even in areas where they did not go: through, for example, exterminating elephants and other great herbivores, or through altering the fire regime. One begins to doubt whether, in historic times, there have been any extensive land areas entirely without cultural influences, except for islands that never had human inhabitants.

Although the very idea of wilderness, narrowly defined, begins to look unrealistic, in a broader sense it is still valid. An area which has had no human impact, or only certain limited kinds of human impact, for a thousand years or so is likely to be of high conservation value. But so are many types of semi-natural and cultural landscape. Increasing research often blurs, instead of resolving, the distinction between the wholly natural and the semi-natural (Peterken 1996).

At the other extreme, historical ecology can help in identifying the unconservable. It is not unusual for certain areas, especially of industrial wasteland, to pass through a stage of succession which is attractive to conservationists (e.g. for a large number of orchid species). It would seem impractical to conserve these temporary, seral stages for more than a few years—unless a means is found of perpetuating them cyclically as stages in a larger-scale activity.

Conservationists are on weak ground when they criticize cultural landscapes for not being primaeval, or when they try to recreate an ideal primaeval landscape which may never have existed. In southern Europe, especially Spain, foresters still plant arid, orchid-rich hillsides with pines in the name of conservation—they plead that they are 'restoring the forest cover' that they think existed even where the annual rainfall is less than

300 mm—and are very hurt when Nemesis strikes 10 years later in the shape of a conflagration.

We may smile at such naive heavy-handedness, but in England coppice-woods are criticized for not being wilderness. Hambler & Speight (1995) have, in effect, called upon conservationists to drop whatever else they are doing and turn all their energies to trying to recreate unmanaged wildwood. (They are not satisfied with the fact that the great majority of coppice-woods are already, for various reasons, undergoing a close approximation to the kind of benevolent neglect which they advocate.) This might be reasonable in a country where there is still wildwood left, but in England history has long ago closed that option. English wildwood disappeared in prehistory, and there is no way of knowing what it would by now have turned into, 4000 years on, had it survived.

Conservationists tend to be interested in sites that meet some pre-determined criterion of what a 'natural area' ought to look like. The Council of the European Community (1992), in its Directive on the conservation of natural habitats, drew up a long and Mediterranean-biased list of 'forests of tall trees'; no doubt a worthy objective, but it excludes forests of short trees, such as the ancient cypresses of west Crete, which are among the oldest and most extraordinary trees in Europe (Fig. 8.7). The list ignores savanna (grassland with trees), except for a mention of the *dehesa* of Spain, the most extensive and least threatened kind of savanna. Grassland or heath with trees used to be nearly as widespread in Europe as it still is in Africa; its remnants range from the wood-pastures of England to the pine savannas of Greece and the oak savannas of Crete. They are usually of value for both the tree and the non-tree components, and are severely threatened by destruction and neglect. But because European text-books of ecology contain chapters on forests and grassland but not

Fig. 8.7 Cypress *Cupressus sempervirens*, well over 1000 years old, at the alpine tree limit in west Crete, April 1989. (Photo by O. Rackham.)

savanna, and because the Council is ignorant of ecological history, they are excluded.

One meets the argument that developers should be allowed to destroy a site on the plea that they will create an equivalent habitat elsewhere. ('Eurotunnel Moves Ancient Wood', as a headline in a botanical newsletter put it.) A re-creation exercise would be of the greatest value for research into how habitats function, but as a substitute for conservation it raises the most difficult questions. Can a firm of developers promise to remain in existence long enough to see the project through to completion? How can anyone decide when the project has been completed? Is an assemblage of Blogginsley Wood plants the same as the real Blogginsley Wood? Is it possible to assemble the Blogginsley Wood plants on the site chosen for the re-creation, rather than the different set of plants which would now be there had that site remained as woodland, or the plants which would be there in future were it to be allowed to revert to woodland naturally? The reality with woodland and other long-term habitats seems to be that each site is the product of a unique sequence of past events which it would be impracticable to repeat.

It is particularly to be regretted that the *National Vegetation Classification*, now one of the systems most widely used by British conservationists for defining vegetation types, makes little use of historical information (Rodwell 1991). The authors divide vegetation into 'plant communities' strictly on floristic criteria. In woodland, especially, they make little use of trees, on the grounds (which they do not substantiate) that trees are a relatively artificial feature of woodland. Instead they rely on woodland ground vegetation, about which there is almost no information for any period except the present. To what extent their categories existed in past centuries is anyone's guess. A classification with no time dimension is of limited usefulness.

8.3 Historical ecology and management

When an area becomes a nature reserve, its history does not stop. Any area which has been a reserve for a long time—Yellowstone National Park in the USA (Chase 1987), Wicken Fen and Hatfield Forest in England (Rackham 1989)—has a history of curiously contradictory management, often because the first people in charge of it were ignorant of its history and thought its management was a simple matter of taste. The City of London Corporation in 1878 solemnly undertook to 'preserve the natural aspect' of Epping Forest on the edge of London. In pursuance of this object, they abolished the pollarding which had been a normal practice for a thousand years and had given the Forest its particular character as an ecosystem; their successors are still having to cope with the consequences of that decision (Fig. 8.8). Not for nearly a century was it discovered, through pollen analysis, what the 'natural aspect' of Epping Forest had

Fig. 8.8 Effects of a century of not pollarding in Epping Forest. The original complex, savanna-like ecosystem has been converted into immense beeches, the remains of dead oaks, and little else (May 1986). (Photo by O. Rackham.)

really been: a limewood which it would be quite impossible to re-create (Rackham 1978). Conservationists who are ignorant of (or, what is worse, think they know) a site's history all too easily throw away its real ecological character in pursuit of what later research shows to be an illusion.

Conservationists are wiser when they continue the existing management of a site until it can be shown to be harmful. This is a cautious and sensible policy: one wishes that the early Epping Forest Conservators had followed it instead of pursuing their own self-willed ideas. But it should be re-considered as soon as the necessary historical research has been done. Hatfield Forest (Essex) is the most complex example of wood-pasture. It combines defined areas of woodland and areas of savanna-like grassland with trees called *plains*. In the 1930s, the National Trust, a conservation body which had acquired the Forest, continued coppicing the woods and grazing the plains, but failed to fence felled areas, so that the cattle got in and ate up the young shoots. The separation of woods and plains, fundamental to this Forest's functioning for at least 600 years, had been overlooked (Rackham 1989).

Management should seek to define the special characteristics of a site, its *genius loci*, the features that make it different from other sites. It is important to maintain the normal character of a site as it has developed, rather than an ideal of how it might have developed had history been different. Many long-continued human activities have come to form part of the environment of a site, and plant and animal communities have become adapted to them. A wood with centuries of coppicing will have coppicing-adapted plants and animals (Fig. 8.9); there is no point in managing it for the benefit of creatures dependent on big hollow trees (unless old coppice stools are suitable).

That is not to say that conservation management should necessarily be purist. As long as the principle of coppicing is maintained, there is no

Fig. 8.9 *Geranium asphodeloides* as a 'coppicing plant' in a recently felled oakwood in the Pindos Mountains, Greece, May 1988. (Photo by O. Rackham.)

objection to varying the length of the coppice rotation (for which there is good historical evidence (Rackham 1980)), or to leaving part of the wood as a non-intervention area, or to having some old or dead trees. Alterations to the historical regime are usually acceptable if they add to, rather than supplant, the range of historical habitats.

Historical research can identify new threats. In England it is very clear from early botanical records that the most severely threatened part of the woodland flora is the plants of permanent unshaded areas in woods: species like *Succisa pratensis, Lathyrus sylvestris* and *Melampyrum cristatum* (Rackham 1980). These 'woodland grassland' plants might be thought a very artificial part of a wood's flora, but some of them have a pollen record going back through the Holocene (and to earlier interglacials) (Godwin 1975; Rackham 1986). They are threatened by increasing shade as rides get overgrown and the tracks round the margins of woods are either overgrown or ploughed up.

Historical evidence also provides a standard against which to check changes in conservationist fashion. Reserves have been through a series of phases: the 'Shoot Carnivores' period (in nineteenth century USA), the 'Do Nothing' period, the 'Grassland Improvement' period (in England), the 'Suppress Fires' period (in the USA) (Pyne 1982), the 'Plant Trees Everywhere' period, the 'Measure Biodiversity' period, and the 'Encourage Public Access' period. Every fashion has taken away something of the special features of a site which is difficult to undo when the fashion passes. (There is also the varying tension between the interests of the site and the interests, or supposed interests, of visitors.) Five-yearly revisions of management plans (in the manner of the Soviet Union) and frequent changes of staff reinforce this instability. Unless continuity is built into a long-term management plan, the future of any reserve, over more than a few decades, is uncertain.

8.3.1 Biological and other kinds of conservation

Conservation is, or should be, a continuum of many things, from the conservation of the Amazon rainforests to the conservation of Benjamin Britten's manuscripts in the Fitzwilliam Museum, Cambridge. The defence of a site is much helped if it can be shown to be of more than merely biological importance: if it is mentioned in a charter of the eighth century AD, or is the scene of a battle, or is the subject of a painting by Constable or a poem by John Clare.

Few biological sites are without historical interest, and most historic sites and even buildings have some biological interest. In England repairs to an ancient building have, by law, to take into account possible disturbance to bats. Church repairers, until recently, were unaware that many churches are first-class habitats for lichens. (Is the present author the first member of a Diocesan Advisory Committee to be also a member of the British Lichen Society?)

Conservationists in different fields, although they have much to learn from each other's methods, do not always talk to each other. I visited a hillfort, an Iron Age earthwork monument, on the Welsh border of England, situated in a medieval (or earlier) limewood. Forest Enterprise (the owner) was converting the limes from their traditional state of coppice to timber trees. My companions and I agreed that this was not an ideal treatment either for the wood or the hillfort: the archaeology would be at risk both from uprooting of the big trees in storms and from machinery used when they were eventually harvested. However, it was claimed that (the wood being very prominent) the public would not tolerate the changes in appearance brought about by renewed coppicing. I was reminded that the archaeological authorities might have recommended that the trees be destroyed, in the belief that trees promote erosion—even though the limewood is an antiquity in its own right, and (while coppiced) has preserved the visible remains unusually well. I speculated on what might then have happened if the limes had withstood attempts to destroy them, as limes often do.

Most of the world's National Parks are, at least to some extent, cultural landscapes. The practice still too often prevails of expelling the original human inhabitants of a national park and replacing them with tourists. This habit is impolitic, for it guarantees that national parks will be opposed. It is also unscientific, for it is precisely in the kind of area likely to be designated as a national park that old-fashioned cultural practices, of kinds that biological conservationists ought to uphold, are likely to survive.

National Parks tend to be presented as if they were wilderness; their historical aspects are played down. The visitor to the Samariá National Park, greatest of the hundred gorges of Crete, learns about its animal and plant life, but not, alas, about the illustrious place of the site in Cretan history and culture; little is made of the antiquities, buildings, chapels, castles, mills, pollard trees and cultivation remains which fill the gorge. The authorities of Yosemite National Park have demolished Yosemite village and are carefully erasing all archaeological traces of its having existed (Olwig 1995).

8.4 Sources for ecological history

8.4.1 Documents

Written archives are of many kinds: they range in date from the Linear B tablets of late Bronze Age Crete to the notebooks of my own visit to Crete in 1968.

For England the earliest documents to describe identifiable pieces of country are Anglo-Saxon charters from about 700 AD onwards. Domesday Book attempts a systematic survey of most of England in 1086 AD; it has never been repeated. Surveys and management accounts of individual estates, including for example woodland, heathland and hedges, go back to the mid-thirteenth century. These, on the whole, reveal a remarkable degree of continuity. About half the wood-lots mentioned in medieval documents were still there in 1900, often with their exact boundaries unaltered; some of the hedges, and many of the roads, in Anglo-Saxon charters are identifiable today (Rackham 1986).

In other countries there are the accounts of travellers, such as Pausanias in Greece of the second century AD, or the succession of travellers, native and foreign, in Spain and Portugal from the sixteenth century onwards. It is essential to go over the ground point by point, and to consider whether the landscape described is consistent with what is there now. This shows, for example, that much of the great extent of savanna in Spain and Portugal is no older than the nineteenth century (though some is much earlier); it is not 'degraded forest', as often claimed, but is heathland which has acquired trees (Grove & Rackham, in press).

Land surveys (somewhat in the tradition of the Anglo-Saxon charters) are a much-used source in the USA. These are concerned with establishing the boundaries of properties bestowed on new settlers, typically in units of 1 mile square. The sides are described and the two trees nearest each corner are named. In middle Texas, for example, it is possible to compare the distribution of forest, savanna and prairie in the 1830s and 1840s (the last Indian period before settlement) with what is there now.

The most useful maps are those at scales of 1:25 000 or larger. For England these exist from c. 1580 onwards; they show every wood, road and hedge, and occasionally individual trees and buildings (Figs 8.10a & 8.11a). From the eighteenth and nineteenth centuries there exist Enclosure Award maps for about half the parishes of England and Wales; where there is not an enclosure map there is usually a Tithe Commutation map of c. 1840. The art of mapmaking reached its zenith with the six-inch and 25-inch

(a)

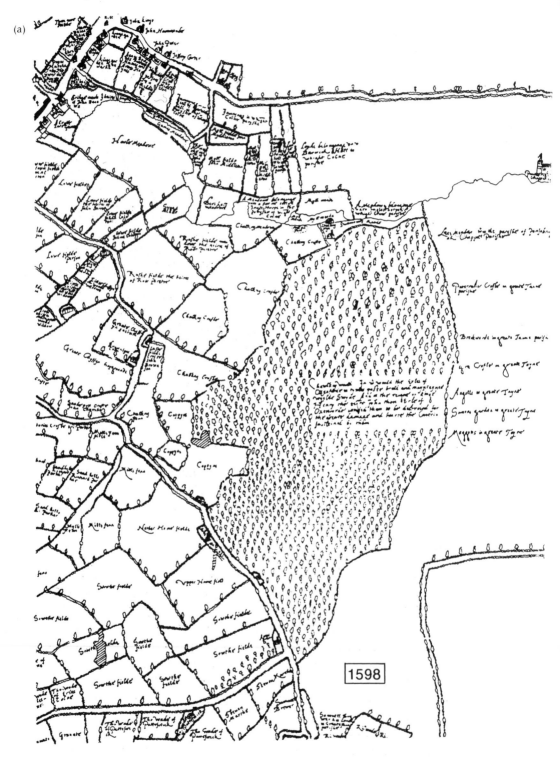

Fig. 8.10 (a) Chalkney Wood in Earl's Colne (Essex, England) and the hedges and fields to the west of it as mapped in 1598. The text in the wood refers to its use as a swine park (for keeping wild boar *Sus scrofa* as semi-domestic animals) up to *c.* 1520. (From Rackham 1990; after original in Essex Record Office.)

Fig. 8.10 *Continued* (b) Chalkney Wood as photographed by the German air force in 1940. The tones within the wood indicate different ages of felling and regrowth. Many of the hedges and fields, as well as the wood's outline, had not altered since 1598. The wood is 1.6 km long. (Original in the United States National Archives, Group 373, GX 10373F/ 1054/SK38.)

(1 : 10 560 and 1 : 2500) Ordnance Surveys of Ireland and Britain between 1835 and 1880; these appear to show every individual non-woodland tree. Other countries with an excellent cartographic tradition include Flanders, Italy and Sweden.

Pictures

The earliest identifiable landscape pictures in Europe are the frescoes of *c.* 1700 BC excavated on Santoríni, which appear to represent the cavernously decayed, multicoloured cliffs of that wondrous Aegean island.

   Landscape pictures are a continuous source from the thirteenth century onwards. Those that are works of art in themselves are often too idealized to be of much value. Better are those that form the background to a human or divine scene, or are topographical views of some definite place, or are illustrations in books of travels. They are not always straightforward to interpret—artists add or leave out features at whim, and remarkably few artists ever mastered the difficult technique of drawing a recognizable tree. Field sketches (such as those of Edward Lear in the nineteenth-century Mediterranean) tend to be better than studio pictures. With good fortune, it is possible to sit where the artist sat and observe the changes. For example,

(a)

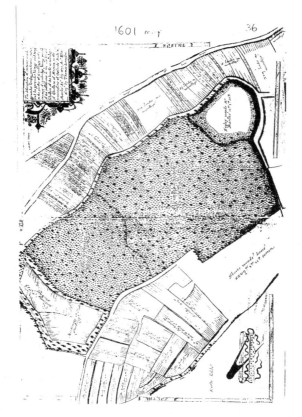

(b)

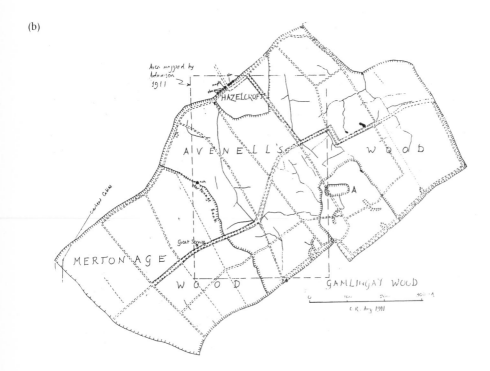

at an estate in a remote dry valley in the chalklands of south England I was able to work out, from a topographical picture of *c.* 1750, changes and stabilities in the rather fluid boundaries between woodland, chalk scrub and chalk grassland. (In this case, as so often, the original viewpoint was obscured by later tree growth.)

Early photographs (e.g. picture postcards) are an important source in many countries. Here, too, it is important to rediscover the exact viewpoint, repeating the photograph with the same focal length of lens and (if possible) in similar light. Of particular note are the immense archives of photographs taken by the Italian Forest Service from the 1920s onwards. Early aerial photographs are another source. The German air photographs of the Second World War for many countries (including England) are in the National Archives, Washington (Fig. 8.10b). The British air photographs, for many countries excepting Britain, are in the Department of Geography, Keele University.

Interpretation of documents

Besides particular accounts, there are contemporary generalizations, such as Evelyn's *Sylva*, the great book on trees and woods in seventeenth-century England. Historians see these as an easy way to avoid the task of piecing together the history of a landscape from the histories of thousands of individual wood-lots, hedges and heaths. But was Evelyn correctly informed, and did he draw the proper conclusions from his data? His pessimistic conclusion about the declining woodland of England was flatly contradicted by Defoe forty years later: who was right? (Defoe.)

Documents are wonderful things, but their limitations need to be understood. With few exceptions, they were not written in order to tell posterity what grew where. Just as historians often write papers on ecological history without identifying the principal trees and other plants and without explaining how they behave in relation to browsing, burning and woodcutting, so ecologists write papers containing a perfunctory paragraph beginning 'Ancient authors say …' without explaining which ancient authors, what exactly they said, and what are the grounds for believing them. In either case what follows is likely to be pseudo-history (Rackham 1987, 1996a).

Fig. 8.11 *Opposite* (a) Gamlingay Wood near Cambridge, England as mapped in 1601. Unlike Chalkney Wood, it was largely surrounded by open-field strip-cultivation. The strip of wood round the field to the southwest represented the enlargement of a hedge. The feature across the middle of the wood is a bank dividing two ownerships. (Original in Merton College, Oxford.) By permission of the Warden and Fellows of Merton College. (b) Gamlingay Wood in 1991. The outline and the internal bank are unchanged since 1601. Although this is perhaps the best-documented wood in England, the earthworks reveal undocumented changes. Area A, containing earthworks which may be Iron Age, lies outside the original woodbank, and only later became woodland. This enlargement of the original wood was too early to be noted by the 1601 map or by any other document. (After Rackham 1992.)

The most obvious limitation, often overlooked by historians, is that documents say nothing about what was going on at times when people were not writing. This is a serious matter in countries like England and Greece, where the most important steps in making the landscape had already happened before the earliest writings.

Historical documents, unless of the most workaday kind, need to be verified. Was the writer in a position to know what he was talking about? Had he been to the place? Was he correctly informed? If a poet or philosopher, was he more interested in getting the metre or the philosophy right than in accurately reporting details? What is the status of the story—historical, mythical, proverbial or a work of pure imagination? (Many scholars are reluctant to accept that ancient authors were capable of writing fiction.)

A pitfall awaits those who rely on other people's translations. Many ancient documents had a precise meaning which is now difficult to reconstruct; scholars guess at words like 'brushwood' (that is, very small twigs suitable for making brushes) which originally meant something different. Difficult passages, like the ancient Greek philosopher Plato's famous account of the (apparently fictional) ecological history of Attica in *Critias*, fall into the hands of smooth-tongued translators who render them in the light of modern views on deforestation and erosion.

Gaps in historical information are not neutral. Felling trees is sudden, conspicuous, profitable and sometimes controversial. Their regrowth, or new trees invading abandoned land, is gradual and inconspicuous, and is far less likely to be put on record. There is plenty of information about people cutting down trees in the Greek and Roman world (how could there not be?), but it does not in itself constitute a history of deforestation. Without knowing about regrowth, we cannot know whether woodland was decreasing. For all we are told, regrowth may have been gaining on felling, as in twentieth-century New England.

Legal documents have a particular fascination for scholars. At any conference on forest history, about two-thirds of the papers deal with the history of forestry laws, the remainder with the history of forests. But laws are not to be taken at face value as evidence of what was happening on the ground, nor interpreted in terms of modern legal practice. Medieval laws were often not intended to stop people from doing things, but to raise revenue from fines. It is records of infringements and penalties that matter.

Official statistics should be treated with particular suspicion. A statement like 'forests cover 54.71% of Ruritania' may seem to be straightforward, but what meaning is to be attached to it in a country where forests do not have nice sharp edges but shade off into savanna or maquis? If the figure at an earlier date was 35.69%, was the change due to the forest getting bigger or to the definition getting broader? How is the figure to be interpreted if, in Ruritanian law, lands listed as forest are liable to seizure by the Forest Service?

Some kinds of information are particularly unlikely to be recorded. The whole corpus of ancient and early-modern writings has remarkably little to say about Mediterranean terracing, although on other grounds it certainly existed. The most usual reason for something not being recorded is that it was too commonplace to be worth mentioning.

Cultural differences among terms for vegetation

North American writers find no difficulty in distinguishing between 'forest' composed of continuous tall trees, 'scrub' of shrubs which are different species from trees, and 'savanna' of trees scattered among grassland or heath. They make much of the distinction between 'old-growth' and 'second-growth' forests. They tend to believe (forgetting the contribution of Native Americans to the cultural landscape, Fig. 8.12) that there exist areas of primaeval wildwood in contrast to 'second-growth' forests which have been felled since European settlement.

In Europe things are different. There is no equivalent to the date of settlement (except on outlying islands such as Crete) and no distinction comparable to that between settlers and Indians. Roman Britain and Homeric Greece were already lands with centuries of civilization behind them. Virtually all forests are umpteenth-growth; wildwood (with the rarest exceptions) is not recorded or remembered, and has to be re-constructed from pollen analysis. Depleting forests is an entirely different process from destroying them. There have been several advances and retreats of civilization, and inversely of woodland. Woods which have existed all the

Fig. 8.12 Savanna in Texas. This spot was the scene of a Wild West gunfight in 1831 and has changed little in appearance since. The trees, *Quercus fusiformis*, grow in clonal groups called motts, probably the result of burning (see Fig. 8.1), which have lived through four different human cultures. But this is a cultural savanna—if not grazed or burnt it turns into a forest of *Juniperus ashei* (near Menard, Texas, January 1996). (Photo by O. Rackham.)

time (however often they have been felled) may be different from those that have sprung up on ex-farmland.

In Europe the distinction between trees and shrubs is weak. In the Mediterranean it disappears: shrubs are trees reduced to a shrubby form by some combination of browsing, woodcutting, burning or drought. If these factors are removed they grow up into trees again. The mention of trees, therefore, does not imply either forests (in the American sense) or timber. To an English writer, *wood* or *woodland* does not imply tall trees: whenever a wood is felled, it retains its identity and name and continues to be a wood throughout the ensuing regrowth cycle. In historic England, as in modern Scotland, *forest* meant a place of deer, not necessarily of trees— Sherwood Forest (Nottinghamshire) was mostly heath, Dartmoor Forest (Devon) was entirely moorland* and Epping Forest had trees in the form of savanna.

Many Europeans do not notice the distinction between forest, scrub and savanna. In Spain and Greece the normal words for 'forest', *monte* and δασος, are used indiscriminately for forest, coppice-wood, maquis (trees in the form of shrubs) and savanna. *Monte* can mean anything from 'mountain' (its original meaning) down to phrygana (*Thymus, Cistus* and other undershrubs which are not potential trees). In much of medieval Europe, scholars refer to 'forests', in various languages, as constituting grazing land for animals. If one asks what it was the animals were eating, one gets an evasive reply. The answer must be that much of what is referred to as 'forest' was really more like savanna (Rackham 1996b).

Europeans do, however, distinguish between timber and wood. In English, French, German, Portuguese, Spanish, Italian or Latin, *wood* produced by coppicing or pollarding, or from the boughs of felled trees, is a different word from *timber* big enough to make beams and planks. In North America this distinction largely disappears, so that among American writers trees, timber, forest and woodland are almost synonymous. ('Shooter in hand, the Kid urged his mount through the belt of *timber* at the edge of the prairie.')

8.4.2 Archaeology

Archaeological evidence documents the march of civilization and the retreat of the primaeval landscape (or the retreat of civilization and the re-establishment of the wild). Sometimes this is obvious from the mere density of visible remains. The vast numbers of Iron Age monuments in Ireland, Sardinia and parts of south Sweden prove that a continuous cultural landscape existed by the late centuries BC and there was no room for extensive remaining wildwood. (In Sardinia and Sweden these monuments continue into what are now great forests.)

---

* Vegetation of grasses, Ericaceae and *Sphagnum* on peaty soils, characteristic of mountains with an Atlantic climate.

Cultures which did not leave conspicuous monuments have to be reconstructed from archaeological surveys. In Crete, recent surveys show that by the end of the Neolithic settlement had spread throughout the island, including the west (hitherto thought to be empty) and to the high mountains (Rackham & Moody 1997). In England, Bronze and Iron Age and Roman sites are to be found throughout the country, including most of what later became the big wooded areas of the Middle Ages. Archaeological surveys can never discover all the settlement that there was, and in some types of terrain (especially where there is now dense vegetation or active erosion) can find only a small fraction of sites.

Some archaeological evidence is related to the nature and use of surviving roughland. In England, medieval woods are surrounded and sometimes subdivided by banks and ditches, which are evidence of the great importance attached to woodland conservation. These earthworks also record the exact extent of the wood and any changes in its boundaries (Fig. 8.11b) (Rackham 1980, 1990). In many countries charcoal hearths record the use of woodland for making charcoal as an urban and industrial fuel.

Discrepancies should be looked for. An area of ridge-and-furrow (earthworks produced by medieval or sub-medieval ploughing) within a wood proves that the wood was once smaller (Fig. 8.13). Charcoal hearths extending beyond a wood on to moorland (provided they were for wood, not peat, charcoal) prove that the wood was once bigger.

Archaeological evidence, unlike documentary, tends to be precise as to location but inexact as to date.

8.4.3 Vegetation

Old trees often afford an independent date in their annual rings. These record not only the age of the tree but the various vicissitudes that have happened to it—pollarding (Fig. 8.15), fire scars, the encroachment of neighbouring trees, etc.

In Europe, obvious old trees (up to 1000 years old, rarely more) are a feature not of woodland but of wood-pasture, savanna and as hedgerow and field trees in farmland. They are particularly useful in England and Greece. Here too, discrepancies should be looked for. A big, spreading old tree, surrounded by densely packed younger trees in a forest, is evidence of a savanna history (Fig. 8.14).

Within woodland, ancient trees usually take the form of coppice stools. If a suitable tree (e.g. lime, maple, ash, chestnut) is cut down every 10–30 years, it lives indefinitely; over the centuries it grows into a massive base, 2 m or more across, from which an indefinite succession of poles arises (Rackham 1990). I have seen ancient coppice stools in England, Wales, Scotland (rare), Ireland (rare), France (local), Norway, Sweden (rare), Spain (Alpujarra), Majorca, Italy, Sardinia, Greece, Crete and Turkey. (In America,

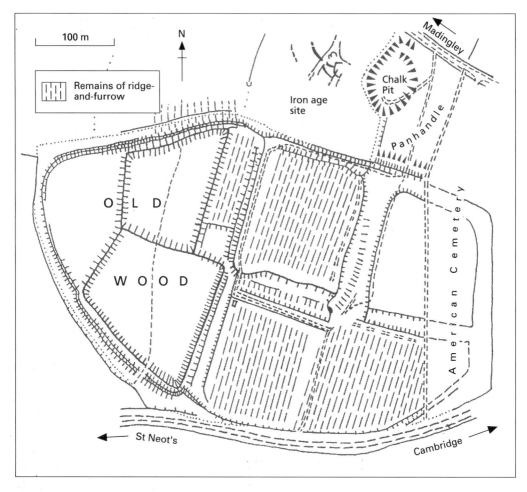

Fig. 8.13 The complex earthworks of Madingley Wood near Cambridge, England. The eastern part of the wood was grubbed out, probably some 350 years ago, and spent some time as four arable fields before again becoming woodland. Areas of ridge-and-furrow earthworks are relics of this period. The 'old wood' has been woodland at least since the Middle Ages, but contains earthworks which are probably relics of Iron Age fields, some 2000 years ago. Plant records for this wood go back to the seventeenth century. (After Rackham & Coombe 1996.)

ancient stools can exist in wildwood, because certain trees, e.g. *Tilia americana*, are self-coppicing.)

The presence of particular species often goes with sites that have particular kinds of history. In England, plants such as *Paris quadrifolia, Crataegus laevigata, Anemone nemorosa, Primula elatior* and *Sorbus torminalis* tend to grow in ancient woodland; they may be missing from newly established woodland even after a lapse of centuries. Where cultural practices are long-established, plant communities develop in response to them. Examples are the flowers of ancient meadowland, or the 'coppicing plants' that bloom in profusion in ancient woods every time the wood is felled (Rackham 1980, 1990) (see Fig. 8.9).

Fig. 8.14 Stump of a deciduous oak, showing cycles of narrow and wide annual rings produced by pollarding; these can be traced back at least 600 years (South Macedonia, Greece, April 1990). (Photo by O. Rackham.)

Fig. 8.15 Old, possibly pollarded, yellow birch *Betula alleghaniensis* now in a forest of American beech *Fagus grandifolia*. It cannot have grown up in this environment, and is probably a relic of cultural savanna (Great Smoky Mountains, Tennessee, May 1981). (Photo by O. Rackham.)

8.4.4 Pollen analysis

Pollen analysis depends on having a stratified deposit, in which pollen grains are preserved and which builds up from year to year. This usually means a permanently wet place, such as a peat bog or the mud at the bottom of a lake. Sites are particularly abundant in peaty areas like Ireland or Minnesota.

Pollen analysis is the major, often the only, source of information about wildwood and the early stages of the impact of civilization. It records chiefly those plants that produce pollen or spores. Trees such as lime, which shed little pollen, are underrepresented, as are all insect-pollinated plants. Trees whose pollen production is suppressed, either temporarily because they have recently been felled or permanently by the shade of taller trees, will not appear in the record. Pollen analysis reveals the composition of forests and other plant communities, but says little directly about their structure (e.g. the height and spacing of the trees). (A sample dominated by hazel *Corylus* pollen is unlikely to have consisted of tall trees, because hazel is of shortish stature and produces no pollen when shaded.)

Pollen analysis becomes more difficult as civilization advances, because the landscape becomes increasingly complex and sorting out its various components becomes problematic.

In semi-arid lands such as southern Europe pollen analysis is much more difficult. There are few of the waterlogged deposits or acid soils in which pollen is preserved. These have often lost their top layers through drainage. Of necessity, they record the vegetation of unusually wet places in a dry landscape. Many important pollen types are difficult or impossible to distinguish (e.g. the various ecologically distinct oaks). Many common plants are insect-pollinated and therefore poorly represented in the pollen record. The presence of tree pollen is traditionally taken as implying forests, but this is not so. Trees such as *Quercus coccifera* can be bitten down to shrubs less than 1 m high and still produce pollen. Many south European pollen diagrams, on closer examination, contain the pollen of non-shade-bearing herbs and undershrubs such as *Cistus* and *Asphodelus*. Although small in quantity, these come from insect-pollinated plants and are highly significant—they prove that the vegetation was not forest but savanna or maquis (Grove & Rackham, in preparation).

8.4.5 Off-site archaeology: building timbers

The timber and underwood used in buildings give information about the species of tree, their ages at felling, sizes, growth rates, competition from other trees, etc. They thus reveal the composition and management of woodland and sometimes non-woodland trees (Rackham 1990; 1997).

This information is most useful for medieval buildings: for instance in fourteenth-century England oak trees with a usable length of more than 8 m were rarer and more valuable than ever before or since. It can be extended back through excavations in waterlogged sites in which timber and wood are preserved. This gives the earliest evidence of woodland management, from the Neolithic of the Somerset Levels in England.

Timber was not necessarily local: for instance the timbers of ancient buildings in Spain tend to be pine, brought from distant sierras, the local oak not being regarded as a timber tree. Boards in medieval England tend to be from straight-grained, slow-grown trees, brought from eastern Europe.

Here, as always, discrepancies are instructive. In Sardinia the preferred house timber, from the thirteenth to the eighteenth century, was juniper, which was often 5 m long. There was evidently something which permitted a continuous supply of this light-demanding, non-forest tree. In modern Sardinia, juniper is uncommon and rarely reaches this size; the increasingly dense forests are hostile to it.

8.5 Conclusions

In historical ecology it is essential to make use of all the information, including information derived from fieldwork. Different sources of information are more than merely additive. Few of them are conclusive in themselves, with no problems of either sampling or interpretation. A story told from documents alone—whether in Classical Greece or World War II England—is almost certain to be wrong. It becomes convincing if it is corroborated from archaeological survey or the evidence of old trees. The best-documented wood in England, as far as is known, is Gamlingay Wood near Cambridge; yet this has many features on the ground, including a change of boundary (Fig. 8.11), which are not recorded in the documents.

One of the most fruitful sources of factoids is the habit of historians and archaeologists to claim to know all the answers, even where there is not enough evidence. For example, a consistent change in ancient woods in England at present is the increase of ash *Fraxinus excelsior*. If I, before whose eyes the change is taking place, cannot offer an explanation, can we accept the explanations offered by scholars for similar changes in the past known only from palynology?

Direct observation of historical changes has always been difficult, for anthropological reasons: ecologists die or are promoted, and the question they were studying becomes unfashionable and is neglected by their successors. This gets worse as ecology becomes increasingly professionalized and dependent on expensive equipment, and is thus restricted by careers, employment practices, research grants and even safety codes. (English county wildlife trusts can be fickle employers, sacking staff for administrative reasons regardless of work left unfinished.) It used to be difficult to understand woodland because of the long timescale. This difficulty now extends to grassland—research needing more than 2 years' observations tends not to be done.

Acknowledgements

Among the many people who have contributed in various ways I would particularly mention Dr S.P. Bratton, Dr J. Dickson, Mr A.T. Grove, Dr J.A. Moody, Dr D. Moreno, Mrs A. Parker, and the late Mr C.E. Ranson.

CHAPTER 9
Selecting areas for conservation

Colin J. Bibby

9.1 Introduction

The scale of the crisis facing biodiversity imposes a very strong need for the establishment of priorities for conservation effort and expenditure. There is too little knowledge, too little time and too little information for conservation to proceed efficiently on a basis of saving species one by one. It was once common to believe that a focus on charismatic species would, additionally, safeguard ecosystems and other less well-known or attractive species. Much early conservation legislation concentrated on species. There is, however, growing evidence that this is not an adequate basis for conservation planning.

A high proportion of expenditure on biodiversity conservation is geographically targeted. Protected areas safeguard habitats and their constituent species in places where relatively high priority is given to conservation. Legislation and policies influencing land use are also restricted in their geographical area of influence.

This chapter addresses the question of how areas for conservation are selected. What might then be done in the selected areas is the subject of other chapters.

Approaches differ with scale. This is partly because the kinds of decision that are made at different scales vary, and thus have different information needs. It is also because the costs and difficulty of assembling and analysing relevant data rise with increasing area.

At the largest scale, resources are allocated between countries. This is particularly true because of the huge funding needed for biodiversity conservation in developing countries. Nations carry an ultimate responsibility for conservation of their biodiversity and relatively large flows of international funding contribute to such endeavours. Many international aid and development programmes include a component of support for biodiversity and the environment. The Convention on Biological Diversity calls for expansion of such financial support.

The terrestrial boundaries of most nations have no relationship to biological boundaries. Biogeographical regions provide a more rationally justifiable framework for analysis of resource allocation at a regional level.

A biogeographical approach can be applied both between and within nations. A still finer scale of analysis is required to identify and resource protected areas, which might vary in size from less than 1 ha to several thousands of square kilometres.

At all scales, there are huge gaps in the knowledge of the distribution of biodiversity. The data that do exist are often so fragmented as to be difficult and expensive to collate. The search for efficient and practical methods of using existing data and for collecting new data is of great importance. Decisions relating to the environment always have political and economic dimensions in addition to the biological ones. Biologists need to present their information in a clear and relevant manner if it is to receive adequate attention. They also need to take part in processes which combine biological, political and economic information and analysis (McNeely *et al.* 1990; Anon 1992; Munasinghe & McNeely 1994; Johnson 1995). These subjects are discussed in Chapters 11–14.

The output from any selection procedure at any scale usually takes the form of a catalogue of areas whose locations are mapped. Lists may be generated from measurements of attributes such as species richness or uniqueness. Values may be assessed against criteria, or they may be ranked. An important, more recent, approach seeks to identify sets of areas which efficiently achieve some overall objective. Such an objective might be that a population of every species that occurs in a region should occur in at least one protected area. This idea is of sufficient importance as to merit a section for itself.

9.2 Evaluation of nations

A very high proportion of the Earth's terrestrial biodiversity is concentrated in relatively few nations (Mittermeier & Werner 1990). Important decisions about the allocation of conservation resources are made between nations. A question therefore exists as to how efficiently these resources are used, and how gaps might be identified.

It is relatively easy to draw attention to the high-ranking countries for species richness (Table 9.1). At this scale, there is a fair degree of correlation between richness measured by different taxa. Another measure of the importance of a country might be the number of endemic species it contains, for which, therefore, the country has a unique responsibility. Such analysis is somewhat flawed by the accident of how boundaries of areas of endemism relate to national boundaries. Ideally, one would take note of species shared between a small number of countries. This would require a data base of occurrence of species by country which does not yet exist on a global scale, even for the best known class—birds. It is clear from Table 9.1 that there is some degree of relatedness between species richness and richness in endemics. The correlation is, however, by no means perfect. In particular, some countries with high levels of endemism such as Australia, Madagascar

Table 9.1 Rank order of the top 10 countries for richness and endemism in different taxa; only countries appearing more than once are listed. (From data in Groombridge 1992.)

Country	Total spp.						Endemics					
	Plant	Butterfly	Mammal	Bird	Reptile	Amphibia	Plant	Butterfly	Mammal	Bird	Reptile	Amphibia
Colombia	2	1	6	1	6	2	–	3	–	9	8	5
Brazil	1	5	4	3	4	1	?	2	6	3	4	1
Indonesia	7	4	1	4	3	5	2	1	2	2	6	10
Ecuador	9	3	–	5	7	3	10	7	10	–	9	6
Peru	–	2	8	2	10	6	?	6	–	5	10	–
Mexico	4	–	2	–	1	4	–	–	3	7	2	2
India	–	–	9	–	5	7	8	10	–	–	5	9
Venezuela	6	7	–	6	?	?	5	–	–	–	–	–
USA	8	–	7	–	?	?	9	–	5	–	?	8
China	3	–	5	8	–	8	1	–	8	–	?	7
Bolivia	10	6	–	7	–	–	?	8	–	–	–	–
Australia	–	–	–	–	2	10	3	–	1	1	1	3
Zaire	–	8	3	9	?	?	–	–	–	–	–	–
South Africa	5	–	–	–	8	–	4	–	–	–	–	–
Papua New Guinea	–	–	–	–	–	9	7	9	9	–	?	–
Madagascar	–	–	–	–	–	–	6	4	7	6	3	4
Philippines	–	–	–	–	–	–	–	5	4	4	7	–

?, data missing.

or the Philippines, do not feature so highly in species richness. A major flaw in such a simple approach is the influence of size of nation. For some purposes this would not matter. However, for actions such as the establishment of protected areas, which become more expensive with increasing size, the data ought to be corrected by size of country. In other words, investment might have the most impact in a country which is relatively rich in species or endemism in relation to its size.

The real disadvantage of such a simple approach is that it could focus disproportionate resources into the most species-rich countries. Identification of gaps in investment requires measures of current investment and their impact. This is not easy. Perhaps the size of area under protection could be a high-level surrogate measure. The United Nations list of national parks and protected areas (World Conservation Union 1990) does not, however, cover sites of less than 1000 ha in extent. Nor is there information on the degree to which protected areas are, in practice, adequately resourced and protected. The amount of international investment in biodiversity might be another surrogate measure, but such information, surprisingly, has not been gathered together.

In an analysis using data from the World Conservation Union (IUCN) on protected areas and information from the USA and the Global Environment Facility, Balmford & Long (1995) identified some potential priority nations for further action. Their analysis was based on the existence of a smaller area protected or a smaller financial investment than would be expected from a regression of these measures on a biodiversity index.

An indirect consequence of an approach based on the assessment of nations in terms of species richness is that it has tended to focus attention on the habitats most notable for species richness, especially moist tropical forests. This bias has been at the expense of other habitats such as grasslands or dry scrub and forest.

9.3 Evaluation by biogeographical region

The idea of dividing the Earth into biogeographical regions and then seeking to ensure the adequate protection of the species and habitat types comprising each is intuitively appealing. Maps of the major regions (Dasmann 1972, 1973; Udvardy 1975) are used as the basis of UNESCO's Biosphere Programme to select biosphere reserves (UNESCO 1996). The US Nature Conservancy uses an evaluation system where community type provides a first, coarse filter (Noss 1987). Reviews of protected area systems in the Afro-tropical and Indo-Malayan realms have been based on the representation of biogeographical units (Dahl 1986; MacKinnon & MacKinnon 1986a, 1986b).

The practical use of such ideas has run into several difficulties. There is room for ambiguity in defining biogeographical realms or related ideas of biomes, ecological regions and habitats. Specialists in different taxa have brought forward different ideas and results. Even if the definitions are secure, it is not easy to acquire the necessary data. There are problems in dealing with modified habitats. It is not particularly helpful to find lowland Britain in the temperate forest zone when the natural climax vegetation no longer occurs and, indeed, any forest cover is very rare. The general problem is how to distinguish between the mapping of potential vegetation and what actually occurs. Some habitat types, most notably wetlands, fall outside large-scale classifications of biogeographical regions and have to be treated separately. Transition zones, mixtures of habitat types and successional stages are also difficult to represent. They may be particularly important for some species and for the range of genetic variation within species.

A recent classification and assessment of Latin America and the Caribbean (BSP *et al.* 1995; Dinerstein *et al.* 1995) is a good example of assessment by bioregion. The approach proceeds through four steps:

1 Definition of ecoregions.
2 Assessment and ranking of biological values.
3 Assessment and ranking of conservation status.
4 Evaluation.

Ecoregions are described in a hierarchy with five major ecosystem types (METs) divided into 11 major habitat types (MHTs), which are, in turn, divided into 191 ecoregions (Table 9.2). METs are of such a high level as to have little practical significance. MHTs group regions similar in general structure, climate, ecological processes and species turnover with distance (beta diversity). Ecoregions divide MHTs geographically into units sharing a high proportion of their species as well as environmental conditions. Some ecoregions include mixtures of habitats, such as gallery forests and grasslands. Ecoregions are identified on the basis of their original natural vegetation, and so are substantially influenced by climate and topography and may not reflect actual present vegetation at any one place.

The advantage of a hierarchical approach is that the values of ecoregions can be assessed within their MHTs on the basis of their species richness and endemism as well as their size and the scarcity or abundance of eco-regions within an MHT. This avoids the inappropriateness of comparing very different formations such as tropical moist forests and montane grasslands directly. It makes it possible to identify the biological value of ecoregions at global, regional, bioregional and local levels. Some of the judgements had to be made on expert opinion rather than hard data because of a lack of data.

The evaluation of human factors is a separate dimension. It can include both measures of damage and potential damage, which are equivalent to threat, and assessment of potential to secure conservation. The measures used by Dinerstein *et al.* (1995) were:

1 Total habitat loss—which is measurable from satellite imagery, given some definitions of the degree of disturbance to be assessed as total loss.

Table 9.2 A hierarchical division of Latin America and the Caribbean into ecoregions as a first step in a regional conservation evaluation. (From Dinerstein *et al.* 1995.)

Major ecosystem type	Major habitat type	Number of ecoregions	Mean ecoregion size (km²)
Tropical broadleaf forests	Tropical moist broadleaf forests	55	149 350
	Tropical dry broadleaf forests	31	33 660
Conifer/temperate broadleaf forests	Temperate forests	3	110 770
	Tropical and subtropical coniferous forests	16	48 180
Grasslands/savannas/shrublands	Grasslands, savannas and shrublands	16	441 160
	Flooded grasslands	13	21 960
	Montane grasslands	12	118 060
Xeric formations	Mediterranean scrub	2	84 370
	Deserts and xeric shrublands	27	84 300
	Restingas	3	11 660
Mangroves	Mangroves	Not divided	

2 Habitat blocks remaining—measured as sizes and numbers with different scaling for different sized ecoregions and different MHTs.

3 Habitat fragmentation or degradation—which again is measurable from geographical information system (GIS)-based data. Some parameters assessed connectivity and existence of core habitat blocks and some assessed losses of species, such as higher plants in grasslands, as evidence of habitat degradation.

4 Habitat conversion—current rates.

5 Degree of protection.

All these attributes are sufficiently measurable, rather than a matter of opinion, as to be capable of being challenged and corrected, and being tracked over time. Dinerstein *et al.* combined them with an arbitrary weighting for each. The final evaluation of priorities for action made a two-dimensional model giving higher priority to the most important and threatened (Table 9.3).

This study is a good example of evaluation at a level which could potentially be applied globally. Amongst its positive features is its GIS basis, so that it is capable of evolution and of being overlain with other data, especially social and economic. The ecoregion classification is hierarchical and this rationality could be developed at finer scales. Most of the measures

Table 9.3 Conservation priorities amongst ecoregions in Latin America and the Caribbean derived from assessment of their biological distinctiveness and conservation status. Each status type shows the numbers of ecoregions included and, in brackets, the proposed priority for conservation investment from I (highest) to IV. (From Dinerstein *et al.* 1995.)

Biological distinctiveness	Conservation status						Totals
	Critical	Endangered	Vulnerable	Relatively stable	Relatively intact	Unclassified	
Globally outstanding	9 (I)	6 (I)	12 (I)	5 (I)	2 (II)		34
Regionally outstanding	2 (I)	9 (I)	14 (I)	5 (II)	2 (III)		31
Bioregionally important	7 (II)	19 (II)	19 (III)	13 (III)	2 (IV)	3	63
Locally important	13 (III)	17 (III)	11 (IV)	4 (IV)	2 (IV)	3	50
Totals	31	51	55	27	8	6	178

going into the biodiversity evaluation and conservation assessment are, at least, potentially measurable, which is necessary to make them sufficiently explicit as to be defensible and may also help to generate long-term indication of achievement or failure of biodiversity conservation. The lack of a global agreement on a biogeographical classification system remains a weakness.

In common with many evolutions, this study gives priority to the regions with high biological value and high threat. Increasing evidence suggests that rapid species loss occurred in several regions of the world at the early stages of human impact (Balmford 1996). This observation suggests that more attention might be given to important areas not yet threatened. Such an option could be cheaper because land use conflicts have yet to arise and may offer the best hope for the long-term maintenance of extinction-prone species and ecosystems. Perhaps a better balance needs to be found between fire fighting and fire prevention with a similar rationale to the application of triage when medical services are overwhelmed.

9.3.1 Centres of plant diversity

Several significant initiatives have taken a global approach to seeking areas of outstanding importance for species richness and endemism, based on particular taxa. These share the idea that at this level there are places on Earth where effective conservation (or lack of it) will have a disproportionate effect on preventing or promoting the extinction of species.

Raven (1987) pointed out that forests in Madagascar, western Ecuador and the Atlantic coast of Brazil each had more than 10 000 species of higher plant (from a global total of some 250 000), and had each lost more than

90% of their original area. Myers (1988, 1990) extended this idea to locate 20 hot spots, featuring high concentrations of species, high levels of endemism and exceptional threats of destruction. His first 10 tropical forest hot spots contained 13.8% of all plant species within 3.5% of the remaining primary tropical forest, or only 0.2% of the Earth's land surface.

Practical development of these ideas has been advanced in the Centres of Plant Diversity Project (Davis *et al.* 1994) which aimed, amongst other things, to identify areas around the world which, if conserved, would safeguard the greatest numbers of plant species. The criteria for the selection of sites (some of which were much larger than this term usually implies) were that they must either be:

1 evidently species rich, even though the number of species present may not be accurately known, or

2 known to contain a large number of species endemic to them.

In addition, the following characteristics were also considered:

1 The site contains an important gene pool of plants of value to humans, or that are potentially useful.

2 The site contains a diverse range of habitat types.

3 The site contains a significant proportion of species adapted to special edaphic conditions.

4 The site is threatened or under imminent threat of large-scale devastation.

In mainland areas, species richness was taken to mean more than 1000 species with at least 10% endemism. Islands often have depauperate floras, but high levels of endemism, and were admitted if there were at least 50 endemics, or more than 10% endemism.

A total of 234 sites has been described. The total area of the 234 sites was about 11.5 million km² or about 9.5% of the Earth's land surface. It is not known, unfortunately, how many endemic species these sites contain. The degree of protection varied, with 27 sites having no protected areas, and 152 having less than half their area protected. Some of the areas were so large that it is unclear what an adequate level of protection would be. A total of 74 sites were classed as threatened, and 45 as severely threatened, though the criteria of threat were not explicitly stated.

Centres of Plant Diversity is an important project. Plants are a major determining feature of terrestrial ecosystems and provide much or all of the habitat for many other species. Much of the value of biodiversity to mankind lies in plants or their products, including food, materials such as timber and unknown potential values from genes and biochemicals. The data, quantitative and descriptive, provide a framework from which to develop local actions, and the input of regional and local expertise into the evaluation is only likely to increase the regional sense of ownership and commitment.

The project can be criticized for lack of explicitness in its selection criteria. This is consequent on a rather poor knowledge of the flora of many regions. For many areas described, the numbers of species and of endemics are not known. The results are not based on the naming and listing of species involved. Such knowledge will be essential for significant further development of the approach.

9.3.2 Endemic bird areas

A global study of endemism in birds (ICBP 1992) rests on the idea of identifying centres of concentration of endemism, not confounded with the idea of species richness. Birds were arbitrarily taken to have restricted ranges if their geographical extent was less than $50000\,km^2$ which embraces 2609 species or 27% of the world's birds. Data on these were collected as individually located records of occurrence. Endemic bird areas (EBAs) were defined as areas with two or more species entirely restricted to them, and in ambiguous cases, separated only if they had a greater number of uniquely occurring than shared restricted-range species. Biological evaluation of EBAs was made on the basis of the number of endemic species, after allowing for area, as well as the qualitative assessment of the EBA for endemism in other taxa. Threat to EBAs was assessed by the extent to which their defining species were threatened according to conventional Red List criteria and the proportion of the area given formal protected area status.

The results of this study show that a quarter of all bird species on Earth are completely restricted to just 5% of the land surface. Considering only the more species-rich EBAs, 20% are uniquely confined to just 2% of the Earth's land surface. These areas have been explicitly mapped. Currently, for more than half (133 out of 221) of the EBAs less than 5% of the area is formally protected.

The fact that birds are relatively well known made this study possible and gives birds a practical use as indicators. The EBA analysis is based on explicit data which generate an unarguable conclusion, because the areas identified are, by definition, unique for named endemic species. An important corollary of this study is that adequate conservation of the EBAs would incidentally safeguard a very high proportion of the 12% of the world's avifauna which is threatened. This neatly moves on to the limiting feature of Red Lists—that they are not explicit in pointing out the locations for action required to deal with globally threatened species.

A minor criticism of this study has been the arbitrariness of the $50000\,km^2$ threshold. If a larger range size had been used then, undoubtedly, a few slightly larger areas of endemism would have been found. The study will be more powerful when comparable data are analysed for other taxa and when they can be embraced in the same framework. There is qualitative evidence of congruence with other taxa (Thirgood & Heath 1994) and some similarity with the world map generated from the Centres of Plant Diversity project (Fig. 9.1).

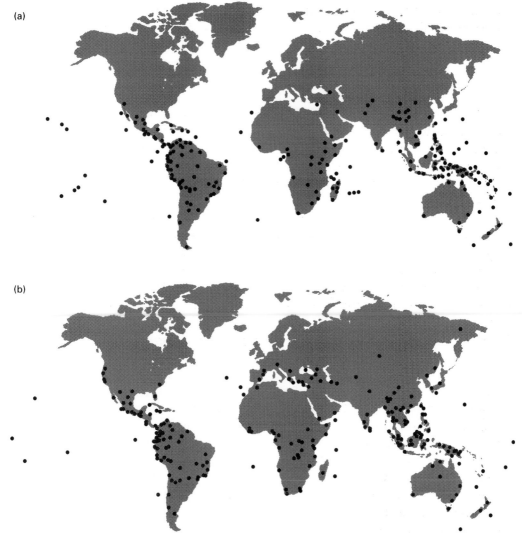

Fig. 9.1 Areas of global importance for (a) endemism in birds (ICBP 1992) and (b) endemism and species richness in plants (Davis *et al.* 1994). Effective conservation in a small area of the Earth has a disproportionate impact. There is a strong qualitative similarity between the two maps. They should be viewed as complementing each other, where they differ.

9.3.3 A global representative system of marine protected areas

In spite of its biodiversity, economic importance and degree of threat, the marine environment has proved difficult to evaluate. A global review (Kelleher *et al.* 1995) was based on criteria developed by Kelleher & Kenchington (1992) and covers the subtidal environment in coastal areas. A first step was the division of the world into 18 biogeographical regions in which separate working groups identified sites of regional and national

priority. Priorities were initially identified on the basis of the following biogeographical and ecological criteria:
• presence of rare biogeographical qualities or representativeness of a biogeographical type or types;
• unique or unusual geological features;
• an essential part of ecological processes (for example, a source of larvae for downstream areas);
• the degree to which the area, alone or with other protected areas, encompasses a complete ecosystem;
• the variety of habitats;
• the presence of habitat for rare or endangered species;
• nursery or juvenile areas;
• feeding, breeding or rest areas;
• rare or unique habitat for any species;
• genetic diversity (diverse or abundant in species terms).

Subsequent attention was given to naturalness, and economic, social or scientific importance. Additional justification for or against selecting a particular area came from considering the probability of establishing and successfully managing a marine protected area. All selected areas were assessed as having a reasonable chance of success as protected areas with respect to social and political acceptability. The resulting inventory covers 1306 areas.

The technical content of this study reflects the relatively poor data available for the marine environment. Its real value will lie in the recommended actions and their implementation as practical projects with institutional and community support. Further scientific information will be needed both to monitor the effectiveness of marine protected areas and to address gaps in their coverage.

9.4 Criteria-based evaluation of sites

Evaluation at an international level gives some context, but further development is needed for most practical decisions which are made locally or nationally, and which affect parcels of land of a size capable of being managed as a single unit, such as a protected area or a forest allocation. Regional or national evaluation may be used to plan protected area systems, for environmental impact assessment, or for management planning at a particular site.

Evaluation systems generally involve the measurement or description of attributes of a series of sites, an evaluation of these measures against a set of criteria, and some method of combining the results. How the series of sites is defined and located may be fairly obvious in a developed country where much of the natural resource has been destroyed. In this case, fragments of natural or seminatural habitat often end at fences or other sharp boundaries, with transitions to a severely modified habitat. It might, therefore, be possible to start with a complete candidate list for evaluation. Where substantial tracts of natural vegetation remain, parcels for evaluation may be delimited by natural features such as water catchments, by land tenure boundaries because of their impact on future options for management, or by prior classification of some kind. In these circumstances, data are often much scarcer than in more man-modified countries. This raises important questions about the efficient design of surveys to collect the most relevant data within cost constraints (Margules & Austin 1991). The spectrum of circumstances is so broad as to have generated important differences in philosophy and methodology of evaluation in different countries.

It should go without saying that the appropriateness of criteria depends, in part, on the purpose of a particular evaluation. This needs to be stated in advance to avoid bias in evaluation. If results are to be defensible, both scientifically as well as politically, criteria need to be explicit and ideally quantifiable, rather than arbitrary or subjective. It is important, therefore, to understand exactly what is meant by a particular criterion, such as rarity, why it is important and how it is measured. Criteria can helpfully be divided into ecological, cultural or planning and management (Smith & Theberge 1986). The most robust criteria are explicit in their assumptions and soundly rooted in the theory or observation of ecology. Many different criteria have been used (Usher 1986) (Table 9.4).

Many of these criteria would make little sense applied across a wide range of site types whose habitats inherently differ in attributes such as species complement. Thresholds of significance, for instance of size, might vary hugely between habitat types. It is normal, therefore, to group sites for evaluation into like types so that grasslands might be compared with grasslands, but forests would be considered separately. Evaluation might be conducted within a higher level classification such as the ecoregions discussed earlier or a more local classification of habitat types.

9.4.1 Diversity

Diversity has been one of the most frequently used measures of ecological value for conservation. It can be measured in terms of species or habitats. Species or habitat richness (numbers occurring) are the simplest and most intuitively simple measures. If population sizes or areas are measured, they can be combined into more sophisticated diversity indices which incorporate another dimension of how numbers are distributed across species. Such diversity indices have obscure practical meaning. Species diversity in a region is manifest at a range of scales (Whittaker 1972). Diversity at a single place is known as alpha diversity. Species occurrence across a set of sites, beta diversity, may be similar or higher depending on how fast species turn over with distance. The total diversity of the region concerned, gamma diversity, reflects species turnover rates on a larger scale. The inclusion of

Table 9.4 The frequency of use of different criteria for the evaluation of sites. Only criteria used in more than one study are included. (From Smith & Theberge 1986.)

Criterion	Number of studies in which used	Type of criterion
Rarity, uniqueness	20	Biotic, abiotic
Diversity	20	Biotic, abiotic
Size	11	Biotic, abiotic, planning and management
Naturalness	10	Biotic, abiotic
Productivity	3	Biotic
Fragility	7	Biotic, abiotic
Representativeness, typicalness	8	Biotic, abiotic
Importance to wildlife, abundance	6	Biotic
Threat	6	Planning and management
Educational value	6	Cultural
Recorded history/research investment	6	Cultural
Scientific value	5	Cultural
Recreational value	5	Cultural
Level of significance	4	Planning and management
Consideration of buffers and boundaries	4	Planning and management
Ecological/geographical location	2	Planning and management
Accessibility	2	Planning and management
Conservation effectiveness	2	Planning and management
Cultural resources	2	Cultural
Shape	2	Planning and management, biotic

beta and gamma diversity in a set of sites in the evaluation framework therefore relates to the idea of representation (see below).

Reasons for valuing diversity include the notion that more diversity is a good thing for capturing more of the range and variety of the natural resource. There are also less tangible aesthetic values in diversity. It has been argued that diversity is an indicator of ecosystem stability, but this notion is not well founded in observation or in theory.

Measuring diversity meets problems with sample sizes and in its inevitable increase with area. Differences in sampling, for instance the number of quadrats measured, can be handled by rarefaction which re-samples the data to estimate the number of species for a fixed (smaller) sampling effort (Simberloff 1972). Area effects can be removed by estimating species richness relative to that expected for area by regression. Counting species is intensive of labour and skill. Habitat diversity may be used as a surrogate if a classification and measurement system exists. Comparing diversity across sites only makes sense within similar habitats, because of the vast range of species richness between habitat types. The appeal of diversity as an indicator has almost certainly led to the underrepresentation of inherently less diverse habitats in many protected area systems.

9.4.2 Rarity

The idea of rarity has emotional resonance in human terms, but is by no means simple in ecology. Rabinowitz *et al.* (1986) identified seven different kinds of rarity. Species may have small (or large) total ranges. Independently, they may have small (or large) habitat specificity. Additionally, they may be scarce or abundant where they do occur. Of the eight possible combinations of these factors, seven would qualify as rare within the range of possible meanings for the term. It is evident that rarity embraces both a spatial and a numerical dimension. For any particular species, some aspects of rarity may be an evolutionary property, such as habitat specificity, small natural range or low natural densities. On the other hand, small range or low densities may be the result of human impact.

Rarity is important to the overall maintenance of the natural resource because many species are rare. Amongst their number is a high proportion of those species at risk of local extirpation or global extinction, either because natural aspects of their rarity render them at risk or because an element of their rarity indicates previous losses as a result of human impact. Loss of rare species may represent a loss of future genetic or other resource values. At a less utilitarian level, the concept of valuing and safeguarding rarity is a fundamental human concern which underpins much of the support for conservation of nature or, indeed, of artefacts.

Inclusion of rarity factors in an evaluation requires data on the range or numbers of individual species (or communities or abiotic features), not only at the sites under study but also at wider scales. Important elements of rarity are scale-dependent. A locally rare species may be regionally or globally rare or it may merely be at the edge of its range.

9.4.3 Size

The meaning of size of sample units is unambiguous in modified environments where natural communities are fragmented and isolated. In more extensive habitats, it is less obvious because the boundaries could have been drawn differently. Larger size might be beneficial because it embraces a greater diversity of habitats and species. In a fragmented landscape, larger blocks may be more likely to contain viable populations of particular species. They may also be more likely to support interior species, intolerant of proximity to the edge of habitats. Finally, size may confer greater management options, such as the ability to resolve usage conflicts or to manage vegetation succession. There has been much erroneous discussion which has handled the issue of size in terms of island biogeographical theory. In some circumstances, very small reserves might be appropriate, for instance in safeguarding plants with minute ranges.

9.4.4 Naturalness

Though popular as a criterion, especially in developed countries where human impact has been greatest, naturalness is not an easy concept to define. Many valued habitats in greatly modified countries, such as grasslands or heaths, are dependent on long-term low-intensity human usage. Naturalness needs careful definition and qualification to separate traditional and indigenous usage from short-term or unsustainable impact. In terms of the impact of alien species or destructive man-induced factors, the notion of lack of naturalness converges on that of threat.

Reasons for ranking naturalness have included the scientific need for sites at which to study natural ecological processes. There are also emotional and aesthetic factors at play in the greater perceived value of wilderness and ecosystems less tainted by man (McCloskey & Spalding 1989).

9.4.5 Representativeness

Representativeness is inherent in the concept of nature conservation aiming to maintain the range of genes, species and habitats for whatever reasons. Two concepts of typicality and of inclusiveness may be separated. Inclusiveness is too fundamental to be a criterion for evaluation of single sites; it only makes sense when applied to a set of sites. Inclusiveness is the assessment of how well a set of sites includes the species and other measures of the range of variation within the study framework. Typicality attempts to embrace the idea of whether a site is central or an outlier within the range of variation of sites of its general type. It thus contrasts with notions of rarity. Typicality can only be measured by classifying or ordinating a set of factors (Austin & Margules 1986; Belbin 1995). Typical sites would lie among the more frequent classes or close to the centre of an ordination diagram. Those with rare features would fall in the rarer classes, or be outliers in an ordination.

9.4.6 Other ecological factors

For various reasons, a range of other ecological criteria has been used in a minority of studies. Productivity is a measure of the rate at which ecosystems capture solar energy and convert it into organic materials. Especially if naturally exported, this may be a valuable function. Examples include coastal wetlands exporting nutrients or providing nursery grounds for fish, or rare productive grasslands fuelling migrating waterfowl in the Arctic. Productivity is not easy to measure in a way with practical utility for evaluation.

Fragility embraces ideas of the scale of response of an ecosystem to external stresses, the speed with which it might return to normal, and the threshold beyond which changes might be irreversible. These concepts are not sufficiently well understood in theory to offer much practical utility in conservation evaluation. To the extent that fragility is about the risk of species or habitats being lost as a result of human impact, the concept could be captured within the criteria of threat and rarity.

Importance often refers to particular species, especially large and charismatic ones. To be useful it requires information on numbers and distribution, both within the study sites and beyond. It has found utility in assessing sites for birds where population estimates exist, and for groups such as waterfowl or seabirds which occur in large concentrations. Some habitats may be deemed as important for their ecosystem functions. Importance may be mandated in law, for example a need to protect migratory or threatened species.

McNeely et al. (1990) have proposed the idea of 'ecologically sensitive areas' based on their resource values, such as for water catchment, or potential medicinal or genetic values. This idea, orientated more to social and economic factors, aims to capture the values which humans might be most likely to conserve. Biodiversity might benefit as a result.

9.4.7 Cultural criteria

Cultural criteria have embraced historical, tribal or other human activity values. Use, or potential for use, for education, recreation, amenity or science has also been included. None of these approaches is relevant to the evaluation of ecological qualities. They are, however, all relevant to the way in which humans place non-consumptive value on sites in a way which might complement rather than conflict with the conservation of their ecological values. Logically then, these criteria should be applied after ecological evaluation, at the point where practical decisions have to be made. Cultural criteria, though politically important, are generally not very satisfactory in being hard to define explicitly and measure objectively.

9.4.8 Combining scores

With some criteria generating numerical outputs (area, species number) and others being only qualitative (naturalness), it is not obvious how to combine them into an overall evaluation of a site. Margules (1986) proposed an operational framework which rationalizes the subject by allowing different kinds of information to influence the decision making at different stages.

Stage 1: Sorting and classification

Sites are classified so that like is compared with like in broad terms. Classification might be by habitat type, by region or both.

Stage 2: Representation

Inherent in stage 1, but stated explicitly by making it a separate stage, is the process of defining bounds and ranges of variation of classes and subclasses which need to be represented.

Stage 3: Threshold criteria

Some sites are eliminated altogether from their class, for example, by virtue of inadequate size, or some threshold level of previous damage leading to unnaturalness.

Stage 4: Ranking

Sites within classes can be ranked by biological criteria such as richness and rarity. While criteria, such as area or naturalness, might have eliminated some sites in stage 3, they could be used again to contribute to the ranking of the survivors at stage 4.

Stage 5: Pragmatic criteria

Practical considerations, such as threat and opportunity, turn the priorities of ecological importance from stage 4 into a shopping list for practical implementation.

9.4.9 Important bird areas and wetlands

An example of selecting sites by criteria is BirdLife International's Important Bird Areas (IBAs) programme. IBAs have been identified and documented for Europe and the Middle East (Grimmett & Jones 1989; Evans 1994). Important wetland sites have been documented for the neotropics (Scott & Carbonell 1986), Asia (Scott 1989) and Oceania (Scott 1993).

   The function of the IBA programme is to identify and protect a network of sites, at a biogeographical scale, critical for the long-term viability of naturally occurring bird populations, across the range of those species for which a site-based approach is appropriate.

   IBAs have the following characteristics:
- They are places of international significance for the conservation of birds at the global, regional or subregional level.
- They are practical tools for conservation.
- They are chosen using standardized, agreed criteria applied with common sense.
- They must, wherever possible, be large enough to support self-sustaining populations of those species for which they are important.

- They must be amenable to conservation and, as far as possible, be delimitable from surrounding areas.
- They will preferably include, where appropriate, existing protected areas.
- They should form part of a wider, integrated approach to conservation that embraces sites, species and habitats.

The criteria have been set in a hierarchy to identify sites of global, regional and subregional importance. At a global level, these criteria embrace the following:

1 Globally threatened species—sites which hold significant numbers of one or more globally threatened species (Collar *et al.* 1994).
2 Restricted-range species—sites which, as a set, include the restricted-range species characterizing EBAs (ICBP 1992).
3 Biome-restricted assemblages—sites which, as a set, embrace the species largely or wholly confined within a biome.
4 Congregatory species—sites meeting the criteria of the Ramsar Convention for waterbirds (Ramsar Convention Bureau 1990) or holding more than 1% of the global population of seabirds or other congregatory species.

Sites of regional and subregional importance relate hierarchically with lower thresholds for congregations and the representation, where appropriate, of regionally declining species. The published wetland directories have been more descriptive than classificatory. They do, however, include many sites that have been, or will be, classified as IBAs.

The IBA programme is possible because birds are popular and relatively well known. The fact that they often concentrate and that they are countable makes appropriate the use of numerical population criteria. Globally threatened birds have been systematically identified (Collar *et al.* 1994) on the basis of the semiquantitative criteria adopted by the IUCN Species Survival Commission (1994). Birds are legally protected by the Ramsar Convention and, for instance in Europe, by the Birds and Habitats Directives. The Ramsar Convention explicitly legislates for the 1% population threshold for identifying an internationally important site for congregatory species. Within Europe, IBAs coincide with the requirements of the Birds Directive and states are legally obliged to protect them (Commission of the European Communities vs Kingdom of Spain 1993).

The main problem with the IBA is that taxa other than birds are nothing like so well known. Many IBAs are known to be important for other biodiversity and there can be little doubt that they are of sufficient importance to merit protection. On the other hand, many sites important for plants or other taxa may not be in IBAs. There is a pressing need, country by country, for the completion of lists of important sites for taxa additional to birds.

9.5 Protected area systems

Evaluating individual sites for protection against a range of criteria has a number of failings, both in practical application and overall impact. Many studies have acknowledged that combining scores across different criteria is at least arbitrary, and at worst inappropriate. Important notions, such as typicality and representativeness, have proved particularly intractable to measurement and to combination.

Protected area systems have evolved over many decades with individual sites being acquired or designated according to principles, processes and perceptions current at the time. Many recent studies have shown that these accumulated sets have not been very efficient in representing the range of biodiversity that could be protected for the same cost in terms of the area within a region given over to strict conservation (Pressey & Tully 1994; Fjeldså & Rahbek 1997).

Rebelo & Siegfried (1992), for instance, looked at the protection of one diverse and representative family, Proteaceae, in the Cape Floristic Region (South Africa), which is one of the world's outstanding concentrations of floral diversity and endemism. Two historical prescriptions for the protection of the flora, and the existing reserves network, fail to protect a significant proportion of the species. Furthermore, the existing network was no better at protecting species than would be a similar-sized network selected totally at random from grid squares in the region.

Another weakness of a site by site approach is that it is hard to justify scientifically and, as a result, is politically vulnerable to competing land uses whose economic attractiveness, at least in the short term, is greater. Many protected areas are located in places where remoteness, topography or low productivity have limited their attractiveness for agriculture, forestry, mining or urban development (Pressey 1994). Evaluation by criteria has, to a degree unwittingly, supported such bias by allowing non-biological criteria, such as scenery, amenity or wilderness, to be included, or even to be dominant. Reserves in northeastern New South Wales are disproportionately located on steep or infertile ground (Pressey 1995) (Fig. 9.2). Surviving fragments in flat and fertile areas largely taken for wheat or pasturage are thus of top priority for future maintenance or restoration. They would not, however, rank highly for naturalness or for size. In Africa, savanna grasslands have been relatively well protected, partly by big game hunting interests, more recently replaced by ecotourism, and partly by virtue of rinderpest or trypanosomiasis rendering them less suitable for cattle and their ranchers (Leader-Williams et al. 1990).

On any one land parcel, conservation designation will have to compete with alternative interests, and may well lose if these interests are economically or politically powerful. Even where protected areas have been established in less contentious regions of little-known alternative value, subsequent changes, such as the discovery of mineral resources, have often been sufficient to remove a conservation designation. Pressey (1994) characterizes traditional reserve planning processes as ad hoc and a further example of Odum's (1982) 'tyranny of small decisions'. Clearly, a

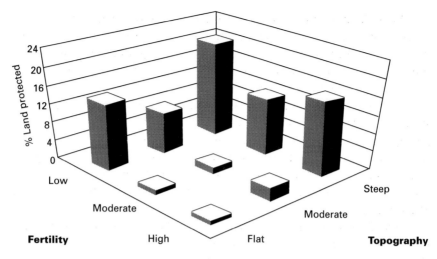

Fig. 9.2 Protected areas in south west Australia are most frequent on the land that no-one else wants and most poorly cover the flat and fertile regions which are in demand for agriculture. (Redrawn from Pressey 1995.)

new model is needed to deal with both scientific and political difficulties inherent in one-off, small-scale decision making.

9.5.1 Explicit aims of a protected area system

The idea of a set of sites which, as a set, meets an explicit aim has emerged and grown over the last 20 years (Pressey 1994). Protected areas, in which the maintenance of species and ecosystems is the major priority, are a keystone of conservation policy (Anon 1992). In the face of competition for land, strict protected areas are unlikely to cover more than 10–20% of most large regions. In some ecoregions, very much less currently survives as being potentially available. How to find a good set of sites to embrace the greatest quantity of biodiversity most in need of formal protection raises questions about the definition and measurement of biodiversity, so that explicit aims can be defined for the set. A process to find the optimum, or at least a good, set of sites can then follow.

This subject has been explored both at a practical level and on a theoretical basis. Practical exploration has mainly been at the scale of a region or habitat grouping, while some theoretical approaches have covered much larger scales up to the global. Theoretical studies aiming to investigate methodological approaches have often used grid-based maps of species or environmental attributes. Some practical studies have used land parcels divided by tenure or physical features such as catchments.

Examples of explicit criteria for selecting sets of sites include the following:
• Include all species at least once or ($n > 1$) times (wetland plants in New South Wales (Margules *et al.* 1988); predicted occurrence of canopy

tree species, New South Wales (Margules & Stein 1989); owls, hawk-moths and tiger beetles in Thailand (Kitching 1996); birds in South America (Fjeldså & Rahbek 1997); Proteaceae in the South African Cape (Rebelo & Siegfried 1992); breeding birds in Britain (Williams *et al.* 1996)).

• Include at least 10% of each environment and 10% of each tree community (coastal hardwood forest in south east Australia (Margules *et al.* 1994)). Environments were determined on the basis of soil, terrain, temperature and rainfall attributes.

• Have a high (95%) chance of including individual vegetation communities at least once (mallee patches in south Australia (Margules *et al.* 1994)). Probabilities of occurrence were estimated by logistic regression of occurrence on environmental variables from a field sampling survey.

• Represent each land system once, or include at least 1%, 5% or 10% (semiarid New South Wales (Pressey & Tully 1994)).

• Represent each plant community at least once or ($n > 1$) times (lakeshore plant communities, Sweden (Nilsson 1986)).

The statement of explicit goals which a set of sites is to achieve will predict the kind of data required for analysis. Species distribution and abundance data might be ideal, but are expensive to acquire. Criteria which include occurrence of all species are not very satisfactory because they are vulnerable to vagrants, weeds or extinct populations selecting inappropriate sites. The representation of viable populations or a stated percentage of populations is more desirable. This raises questions about population viability, which is another aspect of reserve selection and design. Especially in large tracts of vegetation and parts of the world where data abundance is poor, vegetation community or environment types are likely to be needed as a surrogate criterion. Species data can be added where they are available. Important studies in Australia have inferred the occurrence of species or habitats by interpolation of environmental measures. The fact that data on species are poorly available cannot be used as an opportunity for avoiding action. There is no realistic prospect of even describing most species on Earth before most opportunities for protected area designation will have passed. Gaston *et al.* (1995c) propose that measures of the occurrence of families, at least on a large scale, are an adequate surrogate for species richness. This may perhaps have practical utility in guiding fieldwork in parataxonomy.

Although only fairly simple sets of criteria have been illustrated and explored to date, the theoretical approach has substantial capacity for development. In particular, there are prospects for weighting priorities according to endemism to the region, which sets representativeness on a larger scale. There are also prospects for making explicit the decision-making processes handling the competition between different land uses.

9.5.2 Algorithms to define sets

Explicit algorithms can be stated as rules for searching a species or attribute by site data base to find a minimum set meeting the stated objective. A simple example to represent all species might go as follows:

Step 1 Pick all the sites which are already reserved. This is not appropriate in a search for a theoretical optimum design because many existing reserves will not be the most important or necessary for representation of biodiversity. In practical cases, the option of disposing of existing protected areas and starting again is not likely to be realistic.

Step 2 Pick all the sites which have to be in the set because of rare species. If, say, the goal is to represent all species in five reserves, then species with five or fewer locations dictate that all their sites of occurrence must be included.

Step 3 Pick the next site as the one that will make the greatest addition to the set already chosen. There are several possible ways to state this rule which might be broken into subrules. Essentially, it needs to search through the rarer of the underrepresented features. The common features will almost certainly end up well represented anyway. In the case of ties, subsidiary tie-breaking rules might be arbitrary or they might make practical improvements, such as selecting the site nearest another already selected (Nicholls & Margules 1993).

Step 4 Iterate to step 3 until the set is complete and the stated goal has been achieved.

Step 5 Check backwards through the list to see if any unnecessary sites have been included, such that their removal could shorten the list.

This final step in the process responds to the fact that a heuristic algorithm of this kind is not certain to find the optimum result. Underhill (1994) drew attention to the proper mathematical way of handling an optimization problem. It is not at the moment clear whether the more easily understood approach is seriously flawed, so that conservation biologists will need to master optimization theory, or whether the direct comprehensibility of simple but explicit algorithms will prove to be more practically important.

One particular application that has attracted interest is a computer program, WORLDMAP (Williams 1992). This provides the data base software for handling grid maps of species occurrence, a selection of algorithms and colourful graphical presentation of results. WORLDMAP is particularly orientated to the principle of complementarity (see below). It has been used to explore the effects of selection by rarity or richness, ideas of representing biodiversity at the level of taxonomic difference or at the level of variety at family level, and congruence across taxa (Gaston & Williams 1993; Williams 1993; Williams *et al.* 1994b; Gaston *et al.* 1995c). At the global level, with large grid cells, the complete representative set is not very meaningful. Huge grid cells are not all going to be protected and more cells are likely to be needed for complete sets, because more are likely to contain unique features.

Another important application has arisen in the USA under the name of gap analysis (Scott *et al.* 1993). Gap analysis is targeted at the same idea of finding gaps in a representation of biodiversity within protected areas. Gap analysis prescribes a GIS-based approach calling for data layers on vegetation, land use and species distribution, including predicted distributions (McCullagh & Nelder 1983; Austin *et al.* 1984; Gillison & Brewer 1985; Butterfield *et al.* 1994). The strength of the GIS approach is that it facilitates further analysis in landscape ecology, and sets a potential framework for monitoring. It also allows the prospect of integration of other data into a land use planning framework. Gap analysis inherently calls for large-scale collaboration in data collection and synthesis, which will undeniably be necessary for future strategic planning in linking biodiversity and other human activities (Fig. 9.3).

Some central ideas arise from attempts to find sets of areas (Pressey *et al.* 1993).

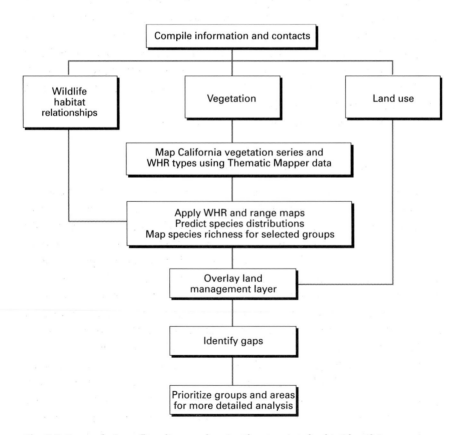

Fig. 9.3 Gap analysis—a flow diagram showing the steps involved in identifying gaps in the protected area coverage in California. (Redrawn from Scott *et al.* 1993.)

9.5.3 Complementarity and efficiency

Given the grid cells or land units subjected to analysis, there will be one or a few ways of picking the smallest sized set that meets the overall criteria. As the set is built with a heuristic algorithm, it is evident that the selection of each new site depends not only on its attributes, but also on the combination of features in the previously selected sites. In this sense, the next site complements the series. The principle of complementarity was apparently first named by Vane-Wright *et al.* (1991), although the idea is much older and clearly related to the notion of representativeness in criteria-based systems. Complementarity also subsumes some notions of rarity and endemism. To build a complementary set, rare and endemic occurrences have to be included. For instance, if all species must occur in the final set, then it has to include all those sites with species of unique occurrence.

The idea of complementarity gives rise to a minimum-sized set which meets a stated goal. *Ad hoc* approaches or procedures based on the application of criteria to single sites will usually produce a larger series often with no way of knowing when it is complete. In practice, existing protected area networks are often larger than they could be to achieve the same level of representation of biodiversity, but are also often not very complete in overall representation. This leads to the prospect of defining the efficiency of a proposed or actual set of sites in meeting the stated goal. Efficiency can be defined as $1 - (x/t)$, where x is the number or area of sites designated or required to achieve the goal, and t is the total number or area of sites (Pressey & Nicholls 1989; Pressey *et al.* 1993).

9.5.4 Flexibility and irreplaceability

While there is a smallest sized set that meets the overall goal, real world protected area systems are likely to be larger because they will have to meet other, non-ecological, constraints. In these circumstances, there could be a large or very large number of combinations that meet the overall goal. This is a great help to strategic planning because it gives rise to flexibility. If one site is unavailable for some reason, or is destroyed, then unless it has (or had) unique features, other permutations could still meet the overall goal. Flexibility also makes it possible to include additional constraints and considerations into a real network. Nicholls & Margules (1993), for instance, showed how a rule to pick continuous blocks, if possible, only led to quite a small reduction in efficiency. On the plus side, such a rule would lead to greater practical ability to manage the sites, and greater prospects of maintaining viable habitats and populations of species in the longer term. Flexibility is continuous in time, allowing response to changing circumstances as sites come onto the market, or knowledge or the occurrence of species changes.

Another feature of flexibility is that all sites without unique features do not have to be included in the final series. Such sites are replaceable. Replaceability comes in degrees; some sites are more replaceable than others. A possible measure of site replaceability could be the proportion of all possible fixed-sized combinations of sites that meet the overall criteria, and include the focal site. If the site occurred in relatively few combinations, then it is fairly replaceable. A site with unique attributes would occur in all combinations and is totally irreplaceable. In practice, this analysis is

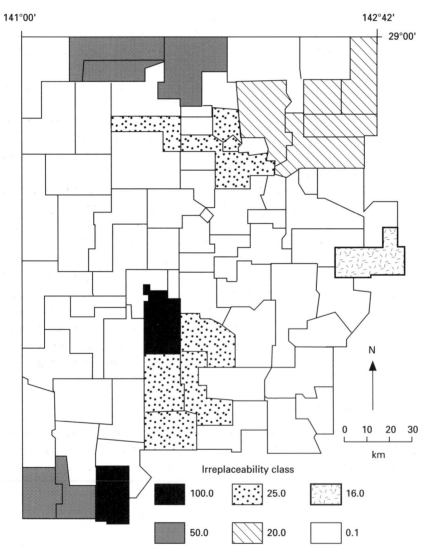

Fig. 9.4 Spatial pattern of predicted values of irreplaceability for individual properties in a region of New South Wales. Higher values indicate the potential contribution of each land parcel to a reservation goal and the extent to which options for a representative reserve system would be lost with the destruction of that parcel. (Redrawn from Pressey *et al.* 1994.)

not easily done. There are about 2×10^{27} ways of picking a combination of 20 sites from a total of 200. This might be quite a small real world example, but it is an implausible computational challenge to examine all combinations. A solution to this problem may be known to the appropriate branch of mathematics, or may be readily findable within theory. Pressey *et al.* (1994) made a pragmatic proposal for a way to estimate irreplaceability and tested it on a small data set. Their suggestion was that irreplaceability can be predicted from the rarest feature of a site not represented in the sites with higher values.

The idea of measuring irreplaceability is very important. It could be a single measure of conservation value that is defensible. The loss of the more irreplaceable sites would close options of ever achieving a representative protected area system. Replaceability is a flexible idea, in that it changes as circumstances change, and so avoids the rigid determinism of criteria scoring methods. It could be used to influence the order of acquisition of new protected areas and the strength of opposition appropriate for proposals that destroy the natural environment. It allows the prospect of generating rational maps of conservation value as inputs to regional planning processes (Fig. 9.4).

The approach of complementary sets says little about the order in which sites might be acquired, notwithstanding the fact that heuristic algorithms of the kind illustrated above pick sites in a sequence. Other algorithms would not necessarily pick sites in a sequence. On the other hand, if the decision algorithm works on rarity then the sequence may be related to replaceability and may thus say something about priority. Questions about sequence are normally constrained by external factors such as budget allocation and land availability. Clearly though, priorities for additions to a protected area set must include the most threatened, and amongst this set will be sites with rare or irreplaceable features where ultimate options for conservation have nearly been closed by previous land use decisions.

Acknowledgements

I gratefully appreciate comments on the draft from Mark Avery, Lincoln Fishpool and Melanie Heath. Sue Squire helped me in the pursuit of references and Elizabeth Wickham in the preparation of the typescript.

CHAPTER 10

Managing habitats and species

William J. Sutherland

10.1 Introduction

Our perceived need to manage habitats or species is an admission of failure. In near-pristine habitats, such as most of the Antarctic, there is little reason to interfere and conservationists simply need to ensure that access, exploitation and environmental degradation do not affect the area. However, over much of the planet, human influence has so radically changed the ecology that it is naive to ignore its implications for conservation policies. It is no coincidence that Britain and the Netherlands, who both destroyed their pristine habitats centuries ago, are leaders in habitat management and restoration.

There are growing calls for a decrease in human intervention and increasing interest in attempting to restore natural processes. These include allowing straightened rivers to create their own course after digging to restore meanders, tolerating erosion within sand dunes as part of the natural process, and removing coastal protection (where safe to do so) to restore the processes of erosion and deposition.

The Wildlands Project in the USA is taking this idea further with ambitious plans to recreate extensive wild landscapes by restoring natural ecosystems and removing or downgrading roads (Noss 1992). The intention is to restore degenerate habitat in order to link reserves, areas of wilderness and areas of conservation interest.

Soulé (1992) suggests that the vision of establishing large interconnected landscapes is most likely to be achieved with a long time perspective. For example, individuals are more likely to donate their land if they pledge to hand it over after a delay, such as after the death of their children. The donation could also be part of a tax deal. Soulé also suggests that the condition that a piece of land is eventually incorporated into a wilderness area could be considered in negotiations for compromises between conservationists and others, such as foresters.

10.2 Why manage?

If withdrawing human intervention is considered to be so admirable, why are many reserves, especially in Europe, so heavily managed? One simple reason is that the ecosystems have usually been greatly altered and have experienced considerable changes in hydrology, soil characteristics or in the distribution and composition of the communities present. Any attempt to restore a more traditional state requires considerable and perhaps perpetual work.

A major reason for the need for intervention is that many species of conservation importance are dependent upon uncommon, early successional habitats which need to be maintained at an early ecological stage by cutting, grazing, burning or disturbing. The problem is confounded when these early successional sites are small, isolated fragments. In an ideal world, a range of successional stages might be maintained in the landscape by occasional natural catastrophes and by a mosaic of small-scale disturbances. For example, field wormwood *Artemisia campestris* is now a rare plant in Britain whose main location is a 0.11 ha reserve within the Brandon Industrial Estate (Fig. 10.1). Management of the reserve for this species involves preventing succession by disturbing patches of soil through regular rotavation and removal of trees and bushes to prevent succession. The surrounding landscape (known as Breckland) was once a large area of inland arid heaths, shifting cultivation and mobile sand dunes (Dolman & Sutherland 1991, 1992) and the species was then much more widespread. Now that this landscape has all but disappeared, we are unfortunately left with the need for continual management to ensure that sites, such as Brandon, are retained at the same successional stage.

Natural processes, such as fires, floods or storms, may be an essential component of the ecology (Section 10.6) but intolerable to the public because of the risks to property and life, and the perception that such natural

Fig. 10.1 Field wormwood reserve in Brandon Industrial Estate, UK. This isolated reserve has to be artificially maintained at an early successional stage for the field wormwood. (Photo by W.J. Sutherland.)

processes are disasters that should be prevented. Thus, although fires are a natural and important component of the dynamics of America's Yellowstone National Park ecosystem, the plan to not extinguish fires unless they threatened human life was highly controversial. If floods, fires and storms are not allowed to create early successional stages, then these may need to be created by more artificial means such as scrub clearance, grazing or soil disturbance.

The conflicting views between the general public and ecologists on the role of natural disturbance is illustrated by the 1987 storm in southern Britain which resulted in large numbers of trees being blown over. Many conservationists consider this to have been, on balance, beneficial by creating gaps within woods. However, it was perceived as a national disaster and large sums were donated by the public to conservation bodies for replanting—even though natural regeneration has usually taken place and speeding up this process is probably often detrimental to the conservation interest. Some organizations were both obtaining money to fill in gaps with tree planting and to create open areas by coppicing!

Many habitats around the world result from a long history of human intervention and this provides a further reason for active management (Chapter 8). Some species will benefit, others will be indifferent, while those species that cannot withstand the human intervention will have gone extinct. Stopping management may simply result in a loss of those species that benefit from intervention, while those species characteristic of unmanaged areas may fail to re-colonize. For example, traditional woodland management in Britain consists of leaving a few trees, known as 'standards', for timber while cutting other trees or shrubs at the base (coppicing) to provide poles on a regular rotation of 10–30 years (Chapter 8). The periodic clearance of the dense shade by coppicing results in a burst of growth and flowering by woodland plants followed by a rapid increase in the populations of butterflies. The reduction in coppicing as a result of the collapse in the demand for coppice products has resulted in a massive decline in these woodland butterflies (Thomas 1991).

A history of human intervention may have allowed species to persist even when the climate became unfavourable. During the earliest human clearances in Britain around 3000 BC, the climate was 2–3°C warmer than it is now. Thomas (1991) suggested that a number of butterflies that would usually be dependent on the warmer climate could still persist in Britain in the cleared areas of disturbed wood, grassland or heaths as the microclimate in such cleared patches is markedly warmer. These cleared areas persisted until the twentieth century when many seminatural habitats were destroyed or degraded for these butterflies either by intensified agriculture and forestry management or by complete abandonment.

Management may mimic natural processes, but it needs to be recognized that, in practice, they are often very different. The 1988 fire in Yellowstone National Park that burnt 400 000 ha resulted in a mosaic of patches of different burn intensity due to factors such as chance, soil moisture and aspect (Christensen *et al.* 1989). Clearfelling blocks of woodland may seem similar but results in a very different habitat (Noss & Cooperrider 1994). Similarly, cutting is often considered as a substitute for grazing but is much more uniform and omits the trampling and faeces that can have considerable consequences for maintaining diversity (van Wieren 1991).

Previous chapters in this book have described the main concepts in conservation biology. The management and restoration of species and habitats provide a rigorous test for the validity of these ideas. Indeed, failure in management and restoration can improve our understanding. One classic early example was Aldo Leopold's failure to create prairie habitat until regular burning was incorporated into the restoration process (Jordan *et al.* 1987), and it then became clear that this was an essential component of prairie ecology (Curtis & Partch 1948). Another example, that is repeated far too often, is the protection of a population of rare early successional plants by a fence in order to prevent the grazing or trampling that has damaged some individuals. The usual result is that the population flourishes for a few seasons and then enters a long-term decline towards extinction as succession proceeds.

10.3 The need for objectives

Conservation managers often lack a clear set of objectives and thus flounder unnecessarily. In practice these objectives should direct the work. It is naive to simply ask which is the best management practice. Practically every technique, such as grazing, burning, control of predators or unrestricted access will benefit some species and be detrimental to others. Possessing clear objectives can make it easier to decide which management practices should be carried out.

If maintaining an area, start by writing a management plan (see Hirons *et al.* 1995 for one format), or if responsible for managing a species, write a species action plan (see Williams *et al.* 1995). This approach may seem excessively bureaucratic but is probably essential when a number of individuals or organizations are involved. It is also useful in ensuring all the information has been collated, in assessing priorities, in clarifying thinking and in ensuring all parties are informed and involved in the decision-making process.

Management plans and species action plans are usually of little use unless those who are responsible for implementing them are also involved in writing them. There is a long tradition of management plans being written by outsiders with little or no consultation with local decision makers; these plans are then usually ignored and forgotten.

One common mistake is to be too parochial. Thus a manager of a large grassland reserve in a region famous for its grasslands may be tempted to diversify the range of habitats and species (for example by digging a lake) within the reserve. However, from a national or international perspective, maintaining the entire area as grassland may well be more important. Clear objectives should help prevent this problem.

People care most about the conspicuous and glamorous species, especially large mammals, birds and flowering plants, but invertebrates may be the group that is declining most rapidly. For example, Thomas (1991) showed that in the county of Suffolk (UK), between about 1850 and the 1980s, 44% of butterflies had become extinct by comparison with only 5% of plants, 3% of mammals and 12% of amphibia and reptiles. This suggests that invertebrates thus warrant special attention.

10.4 Myths and research

It is depressing how little we can confidently state about habitat management and how difficult it is to separate facts from hearsay and mythology. For most management recommendations, it is difficult to find any objective evidence to justify them. The fault for this can be divided between professional ecologists, who claim to be answering conservation questions whilst actually studying something else, and land managers, who fail to carry out their management in an experimental way or to document their results.

Land managers should regularly carry out randomized, replicated, controlled and monitored experiments but in practice very rarely do. Most changes in habitat management should be experimental. At the absolute minimum there should be a control (for example the old management technique), the consequences of the change in management should be monitored and the results made available for others to consult. In practice, even this bare minimum is rarely achieved. Instead, the norm is to alter the management (often making a number of changes simultaneously), not have a control area, subjectively decide if the changes have been successful and pass on the opinions verbally. It is thus hardly surprising that there is so much mythology or that it is often impossible to find the origins of many statements in habitat management.

10.5 Area and isolation

A crucial issue is deciding the priority areas for habitat management. Small patches are likely to contain small populations and, as described in Chapters 6 and 7, small populations are likely to go extinct as a result of genetic, demographic or stochastic processes. A further reason for high probabilities of extinction in small areas is that emigration may be too frequent. Thomas & Hanski (1997) showed that, for butterflies in small patches, proportionately more individuals leave and fail to return than in larger areas. Populations in small patches are simply unable to sustain such high levels of emigration. Butterflies may then require a minimum area for the population to persist and it follows that highly mobile species are likely to require larger areas than highly sedentary species.

Isolated patches are less likely to be colonized or re-colonized if the population goes extinct. For example, some species are very reluctant to cross unsuitable habitat. Thomas (1983) showed that the adonis blue butterfly *Lysandra bellargus* readily flew 250 m over open calcareous grassland, but avoided 100 m gaps of improved grassland.

Permanent herbaceous communities tend to be dominated by clonal plants that mainly disperse by means of rhizomes, stolons or bulbs. For some of these species, successful germination from seed is rare (unless an area is cleared of vegetation). Thus dispersal is re-stricted to the few millimetres or centimetres resulting from the growth of the rhizome, stolon or bulb and thus even narrow barriers of unsuitable habitat are difficult to cross. Once populations of such a species go extinct, re-colonization is unlikely. A consequence of this is that ancient habitats can be recognized in each region by a suite of poor-dispersing indicator plant species.

Patches are thus more likely to be unoccupied if small or isolated. In practice these factors interact. For example, Fig. 10.2 shows which patches are occupied by silver-spotted skipper butterflies, *Hesperia comma*, in relation to their size and area (Thomas & Jones 1993). Small isolated patches are unlikely to be occupied, while large patches near to others are likely to have a colony. The pattern probably applies to most species from microbes to large mammals, although the scale of the area and isolation will vary enormously.

The importance of patch size and isolation for population persistence has a number of consequences for habitat management. It shows how important it is to ensure that there are blocks of habitat of sufficient size. This is the justification for enlarging areas through management or restoration or for combining nearby areas to create a single block. Even if the restored areas are not of the same quality as the core areas, they often play a useful role in allowing populations to expand. This relationship between area, isolation and persistence shows the crucial importance of ensuring that blocks of habitat are sufficiently connected so that populations can persist. For example, the heath fritillary butterfly *Mellicta athalia* is dependent on recently cut coppice woodland and is also a poor disperser (Warren 1987a, 1987b, 1987c). Its persistence within woodland thus de-pends upon ensuring the continued presence of suitable habitat patches, located so as to facilitate colonization.

The giant panda *Ailuropoda melanoleuca* has declined steadily in numbers in China over the last 25 years. Every 40–60 years the arrow bamboo, which is the main food item, undergoes mass flowering over large areas and dies. It takes another 15–20 years before it has regenerated to the size that the pandas can eat (MacKinnon & De Wulf 1994). After mass flowering, starving pandas sometimes leave the forest and are found on village lands where they used to be captured and used for captive breeding.

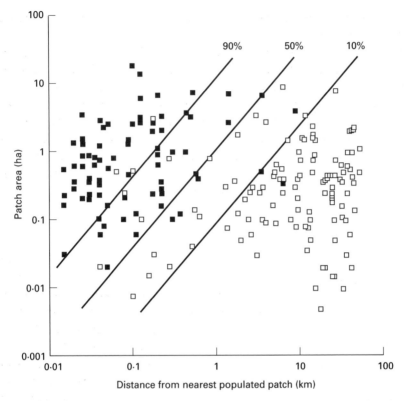

Fig. 10.2 Silver-spotted skipper butterfly distribution in relation to patch size and isolation. Occupied (solid) and vacant (open) patches are shown. The lines show the combinations of area and isolation which give 90%, 50% and 10% probability of patch occupancy. (Adapted from Thomas & Jones 1993.)

As a result of an analysis of satellite images, the panda conservation priorities have changed (MacKinnon & De Wulf 1994). The areas where the pandas had become extinct since the last mass flowering were the isolated and fragmented forests. It is now appreciated that the wandering pandas are simply seeking patches of non-flowering bamboo. They suggest the need is to ensure that the blocks of habitat are connected, and the Chinese authorities are now redesigning their reserve network.

Many insects have an annual or shorter life cycle and thus if unable to breed in a year the population will go extinct. The failure to ensure that suitable habitat is always present is probably central to the local extinction of many invertebrates. Cutting, grazing or flooding a site may result in only gradual changes in plant species composition, but may eradicate certain invertebrates. For example, cutting an entire site may remove all the stems, flowers or seed heads required by a range of invertebrate species. Small-scale management, in which some areas are cut while others are left, will prevent this. The conflict is that it is more convenient and often cheaper to carry out all the management procedures, such as clearing ditches or cutting hedges, at one time, while for invertebrates it is better if small sections are managed each year.

The simple and attractive idea of creating corridors to link habitat patches follows logically from these ideas. By linking adjacent patches, a population can be re-established if it goes extinct and reinforced if it declines. It is, however, as yet far from clear that it is better to create or protect a corridor than it is to put the same effort into enlarging one of the existing patches or improving the management of these patches. Other problems with corridors include the fact that there is often little evidence that they are used and that they may also act to spread disease, introduced predators or competitors (Hobbs 1992; Simberloff *et al.* 1992).

10.5.1 Patches within patches

Much of the concern of conservationists has been about the size of areas of wood, heath or grassland, despite the fact that many species only use very restricted parts of these habitats; this is particularly true for invertebrates. Many invertebrates have extremely particular habitat requirements, such as the thysanuran known only from the guano floating on the underground lake in Dragon's Breath Cave, Namibia (Irish 1989) or the sheep nostril fly *Oestrus ovis*. A major finding from the studies of butterflies is that the adult females often have very precise habitat requirements for egg laying which are related to the conditions under which the caterpillars can survive. In many cases, although the host plant is present in profusion, the butterfly Plebejus argus has gone extinct because the host plants are too long, too short or too shaded (Thomas 1991). This specificity has important consequences for conservation as illustrated by Webb & Thomas' (1994) study of three invertebrate species occurring on lowland heath in southern England. The three species range in habitat specificity, from the silver-studded blue butterfly *Plebejus argus* with a reasonably broad requirement (dry or humid heath facing in any direction which has been burnt or cut in the last 10 years) to the ant *Myrmica sabuleti* with the fairly specific requirement of dry south-facing heath, with a grass-dominated community which has been disturbed in the last 5 years. The third species, the heath grasshopper *Chorthippus vagans*, is intermediate in its specificity.

Recognizing the specific requirements of species is important. For example, one heath which conservationists would typically consider to be a single large patch actually contained 29 separate patches suitable for the silver-studded blue, each isolated by a greater distance than they regularly fly. These patches would thus be considered as separate islands from the butterflies' perspective. For *M. sabuleti*, the number of patches declined from 42 in 1978 to six in 1987. Furthermore, the mean distance from each patch present in 1978 to the nearest one present in 1987 was 9.5 km. It is hardly surprising that *M. sabuleti* and other species with similar requirements are rapidly declining.

The detailed study by Webb & Thomas considered only three invertebrates. It is clearly impossible to manage an area incorporating the habitat
requirements of every insect present. Webb & Thomas suggest that this is
not the insurmountable problem it seems; for example, the requirements
of *M. sabuleti* are very similar to those of many other ground-dwelling
thermophilous species and thus suitable management for *M. sabuleti* will
be suitable for them all. They suggest that all the heathland insects of
southern England can be assigned to one of about 10 different habitat
requirements.

The reserve manager usually has to maintain populations of species
whose requirements are unknown and, in the case of many invertebrates,
where the reserve manager does not even know that the species is present!
Again, in most cases, the best approach is to ensure the continuation of
the range of habitats.

10.6 Disturbance

A surprising number of species are dependent upon open ground. These
areas can only stay open as a result of poor growing conditions (low
nutrient, heavy metals, arid or very cold conditions) or regular disturbance.
Grazing, cutting or burning may prevent succession and create bare ground.
As explained in Section 10.2, the ideal is to have a sufficiently large natural
area that these early successional stage communities persist as a shifting
mosaic with gaps created by floods, fires or trees falling in storms. The life
of these gaps may be extended as a result of grazing by natural herbivores.
In reality, however, many reserves are too small and too protected for
such a shifting mosaic of disturbance to be important. If disturbance and
grazing have been responsible for providing suitable conditions, then it is
clearly necessary to ensure that similar processes persist.

10.7 Nutrients

It seems paradoxical that, while gardeners routinely fertilize to encourage
plant growth, highly fertile conditions are often a problem to plant communities of conservation importance. One reason is that, prior to the invention
of artificial fertilizers, almost everywhere would have been of low fertility;
hence most species are adapted to such conditions (Sutherland 1995). Only
areas such as river flood plains would have been of high fertility. It tends
to be the plants from these fertile communities that have flourished with
human influence. The answer to the above paradox is that, while almost
all species benefit from extra nutrients in the absence of competition, they
lose out to a small number of tall, competitive, fast-growing species when
in competition. Eutrophication is a similar and considerable problem in
water bodies in which the phytoplankton bloom resulting from the nutrient
addition results in deteriorating conditions for the waterweeds (Sas 1989).

The usual conservation priorities are to identify the nutrient source and find ways of reducing the input. This may not be straightforward, for example if the nutrient is airborne. The increased dominance of vigorous grass species to the detriment of the rest of the community, on both heath-land and calcareous grassland in the Netherlands, has been attributed to high nitrogen deposition resulting from increased anthropogenic emission of nitrogen oxides and ammonia (Heil & Diemont 1983; Bobbink & Willems 1987). It can also be difficult to reduce the existing nutrients. Solutions include stripping off surface layers, adding nutrient-poor substrates, soil disturbance to expose nutrient-poor lower layers and, for water bodies, pumping out the mud.

10.8 Grazing

On many conservation sites, mammalian herbivores are important in influencing the structure of the site. Frequently, an important role for land managers is to ensure that the herbivores are neither too rare nor too abundant. Controlling grazing may entail manipulating the native species, introduced species or stock.

Herbaceous (dicotyledonous) plants grow from meristems in the stem and grazing will remove many or most of the meristems and thus severely hinder growth. By contrast, monocotyledonous species, such as grasses, grow from meristems at the base of the plant and thus are less affected by grazing. The consequence of this is that grazing will tend to increase the proportion of monocotyledonous plants. As conservationists frequently (but by no means always) are trying to maintain the abundance of herbaceous plants, high levels of grazing can be considered detrimental.

Grazing can either increase or decrease species diversity. The general rule is that if the dominant species is palatable then grazing will increase diversity by reducing the abundance of the dominant species, while if the dominant species is unpalatable then grazing will reduce diversity by making the dominant species even more common.

Moderate grazing intensities can increase the structural complexity. Grazing obviously reduces the biomass and thus favours those species that flourish in less competitive surroundings. In The Netherlands, the objective for grassland management is to have approximately equal areas of short grass and tall grass (Van Wieren 1991). Fertile sites have to be grazed more intensely to achieve this. If the grazing pressure is too high, the structural variation starts to disappear (Looyen & Bakker 1987).

10.9 Habitat restoration and creation

Restoration and creation are undoubtedly important and useful. There have been considerable successes with many superb reserves being created. Wetlands in particular can be created in a short period, presumably as much of the structure often consists of near monocultures of fast-growing species.

A major concern is that restoration can be considered easy and can be used as a justification for destroying habitats by arguing that the destroyed habitat can either be moved or recreated elsewhere. In practice, moving or recreating habitats is often unsuccessful, not least because of the fact that developers usually seek a short-term solution and do not provide the long-term management required.

10.10 Species management

If, as I suggested in the introduction to this chapter, habitat management is an admission of failure, then carrying out species management must often be viewed as an admission of even greater failure. Species management techniques for animals include removing predators, providing breeding sites or providing supplementary food. For plants, the techniques include pollinating, weeding to reduce competition, fencing to exclude herbivores, collecting seed and planting out the seedlings, creating new individuals from tissue culture and moving plants between sites.

Cade & Temple (1995) evaluated the effectiveness of manipulative management of birds by reviewing the outcome of 30 projects in 1993 that had been described in 1977. Forty-three per cent of these were considered to have improved population viability, 23% helped stabilize numbers, 17% were inconclusive and 17% failed (Table 10.1). This suggests that manipulations are often worthwhile.

One of the most dramatic examples of species manipulation concerns the black robin *Petroica traversi* which at one point had a world population of just five individuals, which included only one breeding pair (Butler &

Table 10.1 Summary of the outcome of 30 manipulation projects on wild bird populations. (After Cade & Temple 1995.)

Kind of manipulation	Number	Success/failure rank*			
		1	2	3	4
Nest site limitations	5	2	2	0	1
Alleviating competition, etc.	5	2	2	1	0
Supplemental feeding	4	0	2	0	2
Nesting biology	4	1	1	1	1
Captive breeding	4	3	0	1	0
Reintroduction	7	5	0	2	0
Integrated approach	1	0	0	0	1
Totals	30	13	7	5	5

* 1, Actions resulting in an increased number of breeders; 2, actions serving to stabilize numbers or to slow rates of decline; 3, actions with inconclusive outcomes; 4, failures.

Fig. 10.3 Chatham Island tit feeding two black robin fledglings as part of the programme to save the robins from extinction. (Photo by D.V. Merton.)

Merton 1992). The entire world population was switched from Little Mangere, where the habitat was considered to have deteriorated unacceptably, to Mangere Island. In order to increase the productivity of the parent birds, young were reared in the nests of Chatham Island tits *Petroica macrocephala chatemensis* (Fig. 10.3) in the hope that the adult robins would renest. This programme was a success as a result of careful and imaginative planning, critical analysis and good luck. The recovery of the black robin was largely possible thanks to the breeding female 'old blue' who lived twice as long as any other black robin female and who started productive breeding at an age when most other females were dead. The population now exceeds 100 birds and almost all are descendants of 'old blue' and a male 'old yellow'.

Perhaps the most extreme suggestion for manipulation concerned the only two wild remaining crested ibis *Nipponia nippon* left in Japan (Hadfield 1994). All attempts to encourage the birds to breed have failed and it was suggested that the birds should be shot and stored at −196°C with the intention of resurrecting the species later using gene technology! This suggestion was not carried out.

10.11 Genetic changes in captive breeding

Species are taken in for captive breeding when it is hoped that the population growth rate or the probability of survival will be higher in captivity than in the wild. Adverse genetic change may occur within populations held in captivity as a result of inbreeding depression, loss of genetic variation and genetic adaptation to captive environments. There is strong evidence that captive populations of fish, plants and several fruitfly species (Frankham 1994) show genetic adaptation to captive conditions. For example, a population of *Drosophila melanogaster* doubled in reproductive fitness after a year (eight generations) in captivity (Frankham & Loebel 1992). These adaptations to captivity are likely to be disadvantageous when the captive animals are released in the wild, as shown in experiments on the brook trout *Salvelinus fontinalis* (Lachance & Mangan 1990).

The suggested solutions to these genetic problems include minimizing time in captivity, minimizing mortality, equalizing family sizes and maximizing the generation interval (Frankham 1994). Equalizing family size should reduce genetic adaptation by about a half as it removes the among-family component of selection (Frankham 1995). Making the captive environment as similar as possible to the natural environment may reduce the problems of adaptation.

10.12 Transferring species

> 'I was up in Washington State and the people were so worried about this huge area they wouldn't let them do any timber cutting because of the owls, and I finally asked a relevant question. I said "how many owls are there?" They said "20", and I said "OK, I suggest we send Air Force One out here, transport 'em in absolutely first-class comfort to the nearest national park". Now the owls can live happily ever after in hundreds of thousands of acres in some nearby park, and we can go back to work here'.
> Ross Perot, talking about spotted owls as part of his presidential election campaign, 18 September 1996, San Francisco.

Ever since Noah, transferring species has been a good way of attracting attention and publicity. It is much more exciting and glamorous to allow a population to go extinct and try and re-establish it than it is to maintain the population by sensible habitat management. Unfortunately, the glamorous approach is usually expensive and often unsuccessful. Releasing individuals into the wild can be a useful management tool but understanding why the species is declining and managing the remaining populations properly are often better ways of spending the time and effort.

One major concern with enthusiasm for moving species is that, as described with habitat restoration and habitat transplantation (Section 10.9), it may then seem easy to solve any problem by moving species that are in the way, in the manner suggested by Ross Perot. There is usually considerable publicity when the species is moved, but no publicity when, as so often happens, the transplanted population goes extinct (Anon 1997).

The terminology is confusing and contradictory. The following are now becoming the standard terms:

Re-establishment: the release and encouragement of a species in an area where it formerly occurred but is now extinct.

Introduction: an attempt to establish a species where it did not previously occur.

Re-introduction: an attempt to establish a species in an area where it had been introduced but the introduction was unsuccessful.

Re-inforcement: attempting to increase population size by releasing additional individuals.

Translocation: the transfer of individuals from one site to another (for example to boost the population or save individuals which would otherwise be destroyed).

Re-establishments are often referred to as experiments, although as Soderquist (1994) cynically (but probably accurately) remarks, the term experiment is often used to mean 'all the animals died and it's not my fault'. The usefulness of genuine experiments involving hypothesis testing was shown by the studies re-establishing the bush-tailed phascogale *Phascogale tapoatafa*, a carnivorous marsupial (Soderquist 1994). The re-establishments were designed to ask specific questions. These experiments showed that males disperse considerable distances if released at the same time as females, but much shorter distances if the females were released first and are thus able to establish a home range. Other experiments showed habitat differences in foraging success and greater survival if the release was accompanied by fox control. Not only was this project more likely to be successful as a result of the hypothesis testing, but the knowledge gained and disseminated is likely to be useful for many others involved in re-establishments. Such studies are far too rare.

The history of introductions shows a remarkable lack of the most basic level of monitoring. Beck *et al.* (1994) reviewed the re-establishment procedures and post-release outcomes for 145 distinct projects involving the release of 13 275 295 captive-bred individuals. The total information for all these projects involving 13 million animals filled less than one file drawer! Many even failed to determine whether the species was present afterwards.

Griffith *et al.* (1989) showed that the success in creating new populations: was greater in the core of historic ranges (78%) than either in the periphery or outside historic ranges (48%); was greater for herbivores (77%) than carnivores (48%); was greater with wild-caught animals (75%) than with captive-bred animals (38%); and was greater for release into excellent-quality habitat (84%) than poor-quality habitat (38%). Releases of large numbers of animals were more likely to be successful, although this relationship disappears above about 100 individuals.

Beck *et al.* (1994) carried out a similar, but more detailed, study restricted to captive-born individuals. Their definition of success was if the wild population exceeded 500 individuals or if a population viability analysis predicted that the population would be self-sustaining. Only 16 projects (11%) were successful according to this definition. Surprisingly, there was no evidence that post-release provisioning and medical screening were more likely to result in success. Interestingly, successful projects tended to be those which provided local employment or involved community education programmes, indicating the importance of local goodwill.

The following recommendations have been made concerning re-establishment by the Joint Committee for the Conservation of British Insects (1986) and the Conservation Committee of the Botanical Society of the British Isles.

1 Consult widely before deciding to attempt any re-establishment. Speak to the experts and government officials concerned with that group.

2 Every re-establishment should have a clear objective.

3 The ecology of the species to be re-established should be known.

4 Permission should be obtained to use both the receiving site and the source of material for re-establishment. The effect on the donor population needs to be carefully considered.

5 The receiving site should be appropriately managed.

6 Specific parasites should be included in re-establishment (the logic is that if a species is rare then its parasite is even rarer!).

7 The numbers released should be large enough to secure re-establishment.

8 Details of the release should be meticulously reported.

9 The success of re-establishment should be continually assessed and adequately recorded.

10 All re-establishments should be reported to the relevant recording bodies so that they do not confuse records.

10.13 Cultural change

Behaviour depends upon both genetics and culture, with culture being particularly important in higher vertebrates as becomes clear when re-establishing species after captive breeding. In many cases released animals lack basic survival skills. For example, it proved difficult to persuade introduced Arabian oryx *Oryx leucoryx* to eat native herbs and grasses after they had been raised on hay, and they had difficulty locating water (Price 1989). Within a couple of months of release many of the golden lion tamarins *Leontopithecus rosalia* re-established into Brazil had been killed by predators whilst others had succumbed to plant poisons. Even the rudimentary locomotory skills needed to negotiate vines and branches had been lost in captivity and necessitated training before release (Beck *et al.* 1991). Male hand-raised cockatiels *Nymphicus hollandicus* are poor at raising young while the females often lay their eggs on the ground (Myers 1988). A number of studies show that predator avoidance is innate, but that the young have to be tutored to maintain their proper fear of predators (Curio 1993). Studies of hand-raised parrots and rails show that the fear of humans is not necessarily culturally determined; hand-reared birds are tame while those raised by their parents show fear regardless of whether their parents have themselves been hand raised or parent raised (Curio 1993).

The northern bald ibis *Geronticus eremita* (Fig. 10.4) was once found over

Fig. 10.4 Only about 200 northern bald ibises *Geronticus eremita* survive in the wild, although there are many more in zoos around the world. Release schemes have been suggested. However, such social species with extended parental care learn much of their behaviour from others and extensive training is then necessary before captive-bred individuals can be released. (Photo by C. Gomersall.)

much of Europe, North Africa and the Middle East but disappeared from Europe, probably largely as a result of human persecution: the eggs and young were considered a delicacy. The five re-establishment attempts in Turkey, Italy and Israel (three attempts) were unsuccessful despite 200 birds being released, apparently because the birds had problems orientating and finding sufficient food (Mendelssohn 1994).

Northern bald ibises are a gregarious species with extended parental care and it seems much of their behaviour is learnt. Can the necessary skills be taught to captive-bred individuals? Pegoraro & Thaler (1994) have experimented by hand rearing birds in Austria using human foster parents who taught the birds to find their way to nearby fields to forage. The birds also needed to learn to recognize dangers, but had no alarm call of their own. They were taught to recognize the alarm call of a hand-reared chough *Pyrrhocorax pyrrhocorax*, which could be imitated by the human guardians when danger threatened. When at risk, for example from a car or domestic dog, the humans imitated the chough which flew and the ibises followed. These birds behaved like their wild counterparts and there are proposals to introduce them into the Cabo de Gata-Nijar National Park in Spain using similar techniques. It is not known how well they would breed as in this preliminary trial the birds were returned to the zoo in the autumn. It is clearly important that a release site should be situated away from existing colonies as there is concern that if released birds came into contact with wild birds then they could cause serious social disruption. This study illustrates how difficult and time consuming it is likely to be to re-establish such social species.

In many social species it is likely that predator recognition is learnt from others by the use of alarm calls or mobbing (Curio *et al.* 1978). Recently released naive animals are thus susceptible to predation. One solution is to train animals in captivity prior to release. McLean *et al.* (1994) succeeded in conditioning rufous hare wallabies *Lagorchestes hirsutus* to fear cats *Felis catus* and foxes *Vulpes vulpes* by playing wallaby alarm calls and squirting the wallabies with a water pistol when the stuffed predators were present.

It is not clear how successful such training to avoid predators is in increasing survival once the captive-bred animals are released. Of three studies, one on the bobwhite quail *Colinus virginianus* suggested that survival was increased (Ellis *et al.* 1977), but two other studies (on black-footed ferrets *Mustela nigripes* and New Zealand robins *Petroica australis*) showed no benefits from training although the sample sizes were small (McLean *et al.* 1994).

Migratory behaviour is often learnt in species with extended parental care, such as geese or cranes, in which the young stay with the adults on migration. Individuals bred in captivity from parents that cannot fly do not migrate. It is likely that migratory behaviour is also learnt in many social mammals such as whales. There are plans to teach whooping cranes *Grus americanus* to migrate by imprinting them on ultralight planes.

Migration can be taught by switching eggs so that the young follow their foster parents, for example whooping cranes have been reared by sandhill cranes *Grus canadensis* and have learnt the sandhill cranes' migration route. The problem is then that the young become imprinted on the foster species and are unlikely to breed successfully.

Another example of cultural learning of migration concerns the lesser white-fronted goose *Anser erythropus* (Essen 1991). The population of breeding birds in Scandinavia has declined by 95–99% and this has been attributed to hunting during the winter in southeastern Europe (Fig. 10.5). There is currently a population of free-living barnacle geese *Branta leucopsis* which breed in Stockholm Zoo but winter in The Netherlands. A project has involved taking pairs of barnacle geese to Lapland and giving them eggs from captive-bred lesser white-fronted geese. The lesser white-fronted geese have then flown with their foster parents to The Netherlands. The foster parents returned to Stockholm Zoo while the lesser white-fronted

Fig. 10.5 Hunter with shot juvenile lesser white-fronted goose in Romania. Largely as a result of such hunting in winter, the first-year mortality of Fennoscandinavian birds is 85% and the population is plummeting (Øien 1997). As well as plans to increase protection in the winter range, a project is establishing an alternative migration route to The Netherlands. (Photo by W.J. Sutherland.)

Fig. 10.6 Mountain gorillas *Gorilla g. beringei* have been shown to learn food-processing methods from their parents while infants (Byrne 1993). Re-establishment of such species with complex cultural skills is thus very difficult. (Photo by W.J. Sutherland.)

geese returned to Lapland where some of them have bred. This project is controversial. Some argue that it is unacceptable to tinker in this manner and it deflects attention from the real issues; others argue that it may be the only realistic way of protecting the Scandinavian population from extinction.

The main message is that, for social species with extensive cultural skills such as migration, predator avoidance and food recognition, the release of captive-bred animals will be difficult and time consuming (Fig. 10.6). It is much more cost effective and efficient to try and prevent species declining so that such dramatic measures are unnecessary.

Acknowledgements

Thanks to Nicola Crockford for comments and Phil Riley for secretarial help.

CHAPTER 11
Economics of nature conservation

Nick Hanley

11.1 Introduction

It may seem odd to include a chapter on economic analysis in this book, since it is uncommon for people to put the words 'economics' and 'nature conservation' together (except perhaps in a way that always seems to suggest that economists are the bad guys, who point out the costs of conservation and who promulgate development). Yet economics is highly relevant to many important aspects of nature conservation. For example, conservation decisions are resource allocation decisions, which impose opportunity costs on society when we decide to conserve an ancient woodland rather than allow a new road to be put through it, or when a nature conservation agency must decide how to allocate its finite budget between competing projects. Economics has developed an elaborate, formal and fairly unified way of studying such decisions by, for instance, evaluating and comparing the costs of allowing the wood to be felled with the benefits that would flow from a new road.

Economics is also concerned with understanding the incentives provided by one of the most far-reaching institutions in the world's economies, namely the market. Markets turn out to be efficient allocators of resources amongst competing uses in many circumstances, and provide appropriate 'signals' (prices, profits, rents) to 'actors' within the economy. These actors (firms, households, governments) respond to these signals in predictable ways (for example, the actions of oil companies following an oil price rise), so that an understanding of these responses increases the governments' ability to manage the economy, and our ability to understand its dynamics.

Markets, however, can also fail badly in their allocation of resources, particularly with regard to the environment. This market failure leads to too little environmental quality, insufficient resources devoted to conservation, too much pollution, overexploitation of fisheries and too low a level of biodiversity protection. That is to say, the market system results in outcomes for the environment which are not economically efficient: this provides an economic justification for government intervention, for example in cutting pollution levels.

220

Under what circumstances do markets fail in this way? They fail when property rights are inadequately allocated. This means the market sends the wrong signals to producers and consumers. For instance, no-one owns clean air; thus factories pay no fee (in a free market) to use this resource by discharging pollutants into it, or for the damage lower air quality does to people's health. This kind of effect is known as an *external cost*. Since many people other than the farmer enjoy the landscape that the farmer 'produces' in picturesque rural areas, the farmer is unable to capture all of these benefits by charging for them, and thus has too low an incentive to maintain a beautiful landscape at the expense of modernization. Such resource flows, from which it is impractical to exclude all beneficiaries, are known as *public goods*, which may also exist due to a 'non-rivalness' property, in that the fact that I am happy that blue whales exist does not reduce your happiness that they exist. Public, environmental goods may indeed be global in scale; for example, many people in nations other than Brazil benefit from the preservation of the rainforest in Brazil, but as they cannot charge people for these services, Brazilian landowners have too little an incentive to preserve the rainforest, choosing rather to convert it into products they can sell (such as beef or coffee).

Thus a free-market system generates too many external costs and too few public goods. Governments should intervene to restore economic efficiency by increasing environmental benefit flows. But how, and on what principles? And in what sense are environmental protection benefits 'real' in economic terms?

Any good, service or resource has economic value if it contributes to *utility*. Utility is the word economists use to describe personal satisfaction or happiness. Adding up individual utilities gives an indication of social welfare; although precisely how the adding up should occur is still a contentious issue. We imagine *Homo economicus* as having preferences over different goods, services and resources which conform to certain axioms (such as the axiom of completeness and the axiom of transversality), and which can be described using a utility function:

$$U = U (X_1 \ldots X_m; Q_1 \ldots Q_n)$$

where there are m marketed goods X (such as CDs or holidays) and n environmental resources. Thus Q_1 might be the stock of fish in a local river, and Q_2 the population of orang-utans in South East Asia. The environment has economic value here in three ways:

1 Directly, through those environmental resources the individual actually uses (such as fish in the river if she goes fishing); these kinds of values are known as use values.

2 Directly, through environmental resources such as orang-utans which the individual does not actually exploit, but which he or she nevertheless cares about; these kinds of values are known as existence or passive use values.

3 Indirectly, through its role in supporting the production of marketed goods (e.g. the use of clean water in food products).

Existence values may be motivated in a number of ways, such as intragenerational altruism, intergenerational altruism (wanting to bequeath environmental assets to the future) and pure selfish motives. Summing these elements of value for any one resource (e.g. use + existence values for caribou in Canada) gives that resource's *total economic value*. Existence values have been shown to be statistically significant and to sometimes make up a large share of the total economic value for many environmental resources (such as upland streams, lowland rivers, Environmentally Sensitive Areas in the UK and sea turtles) (Navrud 1988; Whitehead 1993; Willis & Garrod 1995; MacMillan *et al.* 1996). However, it should be noted that values which derive independently of human satisfaction should not be counted as a part of economic value. Such intrinsic values, which may stem from rights-based belief systems, do not share the same theoretical basis as the other components of value discussed above (see Spash & Hanley 1995).

11.2 Economic principles and methods

Space permits only a very brief acquaintance with some of the more important economic principles and methods here: for a fuller review in an environmental context, see Tietenberg (1994) and Hodge (1995).

11.2.1 Optimality

As mentioned above, economics is often most concerned with the allocation of resources. One useful concept in this regard is the economically optimal allocation of resources; for example, the optimal level of output of pollution. In a static model, such an optimum occurs when the difference between the benefits of allowing emissions ($B(e)$) and the costs of reducing emissions ($C(e)$) is at its greatest. It may seem odd to refer to the benefits of emissions, but if the production of consumer goods generates emissions, then zero emissions are only possible with zero output. This position is shown in Fig. 11.1a as e*. In Fig. 11.1b, both the total costs and the benefit curves are redrawn as their marginal equivalents, where marginal cost (benefit) is defined as the change in total costs (benefits) for a small change in emissions. Thus:

Marginal cost = MC = $\delta TC/\delta e$

Marginal benefit = MB = $\delta TB/\delta e$

It may be seen that the static optimum in Fig. 11.1b occurs where MB = MC. This is in fact a general rule for the optimal allocation of many resources: set marginal benefits equal to marginal costs. Note also for Fig. 11.1 that

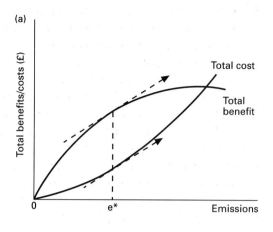

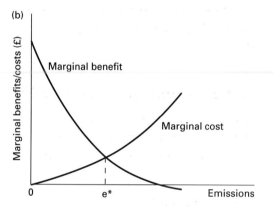

Fig. 11.1 (a) The optimal level of pollution is located where the gap between the total costs and total benefits of emissions is at its greatest (that is, where total net benefits are maximized). (b) Equivalently, it is found where the marginal cost of emissions control is equal to the marginal benefits.

MB is negatively sloped whilst MC is positively sloped; this follows from assumptions about the functional form of the total benefit and cost curves. Also note that it is highly unlikely that all cost and benefit curves have such smooth, continuous forms in reality; in the natural world, discontinuities, non-linearities and surprises may be expected.

This is especially likely when dynamic optimality is considered. Dynamic optimality refers to efficiency over more than one time period. Here, the static rule of MC = MB changes to comparing the sum of the *present values* of costs and benefits. Present values are future values (say, forestry returns in 40 years) adjusted for the effects of time. In particular, economists adjust for the effects of delays in cost and benefit streams relative to now. This is because of a well-established principle known as *time preference* (Chapter 5). This states that benefits will be valued less, the further into the future they are postponed. People are assumed to prefer benefits sooner rather than later due to impatience, the possibility of investing, risk, fear of death and increasing incomes over time. Similarly, costs incurred now are given a higher weight than the same costs occurring at some future time period.

The process by which time-specific costs and benefits are converted into comparable present values is known as *discounting* (Hanley & Spash 1994). If B_t is the future value of a benefit occurring in year t, then its present value (PV) is given by:

$$PV(B_t) = B_t(1 + r)^{-t}$$

where t is the delay from now and r is the *discount rate*, the rate at which people are prepared to trade off present against future benefits. Thus, for example, if a forestry project is expected to produce timber with a value of £5000/ha in 40 years' time, its present value is £5000 × $(1 + 0.05)^{-40}$, or £710 at a 5% discount rate. If instead we had to wait 60 years, then this present value would be only £267. Thus, as the time at which the benefit is received moves further into the future, its present value falls. This would also be the effect of increasing the discount rate. An identical approach is used for costs: thus if global warming costs the UK £100 million in lost output, higher disease incidence and lost species in 100 years' time, this is valued at only £0.76 million now (at a 5% rate); thus it would be rational (on this argument) to spend only £0.76 million now on measures to reduce global warming such that these very large future costs would be avoided.

We can now write down the condition for dynamic optimality. This says that for any activity, it must be the case that:

$$\frac{B'_t - C'_t}{(1 + r)^t} = \frac{B'_{t+1} - C'_{t+1}}{(1 + r)^{t+1}} = \dots \frac{B'_T - C'_T}{(1 + r)^T} \text{ for all } t, t = 1 \dots T \quad \text{eqn 11.1}$$

where B' and C' are marginal benefit and marginal cost respectively. In other words, set the present value of marginal net benefits in each period equal to each other. The choice of discount rate is clearly crucial here. For some resources, we need to adjust this basic formula. For example, for renewable resources such as fish and forests, it is necessary to take account of the natural density-dependent growth rate of the stock. Essentially, this involves recognizing a new cost, that removing fish from the stock forgoes the opportunity of allowing that fish to grow and reproduce (for a formal analysis see Fisher 1980). For non-renewable resources such as oil or bauxite, it is necessary to take account of the finite nature of the stock. This again involves adding a new element to costs, namely that the opportunity costs of mining one more ton or barrel today is that the same ton cannot be mined tomorrow. However, we must also recognize that profits from extraction can then be invested and earn a return (Fisher 1980). Essentially, this approach to the optimal extraction of renewable and non-renewable resources involves treating them as no more than a form of capital asset, the rate of return on which must be compared with that on other assets (such as stocks and bonds). This is a rather unsatisfactory way of thinking about nature conservation, so we will not pursue this line of enquiry further here.

11.2.2 Cost–benefit analysis and the Kaldor–Hicks principle

A very relevant principle which is related to optimality is the Kaldor–Hicks principle. This is one of the fundamental theorems of welfare economics, that branch of economics concerned with the normative aspects of decision making. The Kaldor–Hicks principle underlies one widely used economic technique which is increasingly being applied to nature conservation decisions: cost–benefit analysis (CBA). The Kaldor–Hicks principle addresses the question: how can we tell whether a project or policy increases social welfare? The answer given by Kaldor and Hicks was: only if the gainers could compensate the loosers and still be better off. CBA involves the putting into practice of this principle, by valuing in monetary units all the costs and benefits of a proposed project/policy, and then comparing the discounted sum of benefits with the discounted sum of costs. If positive net benefits are predicted (i.e. if $PV(\Sigma B) > PV(\Sigma C)$), then the project/policy is said to pass the CBA test, and to be efficient (Fig. 11.2 summarizes the CBA process). In modern CBA, attempts are made to include the monetary values of the environmental effects of projects/policies (see below). Thus, for example, if a new hydroelectric dam is proposed for an area of rainforest, CBA would involve identifying all the relevant benefits (such as the output

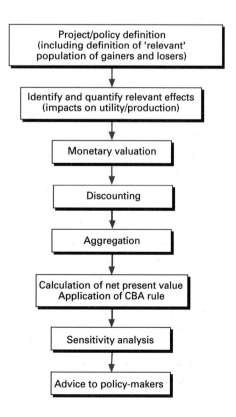

Fig. 11.2 The cost–benefit analysis process.

of electricity) and valuing them, and all the relevant costs. These costs could include the loss of landscape quality, wilderness qualities and wildlife habitat.

11.2.3 Cost effectiveness

In many instances, valuing the benefits of environmental protection programmes is either too difficult, or undesirable for some other reason. In such cases, we may take an environmental target (such as preserving some species at above its minimum viable population size) as given, and seek the least-cost means of achieving this target. An example is the study by Hyde (1989), who compared the costs of protecting the red-cockaded woodpecker *Picoides borealis* in the eastern USA under a number of different forestry management options. These costs may be evaluated in social costs terms, or in terms of the financial costs facing those agents responsible for changing their behaviour. Social and financial costs may differ due to government price support (as in farming in the USA and European Union) or cost subsidization (private forestry in the UK). An example of a cost-effectiveness analysis done on a financial basis is the study by Vickery *et al.* (1994) of options for managing over-wintering Brent geese *Branta bernicla* (Fig. 11.3). Financial-level cost-effectiveness analysis cannot identify the socially efficient (i.e. least-cost) means of achieving the environmental target, but it can give valuable policy advice on how agents (farmers in the Brent geese case) may react to incentives, and on effects on income and employment. Cost-effectiveness analysis can also be used to study habitat protection targets (Colman 1991) and pollution reduction targets (Hanley 1993).

Fig. 11.3 Brent geese grazing on crops have been estimated as costing farmers about 4.2 p (US$0.065) per goose per day in lost yield. A 'human bird scarer' has been calculated to be the most cost effective way for farmers to reduce this loss. However, while farmers are subsidized and overproduce, the geese grazing saves the taxpayers' money by reducing surplus production. This process has been estimated as saving taxpayers 1.9 p (US$0.029) per goose per day (Vickery *et al.* 1994). (Photo by J. Vickery.)

11.2.4 Bargaining and the Coase theorem

Since Adam Smith in the 1700s, the power of the market to produce an efficient allocation of resources under certain conditions has been well known. As was argued above, environmental resources are often not well allocated by the market (between, say, conservation and development) due to a lack of a fully specified and enforced system of private property rights. One suggestion to this market failure problem, according to Coase (1960), is for the government to allocate property rights, and then let the normal bargaining process implicit in markets decide the allocation of resources. This approach, known as the *Coase theorem*, has been roundly criticized in the context of pollution control, due to both theoretical and practical objections (see, for example, Common 1988). However, the theorem may be a useful way of thinking about some nature conservation problems where interdependencies exist between parties, and where bargaining is part of the process of resolving difficulties that arise from this interdependence. One example is the bargaining between nature conservation organizations such as English Nature and landowners in respect of management agreements, where the 'buy' function for English Nature depends on their assessment of conservation values, and the 'offer' function for landowners depends on the cost to them of the management agreement. This interaction is also influenced by the limited budget of English Nature and possible time constraints on the bargaining process (Spash & Simpson 1994).

Another example concerns red deer in the Scottish Highlands (Hanley & Sumner 1995). Red deer *Cervus elaphus* currently number around 300 000 in the Highlands, and may be shot by the landowner on whose land they happen to be (no-one owns the deer, only the right to shoot them). The letting of such shooting is the source of significant revenues to Highland estates. However, deer may roam between neighbouring estates, whilst they may also damage woodland. This externality (woodland damage) could form the basis of Coase-type bargains between a sporting estate and the woodland owner. The woodland owner could offer compensation payments to the sporting estate if it reduced its deer population, if these payments were less than the value of damage done and the costs of preventing damage (fencing). Likewise, neighbouring sporting estates may bargain, either formally or informally, to manage jointly deer populations and thus internalize externalities involved in one estate shooting deer originating on a second estate. The recent trend to establish voluntary cooperation through deer management groups is a step in this direction. The Coase theorem states that such interactions can lead to the optimal management of deer resources (although note that this ignores other externalities imposed by deer, such as damage to conservation interests through the role of deer in preventing the regeneration of native woodlands; only if conservation groups were part of the bargaining process could these externalities be internalized).

11.2.5 Valuing the environment

In much of what has been said above, an implicit assumption has been that it is possible to place a monetary value on environmental resources, or on changes in their condition. To identify an optimal level of habitat preservation, or to conduct a CBA of wilderness protection, it is necessary to value the physical resource changes in monetary terms. For example, what is the monetary value of preserving wilderness? It may be easy to write down the conservation benefits in physical terms: so many species preserved, landscape quality maintained, wilderness recreation opportunities maintained. Yet none of these aspects of wilderness have market prices which reflect their full economic value (use plus existence values). Economists have, since the late 1950s, been developing techniques to put monetary values on such resources: a full review of these methods is available, for example, in Hanley & Spash (1994). These methods are based on the principles of value from welfare economics; namely, that for a good which generates utility, the value of that good can be measured by either the maximum willingness to pay (WTP) of individuals to either have a specified increase in that good, or prevent a specified decrease, or their minimum willingness to accept compensation (WTAC) to either forgo a specified increase in the good or put up with a specified decrease (assuming in all cases that the good is something they derive positive utility from). Thus, the economic value of preserving a beautiful landscape is given by people's maximum WTP or their minimum WTAC. Both these measures combine preferences with ability to pay (income or wealth), and yield estimates of economic values for the environment which are fully comparable with market-valued goods (such as barrels of oil).

How in practice can we measure WTP (we will concentrate on WTP rather than WTAC for the rest of this section) for an environmental 'good'? Two classes of method exist: *direct* methods, which use surveys to measure WTP directly, and *indirect* methods, which infer WTP for environmental goods from expenditure on related market goods. Direct methods include contingent valuation, contingent ranking and choice experiments. Contingent valuation is the simplest and most widely used of these, and has now been accepted for use by many US government and UK government agencies, as well as by the US courts in environmental litigation cases. Perhaps its most famous application was in connection with determining damage due to lost existence values following the Exxon Valdez oil spill in Alaska in 1989. The method involves asking people their WTP for a given environmental change contingent on the description of a hypothetical market. For example, suppose we want to estimate the value of a local woodland threatened with development. People could be told that the wood could be saved if a trust fund acquired enough funds to buy the wood. Their maximum-stated willingness to pay into the fund would measure the utility they would gain if the wood was saved, including both use and existence values. For an empirical example of woodland protection which also compares real with hypothetical payments, see Navrud & Veisten (1996). Nearly 2000 contingent valuation applications exist in the literature now, concerned for example with species preservation in the USA (Stevens *et al.* 1991), landscape quality in the UK (Garrod & Willis 1995), biodiversity protection (Hanley *et al.* 1995) and ecotourism in Brazil (Peters *et al.* 1989).

Indirect methods for measuring environmental values include production function approaches, avoided costs, dose–response models, hedonic pricing and travel–cost models. Travel–cost models are used to estimate the economic value of outdoor recreational resources, such as national parks, public forests and beaches. Like all other indirect methods, the travel–cost model can only measure use values. The basic assumption in the model is that people's expenditure on travel and other access costs must be some guide to the utility they derive from the recreational experience. Statistical relationships are estimated between the (increasing) cost of visiting a resource and the (decreasing) frequency of visits, having controlled for other factors such as the availability of substitute sites and time constraints. In the UK, the method was used to estimate the recreational value of the Forestry Commission estate, showing recreational values to be as important as timber values (Willis 1991). The method has also been very widely used in North America—for an application to fisheries protection see Smith & Desvouges (1986), and for an application of a variant of the method to species habitat protection see Coyne & Adamowicz (1992). Finally, valuation work on the benefits of preserving biodiversity is summarized in Barbier *et al.* (1994) and Pearce & Moran (1994).

11.2.6 Sustainability

Since the Bruntland Commission report in 1987, the issue of sustainable development has dominated environmental policy discussions. Economists, it turns out, had been studying 'sustainability' for many years previous to this, in the context of growth models which tried to identify conditions under which an economy could expect to have continually growing (or at least non-declining) levels of per capita consumption. Much of the work on this in the early 1970s was spurred by a concern about 'limits to growth' imposed by finite non-renewable resource stocks. Since the Bruntland report, much of this work has moved in emphasis to modelling non-declining per capita utility over time, where the environment is not just an input to production but also a waste sink and a direct source of amenity. However, an alternative (and possibly more useful) area of analysis has been to concentrate on the means of achieving utility, rather than on utility itself; in other words, the capital stock which we pass on to future generations. This stock is commonly thought to consist of three components: man-made capital, Km (such as roads and machines), human capital, Kh

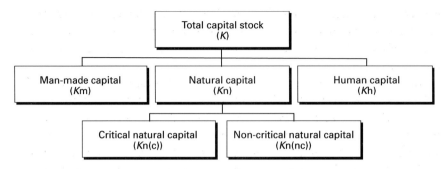

Fig. 11.4 Capital concepts and sustainability rules: weak sustainability—stop K from declining over time; strong sustainability—stop Kn from declining over time; very strong sustainability—stop $Kn(c)$ from declining over time.

(learning), and natural capital, Kn. Kn is defined as all gifts of nature, including renewable and non-renewable resources, biodiversity and nutrient cycles. One school of thought has identified as a sufficient condition for sustainable development (SD) that the total capital stock $(Km + Kh + Kn)$ be non-declining. This, however, assumes perfect substitutability between the three classes. A rival school, however, maintains that Kn is often not substitutable by Km, but that the two may be complements rather than substitutes (Victor 1991). Within the class Kn, other authors (e.g. Turner 1993) have stated that sustainable development does not require holding all of Kn constant, but only some critical part of it, containing 'keystone processes and keystone species'. Figure 11.4 summarizes these capital stock concepts.

With regard to operationalizing any of these non-declining capital stock rules, the key problem is one of measurement and aggregation. How should we evaluate the level of Kn in any time period? Measurement in physical units is confounded by non-comparability (tonnes of ore, numbers of osprey *Pandion haliaetus*, hectares of wetland), unless Kn is compartmentalized and each component is kept constant/non-declining. Measurement in economic value terms suffers from huge practical problems, and also from a conceptual problem, that as a resource gets scarcer its price tends to rise. Thus the value of the Kn stock could be rising even as in physical terms it shrinks. See Hanley *et al.* (1996) for a fuller discussion of sustainability within economics.

11.3 Applying economics to nature conservation: three examples

11.3.1 Cost–benefit analysis of nature conservation decisions

Many nature conservation decisions are concerned with the protection of habitats. Cost–benefit analysis (CBA) is one tool for providing information to decision makers and the community on whether a particular habitat protection decision is worthwhile. The CBA method addresses one criterion only: that of economic efficiency. In particular, it can reveal how the costs of a decision compare with the benefits, over some predetermined population of interest.

One of the best-known conservation issues in Scotland in the 1980s and 1990s is afforestation. With the practical cessation of upland planting of conifers in England and Wales, Scotland became the obvious site for meeting the government's objective of 33 000 ha of new planting each year. Upland afforestation is associated with many environmental costs and benefits. On the costs side, ploughing and draining of peatlands releases greenhouse gases, and leads to the loss of moorland/peatland flora and fauna (such as, famously, the greenshank *Tringa nebularia*). Large-scale afforestation, often in geometric patterns, depletes landscape quality, whilst adding to soil and water acidity. On the plus side, young conifer plantations may be helpful for species such as the hen harrier *Circus cyaneus* and merlin *Falco columborius*, whilst mature forests lock up carbon and provide a habitat for certain birds, such as the crossbill *Loxia curvirostra* and capercaillie *Tetrao urogallus*.

In the late 1980s, much private sector forest planting interest centred on the 'Flow Country' of Caithness and Sutherland, a large wetland area known for its rare breeding birds and unique landscape. This planting was subsidized by the government, yet seemed likely to result in a net loss in conservation interest and very little usable timber. Hanley & Craig (1991) reported on the results of a CBA of allowing any further planting in the Flow Country. The benefits of planting were revenues from clear-felling at the end of the rotation (the area is too windy to allow thinning)—these, discounted, amounted to £543/ha. Planting costs and revenues were evaluated net of subsidies from the government, since these are transfer payments which should not be counted in a CBA. Deducting these planting and management costs from the benefits of felling gave a present value of *minus* £751/ha. The environmental costs of planting were estimated using a contingent valuation survey of 400 Scottish households. This revealed that the present value of mean WTP to prevent the loss of wildlife and landscape quality following planting was £16.79 per household, or £327/ha. This, when compared with timber benefits and planting costs, showed that further planting in the Flow Country failed the CBA test, and thus was not economically efficient.

Generally, CBA faces three problems in its application to habitat conservation decisions.

1 Which discount rate to use? The outcome of the CBA process is often dependent on this choice.

2 How can the effects of ecosystem development be predicted? It is very unlikely that complete certainty exists over these effects, or that probabilistic assessments are known. CBAs can be done on a range of possible effects, or magnitude of effects, but this means that we are left with the task of choosing which version of the analysis to place most emphasis on.

3 The valuation of environmental effects. Often this is a complex task, which a simplistic application of the methods noted in Section 11.2.5 is unlikely to solve.

11.3.2 International environmental protection

Many of the most important environmental problems in the world today are international in character, in that their causes and effects spread over many countries. For example, greenhouse gases are emitted by all nations, and global warming may affect everyone in some way. Sulphur dioxide emissions originating in the UK damage lakes in Scandinavia, whilst pollutants entering the River Danube may originate in one country yet affect another. Losses of tropical rainforest in the Amazon affect citizens of many countries, who care about the preservation of wildlife in that area. However, whilst such 'transboundary' environmental problems are important and commonplace, their solution is not easy, since there is no world government which can force country A to, say, reduce pollution emissions which are damaging country B. International negotiations and agreements are the only practical solution to such transboundary problems, and within economics the area of game theory is well suited to the analysis of such processes.

Game theory analyses strategic interactions between 'players', who are all assumed to be largely motivated by self-interest, but whose actions affect each other. Here, the players are countries. A large literature has developed in environmental economics since the late 1980s, applying game theory to problems such as acid rain control in Europe, international agreements over chlorofluorocarbons (CFCs), and the preservation of tropical rainforest (for a survey see Missfeldt 1995 or Hanley *et al.* 1996). Game theory shows how difficult it is to reach international agreements such as the Montreal Protocol and the Framework Convention on Climate Change, when some countries may do better by staying out of such agreements. Agreements usually involve costly actions (such as abstaining from timber harvesting or reducing carbon dioxide emissions), in return for some benefit (such as biodiversity protection or reduced global warming). However, whilst cooperative actions may maximize the net benefits summed across all countries ('world welfare'), each country has an incentive to abstain from the deal if its own net benefits are higher from so doing.

For example, suppose all countries in Europe except one agree to cut their emissions of acid-rain gases. All countries, including the country which 'stays out', benefit from reduced acid deposition, yet the country staying out gets this benefit at no expense since it is all the other countries that are spending money to cut emissions. All countries may thus have an incentive to stay out, and so no deal is struck. Yet deals *are* struck, thus some explanation must be found. Game theory has pointed towards the possibility of 'self-enforcing' international agreements (Barrett 1994), or the possibility of issue linkages (for example, penalizing countries who stay out of the deal with trade sanctions). Finally, if some countries lose out in the cooperative solution, then those countries who gain might compensate them with 'side payments' (Maler 1989): this is what has actually occurred under the revisions to the Montreal Protocol.

11.3.3 Endangered species

For many years, debate has raged over whether trade in exotic game species such as elephants and rhinos should be allowed. The Convention on International Trade in Endangered Species (CITES) treaty places firm restrictions on trade in such species (the elephant was added to Appendix 1 of the treaty in 1990, following the decline in elephant numbers in the 1980s), whilst in many African countries domestic measures are also in force to protect the elephant. Essentially, these measures involve preventing its exploitation as an economic resource, except as a 'wildlife tourism' asset. However, given the very limited resources available to national parks in Africa to prevent poaching, the question arises as to whether the elephant is best protected by banning trade and hunting, or by allowing it (Swanson 1993, 1994). The cases for and against revolve around the issues of property rights and financial returns, and may be summarized as follows.

For trade: by allowing hunting and trading in elephants local people are given an incentive to conserve the resource since it becomes a source of income for them which, given secure property rights (that is, a belief that they will be able to control the resource into the future), will lead to a desire to keep the rate of hunting down to sustainable levels. That is, for any given population level, the rate of hunting will be restricted to equal or less than the growth rate. Since local people can earn a return from elephant exploitation, they will be willing to allocate scarce land resources (e.g. feeding areas) to elephants, up to the point where the marginal returns from elephant 'farming' are equal to the marginal returns from other forms of land use, such as cash cropping. Banning trade will, in any case, result in high black market prices for ivory, which will mean continuing incentives for poaching.

Against trade: due to the insecurity of future property rights, and the wandering nature of elephant populations, local people will not see hunting as a secure future income source, but will instead exploit it as fast as possible. Elephant populations will be the subject of a 'tragedy of the commons' akin to the world's deep-sea fisheries, where open access leads to overexploitation since people know that if they do not shoot an elephant today, someone else will shoot it tomorrow (Chapter 5). This overexploitation will be exacerbated by the relatively low growth rate of the species. The legalizing of trade in ivory products will result in an increased demand for these products, so hunting will increase. Banning trade, coupled with strict enforcement, will safeguard elephant populations, and allow countries to earn income in the form of wildlife tourism, whereby visitors from higher-income countries essentially give the elephants economic value *in situ* and alive. Means must be found of diverting some of this income to rural dwellers to give them an incentive to look after elephant stocks. If elephants can earn income for local people in this way, then again they will be willing to set aside land. Alternatively, governments could compensate them for losses due to elephant grazing damage.

11.4 Limits of economics

In this chapter, we have set out some principles of economics that can be applied to nature conservation, and noted some examples of such applications. Economics offers a powerful set of tools for analysing any resource allocation decision: for example, environmental cost–benefit analysis (CBA) can be used to decide whether a decision to allow afforestation in a wilderness area will generate positive net benefits to the nation. Yet it is also important to be aware of the limits of economics, since it is easy to get somewhat carried away with its power. Space permits only a short examination of these limits, and we can only note some of the more important issues relevant to nature conservation.

11.4.1 Does economics answer the right questions?

Techniques such as CBA provide a very specific answer to a very specific question: does this project/policy, as defined, generate a positive net present value? Whilst this precision has its advantages (such as being consistent with an underlying body of theory), it also imposes limits. For example, only utilitarian values may be admitted to the CBA; thus, analysis may show that it is economically efficient to allow the felling of forests in the Pacific northwest of the USA, even though it is the last local habitat of the northern spotted owl *Strix occidentalis caurina*. Yet people may feel the owl has the right to be protected irrespective of costs and benefits: such intrinsic values are excluded from CBA. If the CBA had given the opposite answer, that protection was efficient, then local people might well object that the beneficiaries were high-income conservationists throughout the USA, whilst the losers were low-income rural households who would lose out on employment from logging. CBA does not cope well with either distributional issues (the relative affluence of gainers and losers) or with unemployment issues. Yet these might be deemed most important by citizens. Finally, a CBA should, as has been noted, exclude transfer payments such as grants and subsidies. Thus it might be that a CBA would show that farmers should adopt more environmentally friendly techniques, such as the abandoning of intensive grazing or the use of buffer strips to improve water quality; yet unless farmers are compensated for the full value of their losses (that is, inclusive of subsidies), they are unlikely to sign up for such conservation schemes.

11.4.2 Discounting

We have already seen the large effects that the process of discounting has for projects with long-lived impacts. Thus, the economic appraisal of projects such as nuclear power, broadleaved forest planting and renewable energy research and development often turns on the choice of discount rate. However, there is a large debate on whether positive discount rates should be used for projects with environmental impacts, and if so whether this rate should be lower than that used in alternative projects (see Hanley & Spash 1994 for a summary of this literature).

Two arguments against discounting in an environmental context are intergenerational fairness and the nature of natural capital. The fairness argument runs as follows. Suppose an economic analysis showed it was optimal to allow an area of rainforest to be felled for its timber, even though this resulted in a loss of biodiversity. The benefits of the project accrue to those around at the present. The costs, in terms of lost biodiversity, accrue to future generations who must live in a world of lower diversity. Moreover, these losses may be non-compensatable, so that even if the revenues of felling had been invested in, say, new roads or housing, future generations would still feel worse off because of the environmental loss they had suffered (for example, residents of San Francisco may still regret the construction of the Hetch Hetchy dam in 1914 and the consequent loss of wild lands). Second, the conventional argument for discounting rests partly on the productivity of capital and its reproducibility. Yet much natural capital may be unique and non-replaceable, whilst some impacts on this natural capital stock may be irreversible. In this light, discounting of environmental costs may be unjustified.

11.4.3 Uncertainty and complexity

Much economic analysis is carried out using highly simplified models of the real world. This is, of course, necessary in order to begin to make sense of the more complex reality, and is common practice in many sciences. However, this may lead to conclusions from such models being extended beyond their remit. For example, the simple comparison of the marginal benefits and costs of biodiversity preservation, in a model, yields an optimal level of preservation (Pearce & Moran 1994). Yet to suggest that in practice this optimal level can be identified is wrong, since we have so little information on the costs and (especially) the benefits of preserving biodiversity. Many economic models used to produce policy advice are also based on continuous, smoothly increasing or decreasing functions with unique equilibria. Yet in the natural world, pollution damage costs, for example, may be subject to discontinuities and surprises. Again, dynamic models, for example fisheries exploitation, tend to be assumed to have stable properties and to converge to a steady state or at least cycle in some predictable manner or converge to some boundary condition. Yet real-world systems may exhibit chaotic dynamics whereby, although the system variables are bounded, they are not predictable, or else they may be unbounded and 'explode' (Lines 1995).

This section has highlighted a general problem with applying economics to complex environmental problems; namely, that due to this complexity, much uncertainty surrounds the effects of environmental changes. Economics has developed a theory of modelling change under uncertainty, but this tends to be dominated by situations whereby, at the very least, all possible outcomes can be described, and more usually—in addition—where a probability distribution can be attached to this set of possible outcomes. Where the set of possible outcomes is not known, then much of this analysis is no longer appropriate, and economics struggles to say much which is precise. Under this type of fundamental uncertainty, use of the precautionary principle has been suggested; that is, to act in a very risk-averse manner in the face of possible irreversible environmental damage.

CHAPTER 12

Conservation education

Susan K. Jacobson and Mallory D. McDuff

12.1 Introduction

Education programmes have contributed to the conservation of wildlife and wild areas around the world. Yet, increasing human population and natural resource consumption continue to encroach on tropical and temperate forests, savannas, wetlands and marine environments. Future conservation of species and ecosystems will depend upon active public support and participation. Implementing successful education programmes towards this end remains a critical aspect of wildlife conservation and resource management (Fig. 12.1).

12.2 Goals of conservation education

Education at all levels must be designed to enable people to understand the interrelationships between humans and the environment. This is essential if viable solutions are to be found to challenges facing the conservation of wildlife and ecosystems. The clarion call by members of the World Commission on Environment and Development for a 'vast campaign of education, debate, and public participation ... if sustainable human progress is to be made' continues to echo at meetings such as the 1992 United Nations Conference on Environment and Development. Although conservation education alone will not solve environmental problems, effective educational programmes are a prerequisite for better natural resource management, and ultimately for safeguarding the biosphere upon which we all depend.

The goals of conservation education are many and usually include the following:
• Increasing public knowledge and consequent support for the development of appropriate environmental management and conservation policies.
• Fostering a conservation ethic that will enable people to responsibly steward natural resources.
• Altering patterns of natural resource consumption.
• Enhancing technical capabilities of natural resource managers.

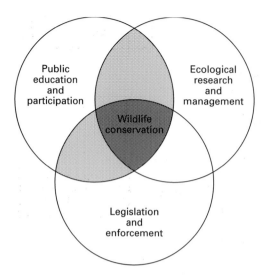

Fig. 12.1 Effective wildlife conservation programmes rely on public education and participation, ecological research and management, and appropriate legislation and enforcement.

• Incorporating resource management concerns into private sector and government policy-making processes.

Education is an essential management tool that recognizes the central role of people in all nature conservation efforts. Indeed, although a conservation goal may be focused on a biological problem, future conservation strategies will more likely focus on communication or education programmes designed to affect people's awareness, attitudes or behaviours towards natural resources. Most conservation education programmes aim to influence long-term behaviour, a goal of great complexity. Debate rages among educators regarding the relationship between environmental knowledge, attitudes and behaviours. Increased awareness of a conservation problem does not guarantee meaningful behavioural changes in support of conservation. Conservation educators initially suggested that the learning process necessary for conservation action passes from ignorance, to awareness, appreciation, understanding, concern and finally to action (e.g. Henderson 1984).

Other researchers have found that many factors affect environmentally responsible behaviour. One model includes cognitive factors, such as knowledge of environmental issues and knowledge of action strategies; action skills; and several personality factors, for example the degree of responsibility and commitment felt towards the environment, one's attitudes towards the environment, and the perception of one's ability to affect change (Hines *et al.* 1986/87) (Fig. 12.2). These factors influence a person's desire and ability to act in accordance with a stewardship ethic. Conservation education programmes must influence not only their audience's knowledge about the environment, but also beliefs and behaviours that can promote environmentally responsible actions in the future. It follows that techniques for conservation education must be multifaceted to influence cognitive, affective, behavioural and skills-related domains.

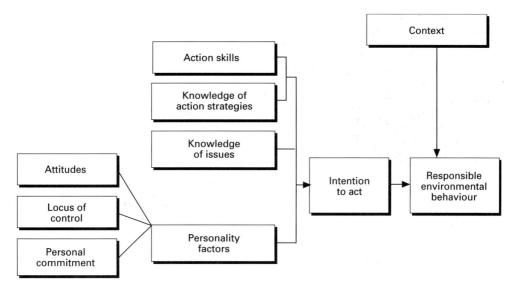

Fig. 12.2 A proposed model of responsible environmental behaviour. (After Hines *et al.* 1986/87.)

12.3 Historical context

The growing awareness of the importance of integrating educational strategies with conservation goals may evoke a false perception of conservation education as a relatively recent phenomenon. The diverse instructional strategies advocated by conservation educators, however, have their roots throughout various historical periods. The lives of early humans, for example, were connected inextricably to nature. These close interactions with the environment served as the primary vehicle for learning about the world. Educational methods used in conservation education also are linked to Greek philosophers such as Socrates (469–399 BC) who advanced inquiry and experiential learning as a path towards knowledge, as well as Plato's (429–348 BC) emphasis on the efficacy of learning by doing. The Romans also contributed to this foundation, through Quintilian's (40–118AD) theory that components of the curriculum should be interrelated, with information presented at the appropriate stage of intellectual development (Freeberg & Taylor 1961). In the formal educational sector such ideas find their manifestation today in support for infusing conservation education throughout a school's curriculum.

Later educational theorists shaped the direction of conservation education—as evidenced by the influence of John Amos Comenius—an early proponent of nature education and interpretation. In the 1600s, Comenius promoted the importance of sensory learning and used a garden as a primary tool for instruction. Referring to sensory learning as the 'golden rule of teaching', Comenius stated: 'It is when things have been grasped by the senses that language should fulfil its function by explaining it still further ...' (Freeberg & Taylor 1961).

Expeditions of naturalists in the 1700s and 1800s increased public awareness of natural history interpretation. From this interest emerged informative expeditions led by nature guides, such as Enos Mills who conducted interpretive walks up Long's Peak in 1889, in the area now known as the Rocky Mountain National Park. Education was promoted within the US National Park Service by the establishment of the National Park Education Committee in 1918. Two years later, the first official nature guides were hired to lead interpretive programmes for the Yosemite National Park in California (Sharpe 1982) (Fig. 12.3).

The importance of conservation education in schools was highlighted in the 1950s, in a report by the National Education Association called *Conservation Education in American Schools* (Freeberg & Taylor 1961). Ironically, this document identified obstacles faced in the formal sector that exist today, such as the reliance on textbooks for learning rather than hands-on experience, and the lack of appropriate training for teachers. This historical context reminds us that challenges in conservation education faced today by teachers in most countries are not a recent phenomenon. Likewise, a broad frame of reference guided by history provides a contextual basis for strategies that have been successful in conservation education, as reflected by the programmes described in this chapter.

In the past three decades, conservation education has continued to diversify and expand. Research on effective techniques, suitable approaches for specific audiences and appropriate methods of evaluation have helped to direct both the science and the art of conservation education. Programme designers now realize that the 'general public' does not exist. Appropriate activities must be developed for specific target audiences. For example, an

Fig. 12.3 An interpretive ranger presenting an educational talk at Rocky Mountain National Park. (Photo by S.K. Jacobson.)

effective educational intervention for hunters may be inappropriate for other groups such as homemakers or politicians. Even among these groups, subcategories based on income, education, gender, etc. all influence the needs of each audience. An understanding of target audiences and their direct involvement in all aspects of programme development is one key to successful programmes.

Effective programmes that guide individuals beyond a general awareness of conservation problems to the commitment, skills and action that eventually will solve these problems have been implemented in a diversity of settings with a variety of audiences. These range from cognitive and problem-orientated programmes in schools, to activities addressing environmental values and attitudes in communities and better technical training for resource professionals. This chapter focuses on informal programmes using parks, nature centres, community organizations, schools, resource agencies and the media to reach both youth and adult audiences. These programmes primarily are instigated in response to local needs and with local participation, where solutions to the challenges of biological conservation and sustainable resource use often can be found.

The programmes described in this chapter demonstrate the extent to which education can lead to the conservation of wildlife and sustainable environmental management. The following examples encompass education for: conserving natural areas, protecting declining species, involving local communities and programming for schools. These innovative programmes enhance our understanding of integrated conservation initiatives that incorporate educational strategies through a systematic approach to programme planning and evaluation. The examples show how education programmes succeed in achieving conservation goals and demonstrate that comprehensive efforts are necessary and worthwhile to help conserve wildlife and manage natural resources. The challenge is to expand these types of programmes so that they reach all people in all places.

12.4 Conservation education for parks: examples from Malaysia and Brazil

Education programmes associated with the conservation of natural areas can potentially have a large impact on natural resource management. Governments have established more than 3500 reserve areas in over 120 countries. Since 1970, nations have expanded the extent of protected areas by more than 80%; nearly two-thirds of these are in developing countries. Audiences for education programmes associated with protected areas potentially span almost all ages, backgrounds and cultures.

Two examples of conservation education initiatives created for national parks in developing countries involve programmes in Malaysian Borneo and eastern Brazil (Fig. 12.4). These programmes used a comprehensive model to guide the development, implementation and evaluation of the

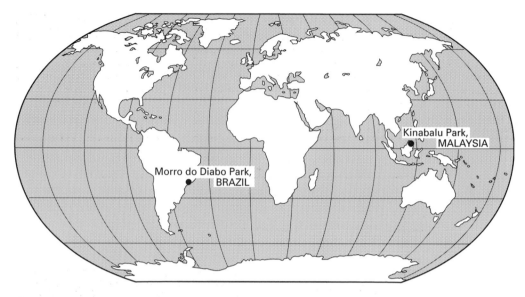

Fig. 12.4 Location of programmes in Sabah, east Malaysia and eastern Brazil.

programme (Jacobson 1991). The model provides feedback from the initial stages of planning through to completion of the programme, allowing for continuous improvement of instructional materials and strategies (Fig. 12.5). This model for the design of conservation education programmes focuses on three phases of programme activity: planning, implementation and product evaluation.

Both programmes used the resources of national parks to target local primary schools (Jacobson & Padua 1995). Kinabalu Park in Malaysian Borneo not only protects a great diversity of plants and animals, from 1500 species of orchid to clouded leopards *Neofelis nebulosa*, but also encompasses the tallest mountain in South East Asia. The programme for Brazil's Morro do Diabo Park protects some of the last remaining Atlantic forests of interior Brazil, home to endangered species such as the black lion tamarin *Leontopithecus chrysopygus* (Fig. 12.6).

The initial stages of planning for both programmes involved identifying the needs of the parks, teachers and students. Needs assessment in the planning phase included an analysis of the perspectives of all stakeholder groups involved in the programme. This included techniques such as interviews with students, parents, teachers, administrators and resource managers, as well as analysis of current educational programmes and government mandates involving education. The objectives of the two programmes emerged from the assessment of needs. Three goals guided the programmes: (i) to familiarize students with the parks and their benefits; (ii) to introduce students to basic ecological principles through outdoor experiences; and (iii) to enhance the interests of students in the environment. Measurable objectives were developed to facilitate later

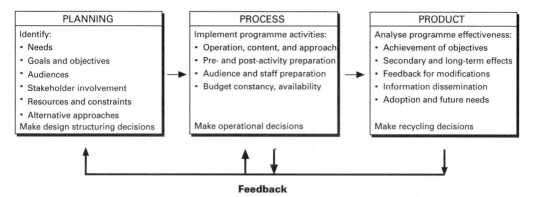

PLANNING	PROCESS	PRODUCT
Identify: • Needs • Goals and objectives • Audiences • Stakeholder involvement • Resources and constraints • Alternative approaches Make design structuring decisions	Implement programme activities: • Operation, content, and approach • Pre- and post-activity preparation • Audience and staff preparation • Budget constancy, availability Make operational decisions	Analyse programme effectiveness: • Achievement of objectives • Secondary and long-term effects • Feedback for modifications • Information dissemination • Adoption and future needs Make recycling decisions

Feedback

Fig. 12.5 Systems model for planning, implementation and evaluation of environmental education programmes. (After Jacobson 1991.)

Fig. 12.6 A school group visits the Morro do Diabo Park on the programme bus which displays the environmental education logo. (Photo by S. Padua.)

evaluation of the programmes' efficacy. After completing the programme, for example, students were expected to be able to describe the ecology of common plants and animals in the park, and to continue activities that promoted conservation in their home and school. Other tasks essential to the planning phase involved identifying the resources, constraints and target audiences for the programmes.

The implementation phase for both programmes involved continuous assessment of the instructional materials and methods as the educational programmes developed. At Kinabalu Park, six school groups participated in a pilot programme (Fig. 12.7). This revealed the need for follow-up

Fig. 12.7 Students explore the montane vegetation of Kinabalu Park. (Photo by S.K. Jacobson.)

activities. A component was added to reinforce student learning and to transfer information absorbed during the park visit to family members through fun activities to complete at home. Both programmes began with a slide presentation, followed by small group activities along nature trails. For the Brazilian programme, follow-up activities involved adult community members through art exhibitions, music festivals and drama. Baseline data were collected through questionnaires and tests used to survey students' opinions and knowledge levels.

The final product evaluation involved assessment of the programmes using a variety of techniques. Students and teachers provided feedback about the programme through written evaluations. Programme planners measured the impact of the programmes by monitoring changes in knowledge and attitudes about the park and conservation, based on formal baseline and post-programme tests. Other measures of assessment included documenting the percentage of students who succeeded in completing voluntary follow-up activities such as a 'Junior Ranger' list of assignments to pursue at home and school. To justify the budget, the staff also kept records of the number of schools and students involved in the programmes. The park administrations used this information from the evaluation as a basis for deciding whether to continue the programmes, which are still in operation.

The results of these programmes expanded beyond the original objectives and included enhanced community involvement in the parks. For example, community members lobbied the government to remove a dump site from Morro do Diabo Park and rescued the park when a fire threatened its forest. The programmes at Kinabalu Park and Morro do Diabo Park present concrete examples of conservation education programmes built through successful partnerships between schools and parks. The evaluation model provides a practical approach for the planning, implementation and evaluation of conservation education programmes that can enhance the conservation of these protected areas, and serve as a general model for designing such programmes.

12.5 Conservation education addressing threatened species: reversing population declines in Canadian seabirds

As we are confronted with the extinction of some 17 000 species annually (Wilson 1988a), educational programmes addressing declining species are especially critical. Populations of nesting seabirds on the north shore of the Gulf of St Lawrence in Canada have been the focus of a seabird management plan that integrates education into the management agenda. During a 15-year period, the Quebec–Labrador Foundation (QLF), in collaboration with the Canadian Wildlife Service, designed, implemented and tested this education and management programme in order to achieve their goal of stabilizing the seabird populations (Blanchard 1995). Due to varying factors of exploitation and management, populations of seabirds, such as the common murre, also known as the guillemot *Uria aalge*, razorbill *Alca torda* and Atlantic puffin *Fratercula artica*, have experienced dramatic fluctuations over the past two centuries. This project provides documented success of an educational programme's influence on decreasing human predation on seabirds and eggs, as well as on changing the knowledge, attitudes and behaviour of local residents. The QLF programme also presents substantial evidence of the importance of collecting baseline sociodemographic data on target audiences and involving them in the programme, as well as collecting data to evaluate the effects of the educational intervention.

The population declines of seabirds precipitated by human exploitation led to a research programme to review hunting practices, as well as the cultural and social impacts affecting the seabird populations. A unique aspect of this research was that its focus was not limited to a biological census of seabird species harvested by humans, but rather incorporated the human variable into the research equation. With its sociocultural emphasis, this study recognized that conservation problems are usually human problems as well. Collection of data included interviews with the heads of households in the area where the population declines were most severe. The survey results revealed a lack of knowledge among local residents concerning wildlife regulations. Responses from the residents reflected widespread illegal hunting, as well as a utilitarian perspective regarding wildlife. These findings were not surprising, given the remote location of the north shore and the role of the seabirds in supporting the semisubsistence lifestyle of the 6000 residents. While half of the residents were employed in cod fishing, subsistence activities such as egg gathering and berry picking supplemented the local economy.

Response to the plight of the seabirds came in 1978 with the collaborative efforts of the QLF and the Canadian Wildlife Service to develop a seabird management plan with a local educational component. Essential to the success of this project was the image of the QLF as a non-governmental organization that had provided community services to residents in the past. Through affiliation with this association, programme planners were able to distance themselves from the regulatory divisions of the government. The goal of restoring the seabird populations while maintaining respect for the local culture resulted in several objectives (Blanchard 1995):
• Increasing the population levels of nesting seabirds.
• Improving local public knowledge, attitudes and behaviour towards seabirds.
• Building avenues for greater local participation in seabird management.
    The bottom line for success of the management plan depended on local participation, rather than outside enforcement of regulations.

The educational programme encompassed four primary strategies: instruction for youths, training in leadership skills, information diss-emination and institutional support. A key activity was a 4-day programme for young people at the St Mary's Islands Seabird Sanctuary which provided local children with hands-on experiences in seabird ecology and wildlife law. The instructors for the programme were local adults and university students who had received training in experiential learning. The objective of the programme at St Mary's was not to convert children to conservation through classroom lectures, but rather to provide fun opportunities to learn about seabirds and the effects of humans on their populations. Other youth activities included theatre productions about seabirds, the creation of conservation clubs and the development of instructional materials about seabirds for use in schools.

Training of local residents was a critical factor designed to enhance and sustain local participation. More than 50 local volunteers and staff received training in instructional methods and field ornithology. The educational materials developed by the project were both practical and relevant for residents, and used mass media such as Canadian radio and television. Some of these materials included a seabird identification poster, a calendar, a nine-part radio series, a film documentary, a citizen's guide to seabird regulations and a school newsletter. Strategies for building support for the programme included study tours for leaders from national and provincial conservation groups.

By 1988, the QLF documented success in achieving the goals of both the management plan and the educational programme. Increases in the populations of seabirds were recorded from 1977 to 1988. In addition, a follow-up survey revealed improvements in residents' knowledge, attitudes and behaviour concerning seabirds. The proportion of heads of households who correctly stated the legal status of selected species rose from 47% of respondents in 1981 to 64% in 1988. While local residents continued to believe that birds should be harvested, the survey results showed a shift in their perspectives to much lower levels of harvest needed for supplemental subsistence of families (Blanchard 1995).

Qualitative measures provided evidence of increased local involvement in conservation, such as increased memberships in conservation clubs, a waiting list for the St Mary's Island Youth Programme, increased local conservation activism and local grant requests for environmental activities. As reflected in this programme and the one described below, sustainable solutions to environmental issues will succeed only on a foundation of local participation.

12.6 Community conservation education: integrated approaches at the Community Baboon Sanctuary, Belize

Education programmes involving community groups are often multifaceted in order to address complex issues associated with conservation and economic development. A community-based conservation programme was developed in Belize that aims to integrate conservation with development. Conservation of tropical forests in national parks and protected areas must be augmented with management efforts geared towards privately owned land. The multilevel conservation and education programme instituted through the Community Baboon Sanctuary (CBS) in Belize integrates private land management with a conservation education and ecotourism programme (Horwich & Lyon 1995). Working with local farmers, the CBS was started in 1985 and extends along a 30 km stretch of the Belize River in central Belize. The primary goal of the project is the protection of the threatened black howler monkey (*Alouatta pigra*), known locally as a 'baboon'. Protection of the monkey's habitat—the semideciduous tropical forest—is the key to protection of the species. To protect this habitat, the CBS targets private land used for subsistence agriculture through a framework of voluntary land stewardship, conservation education and participatory natural resource management.

The challenge for this project was to develop a conservation programme that would integrate the culture of the local farming community, as well as recognizing the constraints placed on local Creole farmers such as limited labour and capital. Ignoring the daily limitations and needs faced by the farmers would have defeated the project. Before planning began, extensive research was conducted with the aid of local people on farming techniques, forest succession and howler monkey feeding ecology in the area (Horwich & Lyon 1995). While many programmes aim to incorporate local participation in daily project activities, few projects actually integrate participatory control into the initial planning process. Programme planners working at the CBS, however, yielded control to the local community and gave veto power to the landowners themselves. Initially, the idea behind the CBS was presented by a local liaison through discussions with landowners and the local political bodies—the Village Councils. Further discussions in the community emphasized that land management plans would be developed through collaboration with each farmer or family. Membership in the sanctuary would involve signing a voluntary agreement to support the management plan. Farmers could withdraw from their pledge at any time with no consequences.

The primary management objective was to establish a continuous forest corridor along the river, thus providing the howlers with adequate forage from the mosaic of successional forest patches. The protection of this riparian land also would improve agricultural land use by decreasing soil erosion along the riverbank. Communicating these benefits to farmers, however, required extensive involvement in the local community in order to cross cultural boundaries. Living within the central village of Bermudian Landing and sharing daily life seemed to engender support from the local people for the researchers spearheading the project. The first formal step in planning involved circulating a petition among the villagers that would allow the researchers to begin investigating this experimental programme. After the villagers signed the petition, physical mapping of the land began with the written support of the community.

Using aerial photographs, survey records and natural landmarks, land ownership maps were developed in consultation with 12 landowners. This information allowed for the development and presentation of 12 individual management plans, followed by the voluntary signing of a pledge to uphold the plan. With the formal and unanimous approval of the CBS given at an annual village meeting, a subsequent expansion included additional landowners in seven neighbouring villages. The key to these management plans was their inherent flexibility and the involvement of the landowners, as it was recognized that sustainability of the management plans would require continual feedback from the changing resources and patterns of land use (Horwich & Lyon 1995).

The CBS was created initially without a formal administrative structure or funding from outside conservation agencies. As the project expanded, a management structure was developed in 1987 coordinated by a Belizean non-governmental organization with a CBS advisory board of eight elected members from the eight participating villages. The core of the programme remains the voluntary participation of the landowners. With more than 120 CBS members to date, no landowner has ever separated from the CBS. Informal conservation education activities, primarily employing word of mouth, are largely responsible for this success.

The educational agenda of the CBS includes four aspects: education for landowners, Belizean school education, Belizean adult education and ecotourist education. The focal point for the educational efforts is the CBS museum, built using local materials and labour to house exhibits highlighting local conservation and cultural themes (Figs 12.8 & 12.9).

Fig. 12.8 The locally developed CBS museum serves as an educational facility for classes and tourists. (Photo by S.K. Jacobson.)

Fig. 12.9 The sanctuary manager, Fallet Young, guides visitors through the museum. (Photo by S.K. Jacobson.)

Another educational product is the textbook, *A Belizean Rain Forest: the Community Baboon Sanctuary* (Horwich & Lyon 1993) which presents information on the natural and cultural history of the area, the CBS project and a guide to the interpretive trail. The text is used extensively in the Belizean schools, as well as by staff members and tourists to the sanctuary.

The impact of this model of community-based conservation has brought national and international attention to the CBS and the communities involved in the project. As a result, other areas in Belize are testing the model. The landowners and villagers living in the sanctuary are responsible for the success of this project and local control in both the planning and the evolving management structure remains the critical foundation for its long-term sustainability, furthering conservation education and land management in the area.

12.7 Conservation education in the schools: the Global Rivers Environmental Education Network

Most environmental attitudes are formed during childhood, thus school children are an important target for conservation education (Fig. 12.10). Conservation education in the formal school system varies among countries in content, scope and disciplinary base, and often is overlooked or ignored relative to traditional subjects. Extracurricular materials and curricular supplements provide the main exposure for students to environmental conservation knowledge and skill formation. The following example demonstrates an ideal approach—hands-on, participatory and interdisciplinary— to involve school children in the complexities of environmental problem solving.

A challenge for conservation educators working in schools is finding instructional strategies that can enable students not only to learn about local and global conservation issues, but also to act in response to environmental problems. To be effective, conservation education cannot be confined to the boundaries of a single classroom, but rather must operate within the context of the environment it seeks to conserve. One programme that has met this challenge is the Global Rivers Environmental Education Network (GREEN), developed by the University of Michigan (Stapp *et al.* 1995). The GREEN project is a water-quality monitoring programme that has been adopted in more than three dozen countries. Through participation in the programme, students explore their local rivers, present their findings to government officials, and exchange data and insights with students in other cultures throughout the world.

Fig. 12.10 Children are a critical target audience for conservation education programmes. (Photo by S.K. Jacobson.)

GREEN has developed into a global communication connection that includes countries as culturally diverse as Bangladesh, Argentina, Kenya and the USA. While the countries differ culturally, many face similar conservation issues concerning the pollution and degradation of aquatic diversity. Through the common link of rivers and watersheds, GREEN seeks to build problem-solving skills among students and increase awareness of similar conservation issues in other countries. GREEN thus works towards three interrelated goals (Stapp *et al.* 1995).

1 To provide students with hands-on experience in chemical, biological and sociological research, by introducing them to the environmental issues of their local watershed.

2 To empower students in problem-solving skills applied to the local environment.

3 To build global and intercultural communication, as well as understanding of cultural differences relevant to environmental concerns.

The evolution of GREEN began in a Michigan high school biology class on the banks of the Huron River in 1984. Student concern about the water quality of the river grew when several windsurfers, including a student at Huron High School, contracted hepatitis A after windsurfing on the river. The students' concern and subsequent testing of the water by the teacher prompted a class at the University of Michigan to develop a 2-week model water-quality monitoring programme appropriate for secondary students. The programme included instructional materials such as maps of the local watershed, a manual outlining standards for performing the nine water-quality tests, material on monitoring water for macroinvertebrates, a slide–tape presentation and a set of water-quality testing kits. The 2-week programme followed this basic schedule (Stapp *et al.* 1995).

1 Discuss water-quality concerns about the river; view slide show (lesson 1).

2 Learn the nine water test parameters (lessons 2 and 3).

3 Monitor the water's quality (lesson 4).

4 Calculate data; interpret the results; derive the water-quality index; incorporate the results into an understanding of the use of the river by humans and animals (lesson 5).

5 Write an action plan and take action (lessons 6–10).

In the Huron River, students found alarmingly high faecal coliform counts of over 2000 colonies per 100 ml of water (counts should not exceed 200 for activities such as windsurfing and swimming). Actions taken by the students resulted in a city-wide evaluation of the underground storm drains which revealed incorrect drain connections as the source of the high bacteria levels. Enthusiasm for the programme from this initial school resulted in the expansion of GREEN to two other high schools on the Huron River. The three high school classes gathered at a congress to share their data and recommendations at the end of the monitoring period, and a model for the programme was published in the first edition of the *Field Manual for*

Water-Quality Monitoring: an Environmental Education Programme for Schools (Mitchell & Stapp 1992). The programme's implementation in other schools resulted in several additions, such as an interactive computer program, the creation of Hypercard packets on watersheds, and the development of international, cross-cultural partnerships.

From these beginnings, the scope of GREEN grew to include an international communication network, allowing students to share their experiences and findings across geographical and cultural boundaries. The University of Michigan faculty and students began planning international workshops in 1989, so that educators from different countries could meet to exchange ideas on watershed programmes appropriate in different geographical areas around the world. Twenty-two workshops were held in 18 countries in Africa, Latin America, Eastern and Western Europe, the Middle East and Australia. As a result of the workshops, some national governments allocated funding to develop programmes and obtain equipment, and committees were formed to prepare curriculum guides for water monitoring. The workshops also revealed that the model developed in Michigan was not appropriate for all countries, as access to both resources and equipment varied dramatically across the globe. Given this constraint, low technology systems based on a biological index have been developed for monitoring in countries without access to the field kits.

Currently, the *Field Manual for Water-Quality Monitoring* has been translated into Bengali, Chinese, Czech, German, Hebrew, Hungarian, Italian, Japanese, Russian and Spanish. With this international interest, the next phase of the development of GREEN focused on the creation of a worldwide network and dissemination of resources for participating schools. The network distributes a semiannual, international newsletter to 2100 educators and government officials in 125 countries. For countries with access to computers, a series of GREEN International Computer Conferences has been established, providing an interactive computer network with an international data base of water-quality data, as well as a source for exchange of ideas. The network now has more than 3000 participants from more than 80 countries. To link classrooms across countries, GREEN has established the Partner Watershed Programme, which matches schools in different countries for communication via mail or computer concerning their experiences with water-quality monitoring.

Based on the enthusiasm for this programme, GREEN participants seem eager to act on real-world environmental problems and to integrate this local action into a global understanding of problems faced by people of other nations. Such dissolution of cultural boundaries serves not only as a model for an effective conservation education curriculum, but also a template for solving global conservation problems in the future.

12.8 Constraints and challenges

Conservation education is needed at many levels and in many forms. Ideally, it should be a life-long process. As these examples demonstrate, a multitude of innovative conservation education programmes can be found around the world. The regions and activities described in this chapter offer insight into the complexity involved in the design of effective conservation education. The interdisciplinary and integrated solutions shared by the programmes provide a broad array of approaches for tackling conservation problems in the century ahead. Key elements that have led to the successful planning, implementation and evaluation of the programmes are outlined in Table 12.1. Factors that impact on programme success include variables such as local participation, clear goals and objectives, the use of existing groups and organizations, the relevance of objectives to the target population, and programme evaluation.

Table 12.1 Elements leading to successful conservation education programmes. (Modified from Jacobson 1995.)

Planning
Have clear goals
Identify measurable objectives
Incorporate an interdisciplinary approach
Determine target audiences and involve them in programme design
Assess participants' social/educational/economic backgrounds
Ensure programme relevance to local populations
Build necessary support (government/community/industry)
Develop a budget and organizational plan
Plan for potential problems and resolution of conflicts

Process
Use existing organizations and groups
Encourage active/voluntary participation
Involve reluctant participants creatively
Be sensitive to the audience
Provide direct contact with the environment/resource
Use key ecosystems/resources/species in programmes
Select appropriate educational media
Focus on economic/cultural values
Provide conservation incentives
Maintain informality/entertainment value of the programme
Be flexible

Product
Evaluate programme components/monitor programme
Use more than one method of evaluation
Collect feedback for programme modification/creation of new programmes
Enhance local control and support of programme
Develop specific long-term plans for sustainability
Disseminate programme results

In a review of documents describing conservation education programmes implemented in tropical countries between 1975 and 1990, Norris & Jacobson (in press) found that the incorporation of systematic evaluation techniques was more strongly correlated with programme success than other programme variables. Unfortunately, evaluation still is not a common component in the design of conservation education programmes due to a perceived lack of time, money or expertise by project staff. Yet the results of evaluation provide accountability for modifying ineffective programmes and expanding educational efforts when success is demonstrated in promoting environmental conservation. Without substantiation of successful strategies, conservation education programmes face the risk of continuing to operate at the periphery of global environmental efforts.

Despite the success stories related in this chapter, conservation education around the world still lacks widespread informal methods for reaching youth and adult audiences. A survey of 13–15-year-old students in the UK, Australia, the USA and Israel revealed that mass media—radio, television and the press—were students' most important sources of information on environmental issues (Blum 1987). Yet conservation educators typically rely on print media and written curriculum supplements (Archie *et al.* 1993). Incorporating advanced mass media tools and innovative advertising techniques into conservation education programmes will lead to increased public awareness of the conservation message. Likewise, more widespread programmes incorporating participatory, problem-solving techniques into project activities will develop and enhance pro-conservation attitudes and behaviours.

Conservation education also has yet to be institutionalized into the formal educational system in most countries. Many constraints limit the efficacy of conservation education in schools. In addition to lack of time in an already crowded curriculum and lack of funding and materials, many teachers are unsure of the scope of conservation education or view it as narrowly pertaining to the sciences. Thus, opportunities for interdisciplinary and holistic approaches are lost. Most teachers receive no training in the content or pedagogy of conservation education, and experiential or action-orientated approaches needed for successful conservation education are not developed. Lastly, many educators do not think conservation education is a priority, and issues such as the sustainable use of natural resources or maintenance of biodiversity simply are not infused into the curriculum. The challenge lies in developing an efficient system to provide educators with the foundation and experience to be effective teachers of conservation concepts.

The examples in this chapter reflect the central role possible for educational programmes in the conservation agenda. Building on past effective instructional strategies remains critical for the growth of future and existing educational programmes. As conservation education programming becomes more systematic and better integrated into conservation strategies, increased public awareness and concern should spark the needed motivation for preserving the planet's wild species and the ecosystems that support us all.

Acknowledgements

Material in this chapter has been adapted from *Conserving Wildlife: International Education and Communication Approaches* by Susan K. Jacobson (ed.) (1995). We thank L. Cronin Jones, S. Marynowski and J. Sanderson for insightful reviews of this chapter.

CHAPTER 13
Conservation policy and politics

Graham Wynne

13.1 Introduction

The fate of the environment is determined by the actions of individuals, local communities and businesses. Much more often than not, however, these actions are influenced by the policies of government and sometimes by the policies of non-governmental institutions. Conservation therefore has to address both people's underlying attitudes to the environment and the legal and other policy instruments that influence their behaviour.

This chapter concentrates on the part played by government policy. It provides an overview of how various elements of conservation policy fit together; some of the improvements that are needed; and how these can be promoted. The model described is inevitably partial and highly simplified.

Some may question the value of concentrating on what governments do. As trade, communications and culture become increasingly global, the autonomy of the individual nation state diminishes. At the other end of the scale, devolution and local empowerment are now real forces in some parts of the world, even if the practice falls short of the rhetoric. Against this backcloth, there is no doubt that too little attention is given to the roles of both international business and local communities, but for better or worse, in environmental matters, national governments remain as very powerful players (Brenton 1994).

13.1.1 The policy challenge

In most countries there is a body of policy whose primary purpose is the protection of the environment. It was assumed until quite recently that most farming, fishing, etc. was essentially benign and that the 'wider countryside' and the marine environment were sufficiently robust to look after themselves. For biodiversity conservation, traditional concerns have therefore been the protection of rare species and regulation of hunting practices; and the protection of areas of special habitat.

There is now widespread awareness that in isolation these activities are wholly inadequate. It is no longer simply rare species and special habitats that are threatened. Much widespread and relatively abundant wildlife is in decline, sometimes rapid decline, and the conservation value of huge tracts of land and sea is deteriorating. The immediate causes are well documented: destruction of native forests; intensification and specialization of farming practices; overexploitation of resources; introduction of exotic species; pollution of rivers and inshore marine waters; drainage of wetlands; habitat fragmentation and destruction by roads and development; and atmospheric pollution. The unpredictable effects of climate change loom on the horizon. And even those areas that are protected as nature reserves are vulnerable to the effects of air and water pollution and changes to the water table that originate from beyond their boundaries.

Underlying the rapid deterioration in biodiversity are the pressures arising from population growth and development (see, for example, McNeely 1995). There are, however, crucial links in the chain which are not so easily summarized. They include a wide array of the social and economic measures that governments (and others) adopt to regulate society. Thus many countries have economic policies and laws to influence patterns of land use and the way farming, forestry and fishing are carried out and these have both intended and unpredicted impacts on the environment. Further complexity is added by energy, transport, trade, aid and development policies, the effects of which are even harder to determine. All of these measures can operate at local, national and international levels (Fig. 13.1).

The greening of government policy can be seen as twin exercises which have to be dovetailed together. The *first task* is to strengthen and increase in scope those policies which are specifically concerned with the environment:
• In only a few countries are even the most important wildlife habitats adequately protected. In some of the richer parts of the world, the rate of habitat destruction has decreased substantially in recent decades, but

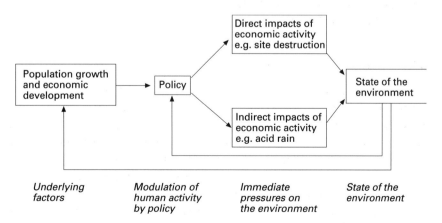

Fig. 13.1 A simple flow diagram to show the role of 'policy' in determining the state of the environment.

damage is still widespread, e.g. through inappropriate grazing regimes. In other parts of the world, destruction of natural and seminatural habitats continues apace.

• Levels of air and/or water pollution that are inimical to many species characterize parts of all countries.

The *second task* is to reform 'mainstream' social and economic policy in order that the negative impacts are greatly reduced and, as far as possible, rendered benign. In some cases, it is possible to go further and achieve both economic and environmental objectives through the same measure(s)—the holy grail of sustainable development.

Of course, little of this is possible unless there is strong public support for conservation—the greatest challenge of all.

13.1.2 Institutions for conservation

Many governments now pay lip service to the need to put the environment at the heart of all decision making. No government has yet evolved an institutional framework which enables this to happen. The first obvious step—the establishment of an environment ministry or agency—has been taken by over 100 nations, compared with no more than 10 at the time of the 1972 Stockholm Conference (Brenton 1994), but this spawns its own problems. Biodiversity and other conservation issues then come to be seen as the sole responsibility of one arm of government, conveniently ignoring the fact that it is underlying economic policy and the policies of the agriculture, forestry and transport ministries which drive much environmental destruction. Initiatives to weave conservation throughout government include cross-departmental strategies, the nomination of 'green' ministers in each department, and the drawing up of environmental accounts to run parallel with traditional financial accounts. All are at an experimental stage.

13.1.3 International milestones

The Stockholm Conference

The United Nations Conference on the Human Environment, held in Stockholm in 1972, was the first major attempt to involve nations in a global response to environmental problems which now clearly transcended political boundaries. The USSR and the East Europeans did not participate and the agenda was heavily driven by northern nations. Current fears amongst developing countries that, through global environmental measures, the north would impose new constraints on their development prospects, were very much to the fore. When the Ivory Coast representative stated that his country would like more pollution problems provided this was evidence of industrialization, he probably gave expression to the thoughts of many. The conference reached agreement around the notion that development and the environment could be managed to the advantage of both, but gave little indication of how (Adams 1990).

However, the Stockholm Conference did succeed in placing environmental problems, particularly pollution, on the international political agenda (Reid 1995) and led to the establishment of the United Nations Environment Programme (UNEP).

The Earth Summit

The United Nations Conference on Environment and Development (UNCED or the Earth Summit) was held in Rio de Janeiro in 1992 on the twentieth anniversary of the Stockholm Conference. As then, the most pervasive feature was the 'division between rich and poor, developed and developing, North and South' (Grubb *et al.* 1993). The formal outputs of the summit (also as at Stockholm) were significantly weaker than had been hoped for, causing Maurice Strong, the Secretary General of UNCED, to refer in his closing remarks to 'agreement without sufficient commitment'. Despite this, the Earth Summit does mark a watershed for conservation. It was the first occasion on which governments worldwide treated biodiversity as a major concern. The inextricable links between the development process and the state of the environment were acknowledged collectively and with more conviction than before—the products of the summit make it abundantly clear that conservation cannot be successfully pursued as a discrete activity. Governments gave a wide range of undertakings that if acted upon would begin to resolve some of the conflicts between economic growth and environmental protection.

The key products of the Earth Summit (Grubb *et al.* 1993) were:
• The Rio Declaration on Environment and Development (the Rio Declaration), which sets out 27 general principles for achieving sustainable development.
• Agenda 21, an extensive programme of actions for the next century to support the Rio Declaration and integrate environmental concerns into a broad range of activities. Non-binding recommendations are addressed to international agencies, national and local governments and non-governmental organizations (NGOs).
• The Convention on Biological Diversity, which is a legally binding agreement signed in Rio by 155 nation states (and the European Union) to conserve biological diversity and bring about sustainable use of resources.
• A Statement of Principles for the sustainable management of forests.
• The Convention on Climate Change, which commits all ratifying countries to prepare national programmes to contain greenhouse gas emissions, and to return emissions of carbon dioxide and other greenhouse gases to 1990 levels by the year 2000.

Progress on implementation of Agenda 21 was reviewed at the United

Nations General Assembly Special Session (UNGASS) held in New York in June 1997. The Rio commitments were reaffirmed but deep divisions between developed and developing countries, particularly over finance, prevented much progress towards new agreements, and continue to overshadow those modest practical steps that are being taken.

13.1.4 Some concepts and principles

Sustainable development

The concept of sustainable development tries to reconcile the objectives of economic development and protection and enhancement of the environment. At least 70 definitions are current (Holmberg & Sandbrook 1992), the most often quoted being '[development that] meets the needs of the present without compromising the ability of future generations to meet their own needs' from the Brundtland Commission (1987). Principle 4 of the Rio Declaration states: 'In order to achieve sustainable development, environmental protection shall constitute an integral part of the development process and cannot be considered in isolation from it.'

   In practice, there is enormous scope for interpretation and the terminology has frequently been used in support of actions which damage the environment and destroy biodiversity (Daly 1992). Many argue that *the conservation of biodiversity is a key test of sustainability*—development cannot be regarded as sustainable if it causes serious or irreversible damage to species, populations or habitats. Making this a central tenet of conservation policy does not overcome definition problems, but it significantly reduces them. The UK Biodiversity Action Plan (UK Government 1994) embraces this definition—the strength of political commitment to it remains to be seen.

Precautionary principle

The precautionary principle states that policy makers should proceed cautiously when intervening in the natural environment so as to safeguard against unexpectedly severe future costs. It explicitly recognizes the existence of uncertainty. This principle is also reflected in the Rio Declaration: 'In order to protect the environment, the precautionary principle shall be widely applied by states according to their capabilities ... lack of full scientific certainty shall not be used as a reason for postponing cost-effective measures to prevent environmental degradation' (Principle 15). Put more simply, if there is a significant element of doubt, the benefit of that doubt should rest with the environment.

Polluter pays principle

The prices of goods and services in the 'free market' frequently fail to reflect the full social costs, including those related to pollution, resource exploitation and other forms of environmental degradation. Mechanisms are needed to bring home these environmental costs to those responsible, whether producers or consumers. (This distinction is something of a nicety—in the end, the consumer pays.) Making the polluter pay encourages more efficient use of resources and reduction in pollution. The principle was adopted by the Organization for Economic Cooperation and Development (OECD) in 1972, although its origins are much earlier. It too was encompassed by the Rio Declaration (Principle 16).

13.2 Biodiversity planning

The Convention on Biological Diversity, through Article 6, calls for signatory parties to: (i) develop national strategies, plans or programmes for the conservation and sustainable use of biological diversity or adapt existing plans, to address the provisions of the convention; and (ii) to integrate biodiversity work into sectoral and cross-sectoral plans, programmes and policies. Conservation strategies are not new but, for nearly all countries, a genuinely planned and integrated approach to biodiversity conservation, with well-defined and monitored biological outputs, would be a major step forward. It is perhaps surprising that conservationists in many countries have not made better advocacy use of this Rio undertaking. Even assuming a particular arm of government or sector of the economy is keen to reform its policies 'to save biodiversity', where does it begin? How does it know what it is supposed to save and how? The forces that generate environmental destruction, often unwittingly, are complex and powerful. In the absence of a clear plan, it is hard to see how the intricate raft of actions that is essential to stem biodiversity losses can be constructed. Early experience of biodiversity planning from 17 countries around the world has been drawn together by Miller & Lanou (1995).

13.2.1 What is a biodiversity plan?

The major elements of a biodiversity plan might include:
- audit;
- objectives;
- priorities;
- measurable targets for species and habitats;
- action plans for species and habitats;
- action plans for policy sectors;
- monitoring and review arrangements.

Audit is simply an assessment of the status and trends of an area's biodiversity. A comprehensive audit of biodiversity does not exist for any part of the world but in nearly all countries some baseline survey work has been done and this, together with 'informed judgement' on trends, is a big advance on planning in an information vacuum. The systematic recording and publishing of what is known also points out where survey and monitoring is most urgently needed.

Objectives describe what a plan is trying to achieve. The most effective distinguish clearly between *ends*—what is wanted in biological terms— and *means*—how to bring it about. Conservation suffers from failure to distinguish between these two, with the result that much attention is given to political and administrative processes, often with too little regard to biological outcomes. The draft Canadian Biodiversity Strategy (1994) includes the following *ends objectives*: 'to maintain wild flora and fauna populations across Canada's diverse ecosystems, landscapes and waterscapes' and 'to restore degraded ecosystems where practical and where restoration will make a significant contribution to the conservation and sustainable use of biodiversity'.

The second of these objectives highlights an important development of recent years: growing recognition that, in the face of the inexorable loss of natural and seminatural habitats, restoration and enhancement must become an ever larger part of conservation actions (see, for example, Adams 1995).

Priorities are invariably necessary because resources are limited and time is short—action must be concentrated where it is most urgently needed.

On a national scale, an appropriate way of ranking species priorities might be as follows:
• Threatened endemic and other globally threatened species (as set out in the *Red Data Books* produced by the International Union for the Conservation of Nature and Natural Resources (IUCN)).
• Species for which a country has a high proportion of the world or appropriate biogeographical population.
• Species where numbers or range have declined rapidly.
• Species which are rare or localized.

The priority categories for habitats might be as follows:
• Habitats on which priority species depend.
• Habitats for which a country has a high proportion of the world resource.
• Habitats rapidly declining in area and/or quality.
• Habitats which are rare, i.e. of limited area.

Measurable targets for species and habitats take objectives to a level of precision which enables action to be planned and taken, and success and failure to be judged. Again, conservation has often suffered from the failure to be specific about what is wanted—this provides governments and others with one of the simplest justifications for failure to deliver. Ideally, targets should be quantitative. Where this is not possible, it is valuable to state whether the intention is an increasing, a stable or a decreasing species population/

habitat area. Targets need to be ambitious, but also realistic and open to revision as knowledge and experience are gained. No government will embrace a targeted approach unless these last criteria are satisfied.

Action plans for species and habitats set down what is to be done to meet the targets. There is little reason to suppose that conservation targets will be achieved without them. They should be produced for as many priority species and habitats as possible, be based on current scientific knowledge and contain, at least, the following:
- conservation status;
- brief analysis of the threats;
- statement of the biological objectives (targets);
- actions needed, specifying who is responsible, timescale, resources, etc.

Action plans for policy sectors describe what policy changes and actions are needed for each sector, such as agriculture, forestry and water, to meet the objectives and targets. They provide a synthesis to help make sense of the process for those charged with implementation in each policy area. Their preparation also brings out underlying principles that cut across many different areas, e.g. the need for environmental impact assessment at strategic as well as project level.

Monitoring and review is the essential feedback loop to assess progress and make changes as necessary. The process should be designed to answer the following questions:
- Are conservation targets being met?
- Are proposed actions still the right ones to meet targets for species and habitats?
- Have the priorities for action changed?

    As actions are designed to meet biological targets, species and habitats have to be monitored. The extent to which actions are carried out also needs to be assessed if their effectiveness is to be judged.

13.2.2 Criticisms of this approach

The approach described above is not dissimilar to that which has been adopted for the UK Biodiversity Action Plan and advocated by a consortium of NGOs in *Biodiversity Challenge* (Wynne *et al.* 1995). The focus on priorities, targets and action plans is not a device to protect just the most threatened species and habitats. While some of these may be saved for posterity by small-scale actions, the future of most depends on substantial policy reform—and this has the capacity to deliver widespread gains for bio-diversity as a whole. Traditionally, conservationists have often been kept at arms length over the 'big' policy issues. A demonstrably rational approach, embracing priorities and targets which are acknowledged as 'valid' by government, makes it harder for decision makers to resist engaging in constructive debate—a plan can act as a key to help access formerly impenetrable bureaucratic processes.

However, some consider that this type of planning is too demanding and unrealistic. Our knowledge of most species, habitats and ecological processes, for most parts of the world, is sketchy at best. Those seeking quick-fix solutions question whether the level of detail involved in the approach is really necessary. Many decision makers dislike processes that clarify what success and failure look like. Even more dislike agreeing to actions which run counter to short-term economic gain, and it is inevitable that a rational conservation strategy will contain such actions. Once agreement has been reached at national level, the strategy has to be implemented through the regional and local tiers of government and non-governmental institutions, so the process is inevitably tortuous.

To avoid these and other difficulties, many conservation strategies and plans confine themselves to a statement of broad descriptions and goals. Politically this is often an essential step and should not be undervalued, but it should not be confused with overcoming the obstacles that confront conservation. Most success in management of the environment has come from identifying and implementing solutions to specific problems; failure arises amongst other things from ignorance and refusal to move beyond generalities. Assembly and analysis of detailed biological and socioeconomic data, and commitment to substantial policy changes and programmes of action, are prerequisites for change on the ground. If political will for this cannot be generated at the plan stage, there seems little likelihood of it existing at the implementation stage.

Participatory approaches can ease some of the problems. If industry, land managers, research institutes and voluntary conservation organizations are involved from the outset with government, the pool of information and analytical skills is increased and the plan has a wider ownership base. Through the translation to local level and implementation, community participation is essential. Many (probably most) action plans and strategies end up being of little use because they fail to inspire, enable or support action by individuals and local communities (Holdgate 1996). The outlook for biodiversity rests heavily on the resolution of this issue. High-level plans are necessary to give the long-term view and to decide and address national and international priorities, but 'top down' and 'bottom up' approaches must be integrated.

13.3 Policy tools: influencing behaviour

Plan implementation rests heavily on persuading people to stop or minimize those actions which damage the environment and to undertake actions which have a neutral or beneficial effect. This is a tough agenda. When stripped back to the basics, the routes that society can take to bring about particular forms of behaviour are few in number. They can be broadly summarized as:

1 *Education*: information and understanding can be imparted to people so that they can make their own informed decisions.

2 *Exhortation/persuasion*: people, businesses, etc., can be asked to do some things and not to do others.

3 *Economic instruments*: sympathetic action can be encouraged with financial incentives and damaging actions can be discouraged by financial penalties.

4 *Legal instruments*: behaviour can be controlled through regulation and law.

These are the essential tools of conservation policy. They can be seen as points on a scale where the freedom of choice to damage the environment is increasingly curtailed. Thus:

Motivation through understanding	→	Persuasion through advocacy	→	Persuasion through money	→	Control by law

In practice, of course, the picture is much less clear. These approaches shade into each other and need to be employed in combination— conservation can only be achieved through a cocktail of complementary measures (McNeely 1993; OECD 1994). Education (see also Chapter 12) and exhortation are discussed very briefly below; legal and economic instruments are discussed in a little more detail in subsequent sections.

13.3.1 Education and awareness raising

It is naive to believe that people will make the right environmental response simply because they are in possession of the facts; it is even more naive to believe they can do so in the absence of those facts.

Environmental knowledge and awareness can be promoted through formal education (in schools and colleges), informal education, provision of technical advice and through information to consumers (e.g. eco-labelling). Although there has been progress in recent years, the environmental education programmes of many countries remain scanty and too divorced from other aspects of environmental planning and action (Holdgate 1996). Even more importantly, neither formal programmes nor informal initiatives are inspiring the level of commitment to wildlife and 'wild' places which is essential if conservation really is to be placed at the heart of public policy.

For the majority of audiences a broad level of understanding is sufficient. For others, such as those involved in land management, a more detailed and practical knowledge is needed. Communication of such technical advice is very often *ad hoc* and inadequate. Most farmers, for example, can justifiably claim that they do not really know what 'environmentally sensitive farming' means, because clear information is not readily available.

13.3.2 Exhortation and persuasion

Programmes promoting specific conservation actions are increasingly common. They include campaigns to dissuade people from taking wild flowers or shooting birds, to persuade consumers to conserve energy and water, and to buy timber and other products from sustainably managed sources. Links with environmental education are vital—people are more likely to respond positively if they understand the nature of the problem and the solution.

Consumer pressure acts as a quasi-economic instrument—if enough people show preference for 'greener' products, companies can be encouraged (in some cases virtually obliged) to improve their environmental performance. There can also be a market advantage for them to do so pro-actively—some people will pay a premium for goods they regard as better for the environment. However, in the USA, this premium was judged to have fallen from 6.6% in 1990 to 4.5% in 1992 (Whitehead 1992). Leaving aside questions of precision, the small scale and trend indicated by these figures is noteworthy. A common attitude remains that the 'polluter pays principle' is fine as long as the polluter is someone else, preferably 'wicked industry', and that green solutions can be found with little knock-on cost to the consumer (see, for example, Holdgate 1996).

Green consumerism is important, but is never likely to be the main driving force behind corporate environmentalism (Cairncross 1995). Companies can of their own free will do a lot (see, for example, Schmidheiny 1992). However, they will not voluntarily set themselves progressively higher environmental standards across the board—the fear (and reality) of losing out to those that ignore such standards is much too great an obstacle. The behaviour of companies, like individuals, will continue to depend heavily on the regulatory and economic frameworks imposed by government.

13.4 Legal instruments

This section highlights aspects of national and international conservation law. For a much fuller account, see de Klemm & Shine (1993). Legal measures are considered here before economic measures because this broadly reflects the sequence in which policy has evolved. Until fairly recently, direct regulation was the almost exclusive strategy to address environmental issues (OECD 1994).

13.4.1 National legislation

The most direct way for society to protect the environment is to pass laws which:

• prohibit or control the exploitation of particular species;

- dictate what activities may (or may not) be carried out in particular areas and habitats;
- control the type and level of pollutants that may be emitted to air or water. (This is not specifically considered here.)

Species conservation laws

Many countries have introduced conservation laws in recent decades to protect a wide array of endangered species, although coverage is patchy at best. Mammals and birds are usually included. Fish are usually excluded as they are dealt with under fishing legislation. Plants, reptiles and amphibians are variably treated. Invertebrates are frequently ignored and microorganisms are always ignored. Protection is often limited to restrictions on the taking of and trade in listed species and even then, for non-game species, enforcement can be difficult.

Species-based legislation can go far beyond protection against taking. In Australia, the Commonwealth Endangered Species Protection Act (December 1992) requires recovery plans and threat abatement plans for federal endangered species and provides incentives for state and territorial governments to adopt similar plans. In Spain, the Act on the Conservation of Natural Areas and Wild Flora and Fauna (1989) requires recovery plans to be drawn up for endangered species; where species are identified as vulnerable to habitat change, habitat plans may also be required. The US Endangered Species Act (1973) requires that critical habitats of listed species be designated and that federal agencies ensure that actions they authorize, fund or carry out are not likely to adversely affect these species.

Integration of species and habitat measures is a prerequisite for the protection of biodiversity, but is fraught with difficulties: it marks the point at which conservation begins to run up hard against other economic interests. In the USA, the designation of critical habitats met with no serious problems as long as only very small areas were concerned. However, the critical habitat of the northern spotted owl *Strix occidentalis caurina* covers substantial areas and the very legitimacy of the Endangered Species Act has been questioned as a result of the associated controversy (de Klemm & Shine 1993).

Protected areas

The protected area is one of the oldest conservation devices, with laws to protect outstanding landscapes first being enacted in the USA in 1864 (Holdgate 1996). Very large areas of land have now been designated as protected (over 20% of Venezuela for example), particularly in the last three decades. However, the 'paper park' is the rule rather than the exception, as the mechanisms, money and personnel needed to bring about protection are frequently lacking. Protected areas are a cornerstone of conservation policy, but to be successful there is a widespread need for them to be better integrated with wider planning frameworks and local economies and to be better resourced. Integrated conservation and development projects (ICDPs), which have the aim of reconciling the management of protected areas with the social and economic needs of local people, are increasingly seen as part of the solution. Success to date is limited, but this does not invalidate the concept. Rather, it points to the fact that in isolation ICDPs are not enough—protected areas can only be effective if the overall policy framework within which they operate is sympathetic to the environment. These issues are explored extensively in Wells & Brandon (1992) and IUCN Commission on National Parks and Protected Areas (1994).

Protected areas are formally established in one of three ways: through public ownership, regulatory measures applied to privately owned land or restrictions self-imposed by private or charitable owners. Although important in some countries, the last approach is not considered further here.

Public ownership can be a powerful conservation tool: once land has been acquired, it will usually be given long-term protection; control of damaging activities may still be difficult, but it is more straightforward than on privately owned land; and positive management measures can be taken as necessary, assuming resources are available.

Protected areas which rely on regulating the activities of private landowners are inherently less secure, but generally less costly for the state. The approach is also more flexible—restrictions can be limited to what is strictly necessary—and is appropriate for large areas where the optimum conservation land use is the maintenance of traditional and extensive farming or other such activities. A further advantage is that restrictions can be applied more quickly than land can be purchased, compulsorily or otherwise. However, problems can arise when positive management is needed—measures cannot generally be imposed by government and depend on a voluntary management agreement being negotiated. The question of costs is also not as clearcut as it might seem as compensation is commonly payable to landowners who are deprived of significant use rights. In the UK, where such a system of *compensation for profit foregone* operates, costs for achieving modest conservation benefits can be very high. Under Swedish law, where restrictions to property rights are particularly severe and result in substantial economic loss, the owner may require the government to purchase the land.

Public ownership and regulation should be seen as complementary systems and deployed according to conservation need. In practice, some countries rely almost exclusively on public ownership because it is difficult to impose conservation land use restrictions on a landowner against his or her will. Others, including Japan and many European countries, rely heavily on the regulation of activities on private land (backed up with financial incentives)—to a significantly greater extent than is justified by the record of success. A more objective-led approach, which also better integrates legal instruments with economic instruments (see below), would help enormously.

Land use planning legislation

Area-based regulatory systems are, of course, not confined to protected areas. Many countries now have quite far-reaching systems to control land use and development, with plans drawn up for all or particular administrative units. These plans typically divide the area concerned into zones and specify where development is permitted, and at what density. They may also specify the type of land use activity that is permitted—for example, commercial, residential or recreational.

The roots of the planning system lie in the need to control urban development and, although land use planning now often extends to the countryside, conservation is normally given little weight compared with economic, social and landscape considerations. Agriculture and forestry are generally excluded from control apart from restrictions that may apply to buildings.

From a conservation perspective, one of the basic requirements is that the planning system fully respects designated protected areas and prevents damaging development—whether inside or beyond their boundaries. In many European countries, although progress is being made in this direction, destruction by major infrastructure projects (particularly road and port developments) is still regularly sanctioned.

More ambitious systems are also evolving. For example, in some regions of Spain, farming, forestry and other operations can be controlled under planning law in zones where they are liable to cause damage to the natural environment.

Environmental impact assessment

Development and land use activities are subject to a wide range of regulations in addition to those stemming from conservation and planning laws. An essential part of conservation policy is to integrate environmental protection into measures which govern activities such as mining, abstraction of water, fishing and forestry.

Assuming an activity requires official approval, under whatever body of legislation, and assuming the decision-making process takes account of likely environmental effects, the authority giving approval needs to know the nature and scale of those effects. To help in this process, an increasing number of countries now have a legal requirement for an environmental impact assessment (EIA) to accompany (significant) development proposals. In the event of this evaluation predicting unacceptable impacts on the

Fig. 13.2 Due to the lack of strategic environmental assessment, infrastructure schemes are still destroying important designated sites such as Twyford Down, UK. (Photo by C. Gomersall.)

environment, the competent authority can either require design changes or refuse consent.

The US National Environmental Policy Act (1969) led the way on EIAs and in one crucial respect remains ahead of measures such as the EU Directive on the Assessment of the Effects of Certain Private and Public Projects (1985). The US legislation requires EIAs for policies, programmes and plans as well as individual projects. This is referred to as strategic environmental assessment (SEA) and recognizes the fact that it is often too late at the project stage to prevent major damage if that project is already part of a bigger approved plan or programme. Major road-building schemes illustrate this point—once the broad route is decided upon, the room for manoeuvre over individual stretches of road is severely constrained, regardless of the environmental impact. This has led to highly charged disputes between conservationists and government in the UK over such schemes as the construction of the M3 through Twyford Down, an outstanding area of chalk downland designated as a protected area (a site of special scientific interest or SSSI) (Fig. 13.2).

The absence of SEA legislation is a weakness in the conservation armoury of most countries, but tentative steps are being taken and some guidance on methodology now exists (Therivel *et al.* 1992).

13.4.2 International law

Given that neither wildlife, nor pollution, nor underlying forces such as trade are confined by administrative boundaries, systems of national law operating independently of each other can only go so far to stem environmental degradation. The environmental interdependence of countries was given some official recognition at the Stockholm Conference on the Environment in 1972. Principle 21 of the Stockholm Declaration re-affirms that states have the sovereign right to exploit their own natural resources, but adds the qualification that they also have the responsibility to ensure that activities within their borders do not cause damage to the environment of other states or international waters.

The dichotomy between national sovereignty and international responsibility remains, with Rio having taken the debate little further forward. However, through a variety of international treaties, states are slowly accepting wider limitations on their sovereign rights over the environment. International law increasingly provides a broad framework within which national laws operate to regulate behaviour.

Regional treaties

At regional level, the most comprehensive treaty on environmental protection is the ASEAN Agreement (1985) between Brunei, Indonesia, Malaysia, the Philippines, Singapore and Thailand. This sets out measures to conserve each element of the environment: air, water, soil, plant cover, forests, fauna, flora and ecological processes.

The two EU directives—the Wild Birds Directive (1979), which requires the establishment of special protection areas for birds, and the complementary Directive on the Conservation of Natural Habitats and of Wild Fauna and Flora (1992), which requires the establishment of special areas for conservation—are amongst the most advanced international measures in that they are underpinned by both sticks and carrots. Legal sanctions can be taken through the European Court against EU member states who permit damage to sites meeting the designation criteria set out in either directive; (limited) community funds are available to help finance essential

Fig. 13.3 The Ramsar Convention has helped bring international attention and assistance to the conservation of Lake Ichkeul, Tunisia. (Photo by D. Pritchard.)

conservation measures for sites with particular concentrations of priority
habitats or species. Even this relatively sophisticated system experiences
regular failure, but it has undoubtedly increased the level of protection
given to important biodiversity sites.

In contrast, early measures such as the Western Hemisphere Convention
(1940) and African Convention (1968) have no governing body, secretariat
or budget and consequently no practical effect.

Global treaties (sectoral)

Four global treaties address specific aspects of biodiversity conservation.
The Ramsar Convention (1971) is primarily concerned with the conservation
and management of those wetlands put forward by signatory countries
and included in the List of Wetlands of International Importance (Fig.
13.3). There are currently 93 contracting parties, with 811 sites on the list,
covering more than 54 million hectares. The only sanctions the convention
institutions can bring to bear on countries which fail to protect their
designated sites are political and moral pressure, but a measure of success
is achieved through these routes.

The Convention on International Trade in Endangered Species of Wild
Flora and Fauna (1973), or CITES, sets out to control trade in threatened
species listed in its appendices. One hundred and thirty-two countries have
become parties to the convention. Its principles are widely accepted, and
it has developed provisions for implementation but enforcement will
always be a problem because of expense and technical difficulties. It can
be argued that it would be more rational to list those species which can
safely be traded, rather than those for which trade needs to be controlled.
However, 'reverse listing' or 'green listing', as this approach is sometimes
termed, is regarded by many governments as too great an infringement
of free trade principles.

The Convention on the Conservation of Migratory Species of Wild
Animals (1979), or the Bonn Convention, remains of limited effectiveness
because of the relatively small number of contracting parties, nearly half
of which are in Europe. Many countries of major importance for migratory
species (particularly birds) and their migration routes are still outside the
convention.

The Convention on Biological Diversity

The idea of a world treaty on conservation was originally put forward
in 1913, and was discussed again in the late 1940s, but there was no real
momentum to take it forward until the early 1980s. Recognition of 'one of
the great extinction spasms of geological history' (Wilson 1992) prompted
arguments that it was time to move beyond the species-by-species approach
embodied in CITES and other agreements. The Convention on Biological

Diversity was finally opened for signature at the Earth Summit in June 1992, came into force in December 1993, and now has 162 contracting parties.

The convention was put forward not to replace existing conventions but to provide a coherent and comprehensive framework for national and international conservation action. The objectives are set out as:

- The conservation of biological diversity.
- The sustainable use of biological resources.
- The fair and equitable sharing of the benefits arising out of the utilization of genetic resources.

Parties are required to establish protected areas or areas where special measures need to be taken to conserve biodiversity; rehabilitate and restore degraded ecosystems; enact legislation for the protection of threatened species; and identify, regulate and manage activities that have, or are likely to have, significant adverse impacts on biodiversity. Parties also undertake to develop national strategies, plans or programmes for the conservation and sustainable use of biodiversity (Section 13.2) and to introduce procedures for EIAs. Emphasis is placed on *in situ* conservation and the maintenance of viable populations of species in their natural surroundings.

As indicated by the third objective, the convention goes well beyond the conservation of biodiversity to encompass such issues as access to genetic resources, sharing of benefits from the use of genetic material and access to technology, including biotechnology. Another striking feature is the explicit requirement for developed countries to make 'new and additional resources' (i.e. additional to existing bilateral and multilateral funds) available to developing countries for the convention's implementation. Without the provision of such resources, developing countries are considered by the convention as no longer bound by their conservation obligations. The financial mechanism for this has been the subject of acrimonious debate, but is currently provided by the Global Environment Facility (GEF), jointly operated by the World Bank, the United Nations Development Programme and the United Nations Environment Programme.

On the down side, the convention lacks detail and none of its obligations are absolute—all are qualified by phrases like 'as far as possible and as appropriate'. This partly reflects deep north–south divisions during the negotiations, with the north very reticent on the financial and technology transfer obligations, and the south equally reticent on the conservation obligations (Munson 1993; Brenton 1994). Concern in the USA, particularly from drug and biotechnology companies, was at first sufficiently strong to prevent their signing the convention. (The Clinton administration has now done so, although their ratification of the convention remains outstanding.) OECD, the United Nations and GEF data, up to and including 1994, provide no evidence that developed countries are fulfilling their commitment to provide 'new and additional' resources for biodiversity conservation. Relevant aid flows from most countries are stagnant or falling, as is aid as a whole (Lake 1996).

Nevertheless, against the odds there is now a global treaty in place, with bold intentions, administrative machinery and a financial instrument. Given its scope, it is unrealistic to expect that it could ever do more than provide a broad framework for other policies and action. The convention's effectiveness will depend in no small measure on how well conservationists use it to lobby at national and international level.

13.4.3 Relationship between national and international law

National and international laws are complementary and interactive. It is often countries with more advanced national approaches to conservation that initiate and facilitate international treaties. Once concluded, other countries are then required or encouraged to improve their legislation to meet the new international obligations. This has been demonstrated in the EU with advances at national level to implement the Birds and Habitats Directives (1979 and 1992, respectively). New national measures may, in turn, lead to the development of international measures with more exacting conservation requirements.

For those promoting conservation at national level, being able to point to international obligations can be a powerful line of advocacy. Many governments strongly dislike being held up as failing in their international duties, even if the legal sanctions are weak, as in the case of the Ramsar Convention and the Biodiversity Convention.

13.5 Economic instruments

Subsidies and grants (incentives) and charges and taxes (disincentives) are economic instruments that can be used to encourage conservation, either as an alternative to or in combination with regulation (see also Chapter 11).

There is a very solid economic rationale for public intervention into markets so that environmentally damaging actions are penalized, and conservation actions rewarded: market prices frequently fail to reflect the full environmental costs of economic activity. Whether these costs are in the form of loss of biodiversity, degradation of soils or pollution of air or water, they are rarely borne fully by the producer, and are therefore not reflected in product prices and passed on to the consumer. These costs are external to the market, partly because of the difficulty of developing markets for environmental resources—who owns them and how can people be required to pay for using them?

Environmental costs do, of course, eventually fall on society at large, but not in a way that reflects who was responsible for or who benefited from the original action. This is economically inefficient and inequitable. Worse, environmental impacts can be so extreme as to be totally irreversible, as when species extinction occurs—or effectively so, because of the very long time period, uncertainty or cost involved in restoration.

Environmental benefits, like the creation of wildlife habitats, also tend to be external to the market, and not captured by market prices. Just as free markets produce too much pollution, they produce too little habitat creation and management.

The importance of the gap between the market price of environmental resources and their worth to individuals and societies is now quite widely recognized, but only modest steps have been taken to incorporate this thinking into policy measures. Some economists argue that if the gap is to be closed, it needs first to be quantified—by valuing environmental costs and benefits in monetary terms to justify the imposition of a tax or subsidy. Biodiversity loss is one of the major costs of economic activity not taken into account by the market, perhaps partly because biodiversity is one of the most difficult resources to give monetary value to. A tropical forest, for example, has *direct use value* (it can provide timber, food, medicinal plants, tourism opportunities, etc.), it has *indirect use value* (through watershed protection or carbon sequestration) and it has *existence value* (people may wish to know that the species that inhabit the forest are conserved irrespective of current or future human use). But many of these values cannot be readily quantified in financial terms. Although some progress is being made with techniques such as contingent valuation and travel–cost methods, the attribution of monetary values to biological resources will remain highly problematic.

It is argued here that this is not sufficient reason to delay the use of economic instruments for conservation purposes—to close the gap between market prices and social values. There are perfectly rational ways of proceeding, particularly if clear environmental targets can be set, based on current knowledge and the precautionary principle. A mix of regulation and economic instruments can then be used to meet these targets in the most cost-effective way, without recourse to monetary valuation. Monitoring the effectiveness of policy against environmental objectives is often more straightforward than monitoring the success of (widespread) government interventions in the market place for other reasons. It is political rather than technical problems that prevent quicker progress— most societies are reluctant to meet the 'here and now' costs of environmental protection and downplay the long-term implications.

13.5.1 Removal of perverse incentives

As noted above, it is difficult for the market to take proper account of environmental considerations without government intervention. Ironically, most current government intervention does exactly the opposite—it discourages conservation and encourages rapid use of natural resources. These interventions are universal and of many kinds. They include subsidies to encourage clearance of forestry for agriculture, and forest revenue systems which encourage concessionaries to pursue highly selective and

Fig. 13.4 The great bustard *Otis tarda* in Spain uses land which is farmed in an environmentally sensitive way as a result of economic incentives to the farmers. (Photo by C. Gomersall.)

Fig. 13.5 A peat bog in Scotland which was ploughed due to perverse incentives favouring forestry, which would otherwise be uneconomic. (Photo by C. Gomersall.)

destructive timber harvesting and provide little incentive for sustainable forestry (Repetto 1988). Most rich countries subsidize agriculture with insufficient regard to environmental consequences such as overgrazing, soil destruction, loss of nutrients and water pollution (Fig. 13.4). Many developing countries, in particular, hold down electricity prices, promoting use and increasing carbon dioxide emissions (Cairncross 1991).

When these incentives cause unintended adverse environmental impacts, they are commonly described as *perverse* (Fig. 13.5). Such measures have a double disadvantage. Not only do they promote damage, they make the introduction of incentives for good environmental management more expensive and difficult to implement. Environmental payments have to be pitched at a higher level than would otherwise be necessary.

Nor do they necessarily make good sense against other criteria. In Latin America, generous incentive schemes to promote cattle ranching have served investors well, but the ranches have failed, both economically and ecologically (Repetto & Gillis 1988). Political pressure can make govern-ments introduce subsidies for social and economic reasons without proper policy analysis and without adequate mechanisms to monitor outcome and effectiveness. Once in place, the same political pressure, inertia and complexity often make subsidies difficult to remove or reform, regardless of whether they achieve their original purpose in a cost-effective manner and regardless of other consequences they may have. These issues are well illustrated by the EU's Common Agricultural Policy which now accounts for £35 billion in public expenditure each year and in the eyes of many commentators represents poor value for money, even before its destructive environmental effects are taken into account.

The removal of perverse incentives warrants much greater attention from the conservation lobby than it has received to date.

13.5.2 Economic incentives

Rather more progress has been made with the introduction of incentives for good environmental practice. Although new subsidies cost the public more than reform of existing policy, there are generally two interest groups in favour—the potential recipients and conservationists. Further, adding to public expenditure may be less controversial than changing the status quo to the disadvantage of a particular section of society. A simple (and expensive) incentive model is one based on *compensation for profit foregone* as noted in Section 13.4.1—landowners are effectively paid for agreeing simply not to destroy particular habitats.

A more positive approach gaining ground in Europe and North America is to reward landowners and managers for carrying out actions which maintain or enhance the wildlife interest. This reflects the fact that, in landscapes with a long history of human settlement, farmed habitats have evolved with high conservation value; and that these habitats depend on land management practices which no longer make economic sense without public subsidy. In effect, society is paying for public benefits which the market will not otherwise deliver. *Positive environmental payments* are generally made against a land management prescription such as a specified grazing, cropping or water level regime. Less explored are models based on *payment by results*, where the results might be units of wildlife produced. If breeding wading birds are the main reason for farming a wetland in a particular way, the reward to the farmer could be according to wader productivity, or at least the number of pairs attempting to breed.

Another approach to conservation incentives is to attach environmental conditions to subsidies which are given essentially for social or economic reasons. Such *conditionality* can again be well illustrated by reference to agriculture policy. Many countries give production subsidies to farmers with little or no regard for the environmental consequences. They are almost always of a different order of magnitude to environmental subsidies.

Particularly where payments are made per unit area farmed, conditions can be attached over such matters as the application of fertilizers and pesticides, the management of field margins, stocking densities or the retention of permanent pasture. Conditions of this type are generally less effective than well-targeted, purpose-designed environmental payments; they are often unpopular with administrators and recipients; and, as production subsidies come under the increasingly close scrutiny of the World Trade Organization, might have a limited life expectancy. However, environmental conditions tied to social and other payments have the potential to deliver conservation benefits across very large areas of the farmed landscape at little or no additional cost to the public.

A similarly wide range of economic incentive schemes has been evolved to promote multipurpose forestry, where the outputs include biodiversity and public amenity as well as timber. Foresters may receive grants to retain particular areas of forest or types of trees (in terms of species and age); to allow natural regeneration rather than pursuing clearfelling and restocking regimes; and to restock with native rather than exotic species. Environmental conditions may also be attached to subsidies for planting or licences for felling.

Conservation based on incentives is not confined to the developed world. There are many schemes in developing countries funded under aid and development programmes to promote biodiversity conservation. They include schemes designed to give local communities a share of the revenue and employment derived from management of wildlife, particularly through wildlife tourism. The underlying argument is simple: whatever the position over land tenure and ownership of the 'wildlife resource', conservation is bound to be even more of an uphill struggle if it brings the local community no apparent benefit. One of the more sophisticated and probably best-documented examples of giving local people a tangible economic interest in wildlife conservation is provided by the CAMPFIRE project in Zimbabwe (Barbier 1992). Safari and hunting revenues fund communal facilities and compensation payments to local residents for crop and animal damage. Meat from culling operations is sold at subsidized prices. As well as providing economic benefit, the scheme is intended to foster community participation in the management of wildlife resources.

13.5.3 Economic disincentives

Taxes, charges and tradable permits to discourage environmental degradation represent the reverse side of the incentive coin. Fines have long been used as a penalty for infringing environmental regulations, but there is now a move to more sophisticated green taxes or pollution payments and resource pricing systems.

Green taxes are an embodiment of the *polluter pays principle*. They encourage firms and individuals to use resources more efficiently and to pollute less. When the state employs tax revenue raised directly to tackle the environmental problems resulting from the activity which is taxed, it is said to be *hypothecated*. One advantage of hypothecation is that the tax (or levy) may be more readily accepted by the public if its purpose is transparent; another is that it may be more effective, as both the tax and the expenditure help meet the objective.

Although the number and variety of green taxes in operation is increasing, with examples given below, it is notable that European governments still raise less than 10% of their revenue from environmental taxes, compared to over half from various taxes on labour (European Environment Agency 1996). Switching taxes from labour to resources and pollution can give a double dividend, by not only meeting environmental objectives, but also helping to create jobs (Tindale & Holtham 1996).

The following are examples of green taxes:

• The USA and Denmark have taxes on chlorofluorocarbons and halons.
• Carbon taxes are levied in Denmark, Finland, The Netherlands, Norway and Sweden.
• France and Germany raise substantial sums from water pollution charges.
• Austria, Sweden, Norway and Finland have fertilizer taxes.
• Denmark has a pesticide tax.
• The USA has a feedstock tax levied on the petroleum and chemical industries.
• Denmark has waste disposal taxes on both landfill and incineration.
• The UK has a landfill tax partly hypothecated to fund environmental trusts, the first tax introduced specifically for environmental reasons in the UK (Fig. 13.6).

Charging systems can also be used to promote the efficient use of resources and to manage demand. For example, switching from fixed water charges

Fig. 13.6 The UK has introduced a landfill tax to reduce excessive waste production. The profits from this will partly go to funding environmental improvements. (Photo by M. Davies.)

to charges based on the volume of water used gives consumers an incentive to reduce use, which in turn reduces the need for water abstraction from rivers, new reservoirs or regional water transfers. It enforces the *user pays principle*. This principle can also be applied to waste charges (volume-based waste collection charges) and road charges (road pricing).

Tradable permits or licences to regulate pollution or resource use are employed by a small number of countries, particularly the USA. The total level of pollution or resource use permitted is fixed by the authorities, and can be reduced over time to meet an environmental objective. Permits can be traded between businesses (and others) to promote cost effectiveness. The main advantage over taxes is the much greater certainty that environmental objectives will be met. However, there are difficulties in developing markets for permits and in taking into account local environmental circumstances.

13.5.4 Regulation or economic instruments?

Some policy makers believe that a greater use of economic instruments should enable a reduction in the amount of regulation that currently

Table 13.1 Comparison of the advantages and disadvantages of regulatory and economic instruments.

Regulation	Economic instruments
Authorities are confident in usage—traditional familiarity	Seen as innovatory and complicated by many authorities
Can be popular with the public if no obvious widespread disadvantage	Taxes and tradable permits can be perceived as selling a right to pollute and can therefore be unpopular with the public
Can be effective at preventing environmental damage and ensuring minimum standards are met	Taxes and incentives can discourage environmental damage, but outcome is less predictable. Tradable permits can guarantee an objective is met but difficult to take account of local conditions
Much less effective at bringing about positive environmental management (such as a particular grazing regime) than maintaining the status quo	Subsidies can be effective at bringing about positive environmental management; if people 'volunteer' to take action they are more likely to get it right
Generally unpopular with those whose actions are constrained and can be difficult to enforce	Incentives popular with recipients; disincentives unpopular with those affected, sometimes more so than regulation
Costs to the public sector include enforcement costs	Incentives can be more costly to the public sector than regulations; but disincentives can raise revenue
Can impose high compliance costs on the target audience	Can be more cost effective for the target audience—greater flexibility allows lowest cost solution

applies to the environment. However, recent research by the OECD (1994) concludes that a shift from a regulatory approach towards one based principally on economic incentives and disincentives 'is unlikely to occur, nor would it be desirable'. Economic measures alone will not deal with environmental problems effectively or efficiently. Combinations of instruments are preferable and within such 'cocktails' economic incentives will have a crucial role to play, in extending the range of conservation that can be achieved and in making regulations work better. As noted previously, there are now several instances of environmental legislation, such as the EU Habitats Directive, being accompanied by complementary economic measures.

Table 13.1 sets out some generalized characteristics of each approach. The choice of instrument can only be decided on a case-by-case basis, but two broad conclusions emerge. Regulation is appropriate to 'guarantee' that minimum environmental standards are met—to prevent gross habitat destruction, pollution, etc. Economic incentives or a mixture of incentives and regulations give a much greater chance of success when the aim is to go beyond minimum standards—to bring about positive management.

13.6 Bringing about change

Governments are not oblivious to environmental problems, the current Vice-President of the USA, for example, having called for the rescue of the environment to become 'the central organizing principle for civilization' (Gore 1992). There is widespread acceptance that, without government intervention, the environment cannot be protected and even some prominent free-market politicians would now sign up to this notion (Cairncross 1991). In many western countries, there is large-scale public support for conservation, with the combined membership of seven US environmental organizations exceeding 10 million and that of seven UK organizations nearing 5 million.

With such breadth of concern, why is the introduction of effective environmental policy so problematic? A major part of the answer is simple. Even a watered-down version of sustainable development—one that accepts that truly green growth is a chimera but that *greener* growth is possible—comes with a price tag. This may be exaggerated and mis-represented by the proponents of *laissez-faire*, but greener growth is likely to mean slower growth (as conventionally measured). Greener growth will certainly mean a redistribution of benefits and those who stand to lose are more likely to be in a position to exert political influence than those who stand to gain, especially as these last include future generations. Politicians and other decision makers are rarely rewarded for putting long-term objectives before present-day expectations. And whatever the political system, they will to some extent be at the mercy of vested interests; they may indeed share those vested interests. The position is neatly summarized by Cairncross (1995): 'Environmental degradation generally occurs because a powerful lobby is getting something for nothing. The forces against change are mighty ...'.

Against this backcloth, it is an exaggeration to say that governments will *only* act in the face of public pressure, but such pressure is enormously important. In western countries, public concern is increasingly channelled through NGOs who over the last 30 years have become prominent advocates. Ironically, their size and power tend to be greatest in those countries whose wildlife and natural resources have been most severely depleted—a fact which the representatives of some tropical countries are quick to point out when being pressed to conserve their forests.

13.6.1 Grabbing public attention

> 'Like most scientists who were new to politics I thought that changes in opinion and action were mainly produced by obtaining facts and arguing logically from them. I was soon disillusioned. I found that there were immense obstacles to implementing the general restrictions we wanted. Facts and logic were only the first phase; success would depend on overcoming economic and political obstacles' (Moore 1987).

Moore's account of his work as a UK government scientist in the 1960s and 1970s on persistent organochlorine insecticides applies equally well today. He and the many others active in this area, on both sides of the Atlantic, eventually produced one of conservation's most clearcut successes—the banning of DDT and dieldrin by the nations of the north with some measure of control in the south. But this outcome was undoubtedly hastened by the public stir generated by Rachel Carson's book *Silent Spring* (1962), particularly in the USA. The scale of impact has rarely been surpassed by any environmental campaign since, although the use of ground-breaking reports and high-profile campaigns to focus public attention on damage and future threats to the environment, and to put pressure on politicians to act, has become commonplace. In isolation, direct lobbying of decision makers on the basis of environmental (or even economic) logic pays very limited dividends.

The pesticides issue illustrates another maxim of conservation advocacy—the chances of success are greatly increased if human health is evidently at risk. The problem of persistent organochlorine insecticides was brought to light through impacts on bird populations—reproduction failure, particularly amongst raptors, leading to massive declines. But it was mainly the implications for human health that eventually forced change. In more recent times, the clearcut risks to human health from the destruction of the ozone layer have done much to facilitate control of chlorofluorocarbons through the Montreal Protocol (1987) and the London Agreement (1990).

When it is 'only' the natural environment that is perceived to be at risk, it is even more of an uphill struggle.

Public interest is most readily generated through the media. As the media has a predilection for the new and the sensational, conservationists courting attention for their subject can be tempted to exaggerate. Sooner or later, this tactic backfires and does damage to the reputation and influence of those concerned. The facts are frightening enough; the trick is to find fresh and powerful ways to communicate them.

13.6.2 Working for solutions

The wildlife conservation movement traditionally put most of its effort into three areas: (i) charting and analysing problems; (ii) bringing these to the attention of the public and government in such a way as to generate action; and (iii) implementing direct solutions such as setting up nature reserves. To these is now added a crucial fourth area—carrying out policy analysis and promoting solutions to government. At the more straight-forward end of the spectrum, this work entails designing and advocating improvements to national wildlife law. Many go further and seek to influence international conventions. Some have taken the next step and include amongst their ambitions the 'greening' of major economic policy instruments, at national and international level.

Few would describe the gains so far from this last class of work as anything more than modest, but the shape of the North American Free Trade Association (NAFTA), the 1992 McSharry reforms to the EU Common Agricultural Policy, the current World Trade Organization debate about trade restrictions for environmental reasons, and the policies of various aid and development institutions have all been influenced to an extent by the work of conservation organizations.

To have any detailed impact on the adoption or reform of policy often means working closely with government. There are sizeable risks to this. It is easy for conservationists to find their energies dissipated by the bureaucratic processes involved; compromises may be made in order to get some result; and contributing in detail to policy development can make later criticism of inadequacies more difficult. However, the balance of advantage for the environment cannot lie in wholesale refusal to engage seriously with those who shape its future. Confrontation is a necessary but insufficient tactic.

13.6.3 Finding common ground: forming alliances

Environmentalists need to work in cooperation with local communities and resource users such as farmers and fishermen—conservation is not possible without their support and in many cases relies on their participation. Common ground is often greater than is immediately apparent and alliances add strength to lobbying. In Western Europe, conservation organizations and farmers have worked together in recent years to generate financial support for less intensive land management practices which deliver both social and environmental benefits. Costs to the public purse may be less than those involved in conventional economic development programmes, while outcomes are more reliable and durable. Even without public subsidy, low intensity land use may be the most profitable, particularly where wildlife tourism or sustainable hunting are realistic options (Swanson & Barbier 1992), but this may need demonstrating to both local interests and those who determine land/resource use policy.

Chances of collaboration are inevitably much less in the many cases where the resource user does not have a long-term interest in the resource—where there is lack of ownership or some other effective form of control, there is little advantage to the user in restraint.

13.6.4 Using the law

Laws to protect species and habitats often fail for one of two reasons: either they are riddled with exceptions (for example, for farming, forestry or public works) or they are quite far reaching, but the statutory conservation authorities do not really dare to implement them (de Klemm & Shine 1993). In some countries, this second failing can be addressed by using judicial proceedings to force statutory authorities to act appropriately, i.e. in accordance with the law.

Conservation organizations in Europe have had some measure of success taking governments to the European Court for failure to protect sites in accordance with the EU Birds and Habitats Directives—for example Sociedad Espanola de Ornitolgia (SEO) vs Spanish Government over Santona marshes or the Royal Society for the Protection of Birds (RSPB) vs UK Government over Lappel Bank. Where, as in the Lappel Bank case, the judgement comes too late to save the site, a measure of compensation in the form of habitat creation may be achievable. More importantly, such actions can deter governments from subsequently ignoring their obligations under conservation law. In the USA, the experience of environmental litigation is much greater, with the northern spotted owl controversy alone generating dozens of actions (Wilkinson 1992).

Legal challenge can be a double-edged weapon—governments are tempted to react by weakening the law. To reduce the chance of such a backlash, conservationists would be wise to avoid gratuitous litigation, but the agenda has to be moved forward. There is little point in having well-crafted protection laws which are flouted. Further, as voluntary conservation organizations work more collaboratively with government, it is important that they continue to challenge—cosy relationships are unlikely to produce dynamic solutions.

13.6.5 Complementarity and other tactics

There is no right or wrong way to bring about policy change. Different organizations, each with their own constituency of supporters, adopt different approaches, from the highly adversarial to the collaborative. Contrasting styles can complement each other, the pressure for change generated by one group's confrontational tactics sometimes increasing the receptivity of decision makers to the policy proposals of others. Some vary their approach according to the stage that has been reached in a particular campaign—proffering solutions to problems which are not recognized, let alone causing political difficulties, is generally a waste of time.

The timescale for significant policy change is invariably measured in years (sometimes decades) rather than months. Long-term commitment to an issue and tenacity pay greater dividends than dilettantism. Equally important, however, is the ability to react speedily to political opportunities—issues can be progressed much more readily at some times than others. As environmental organizations grow and become more bureaucratic, they will need to guard against becoming slaves of their own planning systems.

13.7 Conclusion

Conservation policy has to grapple with some of the world's most intractable problems. Neither individuals nor businesses will consistently respect the environment unless the education, economic and regulatory frameworks put in place by governments send the right signals. Powerful lobbies benefit from environmental degradation and governments are unlikely to act in the absence of public pressure. Ways need to be found to inspire greater public support for conservation, such that major economic policies can be 'greened'.

Regulation without the provision of incentives to promote positive management is ineffective. Equally, the cost of incentives to the public purse means that their use will be rationed—regulation will have to continue to be relied on for much environmental protection. A prerequisite of any conservation measure exerting maximum leverage is the removal of perverse incentives. Selection of the most effective and efficient measures depends on clarity of purpose, including definition of objectives in biological terms.

High-level plans and policies have to be implemented locally. Much more attention needs to be given to the integration of national and local aspirations.

CHAPTER 14

Conservation and development

William M. Adams

14.1 The challenge

There can be no doubt of the importance of either development or conservation. The scale of impacts of human activities on biodiversity and the accelerating pace of species extinction demand attention at the end of the twentieth century as never before (e.g. Heywood 1995) (see also Chapter 2). The moral imperative to tackle human poverty is no less pressing. The *World Development Report 1992* opened by stating 'the achievement of sustained and equitable development remains the greatest challenge facing the human race' (World Bank 1992).

The record of past achievement in development is perhaps particularly depressing. Despite decades of development rhetoric and action, poverty has worsened in recent decades across much of the world, particularly sub-Saharan and North Africa, Latin America and the Caribbean, and there are more poor people than ever before (1.1 billion in 1990). Furthermore, the gap in incomes between rich and poor countries is widening. Between 1960 and 1989 the share of global gross national product (GNP) of the countries with the poorest 20% of world population fell from 2.3% to 1.2% and that of the richest 20% of world population grew from 70.2% to 72.7% (UNDP 1992). The poor often endure degraded environments, and may contribute to their further degradation. Degradation of agricultural land, forests and wetlands is believed to be increasing, and urban air and water pollution are both rising, even in those countries in which economic growth is taking place. Problems such as the lack of clean drinking water are getting more and not less serious. Two million children die of intestinal diseases due to unclean water each year (World Bank 1992) (Fig. 14.1).

14.2 In search of mutual understanding

Debates about conservation and development tend to be dialogues of the deaf. There is enormous scope for misunderstanding between conservationists and developers, even when (as is increasingly the case) they are trying very hard to communicate, and to agree on a common language and agenda for action. The debate about sustainable development that

Fig. 14.1 Children: future conservationists, future consumers? In 1990, 46.4% of the population of low- and middle-income countries in sub-Saharan Africa were under the age of 14 (World Bank 1992). (Photo by W.M. Adams.)

emerged in the 1980s and made global headlines at the United Nations Conference at Rio de Janeiro is both the most exciting product of such essays in collaboration, and a depressing demonstration of the capacity of development and environmental institutions to use new words and phrases to dress up narrow concerns and old ideas.

Notably, the phrase 'sustainable development' has won widespread attention. There are many definitions of sustainable development (Pearce *et al.* 1988), the most commonly quoted being by the Brundtland Commission in *Our Common Future*, 'development which meets the needs of the present without compromising the ability of future generations to meet their own needs' (Brundtland 1987). The success of the phrase, and the bundle of ideas it so effectively co-opts, is chiefly because its flexibility and lack of fundamental meaning has allowed it to be interpreted in many different—and often divergent—ways (Redclift 1987; Adams 1990, 1993; Lélé 1991). Sustainable development has proved its operational worth both to those concerned about poverty and inter- and intragenerational equity in human access to nature and natural resources, and for those concerned with the use of nature itself and the conservation of habitats and species.

Most of those active in conservation planning and action have a core training in the natural sciences, while most of those active in debating or promoting development are trained in the social sciences. It has long been recognized that the disciplines within which people have been educated matter in development studies, giving rise to divergent assumptions and understandings with substantial practical implications (Chambers 1983). The problem is even more marked in the gaps between the fields of conservation and development.

The most fundamental difficulty stems from the common assumption that both conservation and development are unproblematic concepts and that their promoters are agreed in their understanding of the problems, goals and methods of achieving them. For those working in conservation, the unreality of such an assumption is demonstrated by debates within conservation biology or environmental ethics about (for example) the usefulness of *ex situ* conservation techniques, the implications of island biogeography for protected area planning, or the appropriateness of consumptive use as a strategy for conservation. However, the complexity and intractability of these debates are not obvious to those looking in at conservation from the outside. In particular, they are not readily visible to people working in development. They tend to put conservation into a box with a single label, and thereby reduce conservation's complexity to a caricature; for example, assuming that conservation involves an unproblematic commitment to preservation through the establishment of protected areas, the mothballing of potential natural resources, and keeping people and animals (particularly large furry ones) apart.

Development, too, might look as if it has a coherent meaning to the conservationist looking in from outside, but again this masks a highly complex set of debates. The word 'development' is ambiguous. It can be used descriptively (to depict present conditions) and normatively (to present a desirable alternative), and it is not always clear which is which (Goulet 1992). Furthermore, debates about development are both moral (about what the world ought to be like) and technical (about how to explain and direct change in particular directions). They are also extremely complicated. Cowan & Shenton (1995) say that development 'defies definition, but that is not for want of definitions on offer'. This is no place for a detailed account of development theory (see, for example, Corbridge 1995); however, it is important to recognize that ideas have changed significantly over time, and indeed continue to do so (e.g. Schuurman 1993; Crush 1995).

14.3 Development and change

The word 'development' is very powerful. The word, with its associated meaning of unfolding, entered English in the eighteenth century (Watts 1995) and, by the twentieth century, the ideas of development and underdevelopment were both well established in British thought (Cowan & Shenton 1995). Theories of development were and are Eurocentric, and were first imposed through global imperialism. This process was perhaps particularly marked in Africa, which suffered rapid transition with the sudden arrival of colonial rule and the engagement of regional economies and environments with capitalism within a few decades, around the start of the twentieth century. The Colonial Development and Welfare Acts, passed in the British Parliament in 1940 and 1945, rapidly increased the pace and scope of change in British African territories (Low & Lonsdale 1976). Here, as elsewhere, formal 'development' began, focused on the accelerated exploitation and better management of resources, and was pursued through organized state planning. Ideas about change drew on notions of progress, modernization and improvement.

The word 'development' carries two implications about social, economic and environmental change. First, it implies that such change is inevitable. Second, it implies that it will be, broadly, desirable. Thinking has historically been dominated by unconscious assumptions about the inevitability of human 'progress', and the view that western indus-trialized nations are the example of 'success' that all other countries both *should* follow (because of the wealth and welfare that urbanization and industrialization are seen to have brought) and also *must* follow (because of the overwhelming power of the capitalist world economy). The view that development is an inevitable process, a single route along which every country must go, formed the dominant 'modernization' approach to development from the 1950s onwards, epitomized by the new 'Bretton Woods' institutions. The World Bank sought to fund infrastructure dev-elopment in the global periphery analogous to that created to reconstruct Europe after the Second World War (Fig. 14.2). Post-war development built on established western ideologies, and proposed similar methods to pursue similar goals.

The traditional emphasis in development is placed upon economic growth, conventionally measured as GNP. This is the dominant use of the word 'development', and the primary sense used by government finance ministries, particularly those of industrialized economies and the groups they form (e.g. the 'G7' countries or the Organization for Economic Cooperation and Development (OECD)), and by international institutions, notably the International Monetary Fund and the International Bank for Reconstruction and Development (the World Bank). The concern

Fig. 14.2 Making progress on an unsurfaced road near the Tana River in northern Kenya: poor communications limit the opportunities for rural people to market their produce, and the abilities of governments to extend development services or implement conservation strategies. Road construction is a favoured target for development aid. Road maintenance is less popular. (Photo by W.M. Adams.)

to increase the size of the national (and the global) economy has historically been balanced by concern about the distribution of national income and the plight of the poor. There is widespread (although not universal) recognition of the need to balance improvements in the performance of the national economy with a recognition of the need of the poor and the need to distribute wealth actively, rather than hoping it will 'trickle down'. Alternative measures of development include the conventional GNP per capita and the United Nations Development Programme's Human Development Index, which embraces a basket of quality-of-life indices (UNDP 1992).

Development also has a fundamental, but often hidden, cultural component. The political and cultural implications of Eurocentric assumptions, following in particular Said's work on Orientalism (Said 1978), are only now being explored within development studies (e.g. Escobar 1995; McGee 1995; Watts 1995). The industrialized west has been taken as more than a model of how economies should be organized, but also as a model of governance and civil society. Development projects have deliberately involved modernization and the transformation of the economy, society and environment. The desirability of such 'westernization' (with what it might imply, for example, about material consumption, female circumcision or the advent of Coca-Cola) is not self-evident to critics of the western economic model, be they ecofeminist analysts (e.g. Shiva 1988) or advocates of intermediate technology (Schumacher 1973). It is also not self-evident to those people targeted by development, who devise various means of resisting imposed change (e.g. Scott 1985).

A further important dimension of development is its 'ambiguous' character—'a two-edged sword which brings benefits, but also produces losses and generates value conflicts' (Goulet 1992). Gains in material well-being (life expectancy, standard of living, health) or freedom of choice are balanced by the shattering effect that modernization and economic change can have on culture, society and ecology. Furthermore, the gains and the losses are rarely equally shared between rich and poor (whether within one country, the *favelas* and the beaches of Rio, or between rich and poor countries, 'First' and 'Third' Worlds). Adverse impacts of development are often borne locally, and not balanced by local benefits (Fig. 14.3). Furthermore, 'development' can be coercive, where change is imposed by the state, or where affected people are given no effective voice in the nature or rate of change. Development is often something done *to* people (in the name of economic growth or some other aim) not *by* people. From the 1970s, calls for 'bottom-up' or 'participatory' approaches to development have grown stronger, with an emphasis on promoting changes that people themselves locally understand to be necessary and desirable. There has been increasing recognition too of the importance of local or indigenous knowledge and cultural and social norms and institutions, both because of their utility (because experience

Fig. 14.3 Barrage on the Hadeia River, northeast Nigeria. Development project investment can have significant adverse environmental and economic impacts, and these can be distant in space (in the case of dams, on downstream flood plains) and delayed in time. The socioeconomic impacts of development (or conservation) projects can bear unequally on different people, with remote rural communities often bearing the brunt of costs, and distant urban economies enjoying the benefits. (Photo by W.M. Adams.)

shows that if these things are ignored, 'development' projects fail) and because it is seen to be morally right that development should have regard for these things.

It is finally worth noting that 'development' is not something that is confined to those countries variously labelled 'Third World' (in distinction to the industrialized western 'First World' and the industrialized communist countries of the 'Second World'), 'developing' or 'less developed'. Many of these countries lie in the tropics, and most have a recent colonial past (Africa or south Asia, for example, gaining independence within the last half century). It can be argued that they share to some degree characteristics such as small economies, low levels of industrialization, extensive poverty and poor infrastructure development. The labels are, however, unsatisfactory for several reasons. First, these countries vary greatly in their geography, economy and economic history. Burkina Faso or Bangladesh might fit the popular image of a Third World country, but it is less clear that they can be usefully compared with Brazil or Malaysia. Second, giving these countries a special label implies that the processes of economic change within them are somehow different and isolated from those elsewhere. Increasingly, it has been recognized that this is not true. Marxist analysis, the 'dependency school', world systems theory and conventional economics have all come to emphasize the links between First and Third World economies, not least because of the globalization of corporations and the closeness of financial links—capital flows and debt (Corbridge 1992). Development processes therefore span and link industrialized and non-industrialized regions and counties in very different countries, and can be remarkably similar; as can processes of economic change, and the social and ecological impacts that follow from them.

14.4 Development's demands on the biosphere

Homo sapiens has always made demands on other elements of the biosphere. However, the development process increases the scale of those demands, and the rate of rise of demand tends to steepen with time. Furthermore, development planners in the past, as today, have seen the 'natural environment' as a key element in the creation of wealth. Economic historians, for example, have described the importance of staples (furs, forests, minerals) to the growth of Canada, and the World Bank sees sub-Saharan Africa's comparative advantage as exploitation of its renewable and non-renewable natural resources (Watkins 1963; World Bank 1989).

The most ancient impacts of human activity on the biosphere derive from consumptive use of animal and plant species. In many areas, these impacts seem to have remained small until relatively recently. Evidence of human clearance of post-glacial forests appears in the palynological record of Western Europe from 5000 BP, and vegetation was extensively transformed by the dawn of written histories of such places (Rackham 1976, 1986; Birks 1986; Vincent 1990) (see also Chapter 8).

There is a widespread romantic tendency to assume that in preindustrial societies such use was sustainable. However, there is little evidence for this, and Pleistocene extinctions in the Americas are quite obviously linked in time to human occupation. The notion that areas 'discovered' and annexed by European explorers from the fifteenth century onwards were either unoccupied (the legal fiction of Australia's *terra nullius*) or were barely influenced by indigenous human action is increasingly being seen as a romantic fallacy. The political reality and the ecological complexity of the pre-Columbian presence in the Americas, with its culture and cosmology, its fire and its hunting, are now increasingly being recognized (Nabhan 1995). The idea that, even in the nineteenth century, places such as tropical Africa were unspoilt Edens, where human impacts were slight and nature ruled, is clearly quite wrong, although the influence of such ideas has been very considerable (Anderson & Grove 1987).

Conservationists have been rather too eager to adopt 'Edenic' ideas of this kind, and to bemoan contemporary changes. Garret Hardin's 'tragedy of the commons' is endlessly cited as both proof and explanation that non-privatized resources will suffer degradation and exhaustion (Hardin 1968). This argument is the primary platform for attempts to 'conserve' the biosphere, for example protecting land and vegetation resources against the perils of overgrazing (e.g. Bourn 1978; Jewell 1980). The flaws in this simplistic narrative are now widely recognized (e.g. Homewood & Rodgers 1987; Horowitz & Little 1987). Hardin's description referred to open-access resources, whereas research is now widely demonstrating that indigenous

natural resource management systems are based on common property or open-pool resources, where specific groups of people have rights in and exercise control over specific resources. There is increasing research interest in the norms and institutions by which such property management systems function, and indeed a whole field of study, new institutional economics, devoted to it (Berkes 1989; North 1990; Ostrom 1990). Thus most recent studies of pastoralism turn older accounts, which suggested that open-access grazing systems threatened to destroy equilibrial savanna ecosystems, on their heads. They argue instead that indigenous human management is skilful, knowledge based and a flexible response to disequilibrial systems (Western 1982; Behnke & Scoones 1991; Scoones 1994).

There is little doubt, however, of the impact of the development process on the biosphere. The industrialization of consumptive use has rarely been sustainable, the record of capture fisheries being particularly grim (Cushing 1988). The impact of the forestry industry has been similarly drastic, for example in North America, and more recently in the tropics (Williams 1989a, 1989b). The conversion of ecological systems through agriculture has also become industrialized, and while the amount of usable output has been maximized with great efficiency, it has not been without cost. From the nineteenth century, what has recently come to be called 'organic' farming gave way to increasingly controlled crop environments, with artificial fertilizers, mechanization, pesticides and irrigation, in turn transforming agricultural systems in North America and Europe. Crops and domestic animals were developed through selective breeding, and this process is now being further accelerated through genetic engineering. In the Third World, international public funding of agricultural research led to the so-called 'green revolution' technologies, integrated packages of high-yielding crop varieties, fertilizer, pesticides and irrigation, that transformed food production, particularly in south and east Asia (Lipton & Longhurst 1989).

The result of this streamlining of terrestrial ecosystems has been a rising appropriation of the products of global primary production by people. Vitousek *et al.* (1986) estimated that organic material equivalent to 40% of terrestrial primary production was being co-opted by humans every year, either directly or indirectly. This has obvious implications for ecological change and for species extinction. It is one measure of the human impact on the biosphere, and of the threat of species extinction (Wilson 1992). Globally, Hannah *et al.* (1994) estimate that natural habitat has been disturbed by human activities over almost three-quarters of the habitable surface of the planet (90 million km^2 is undisturbed, 52% of the total surface, but much of that is uninhabitable).

The extent of the ecological transformation is already locally large (Fig. 14.4). In the UK, for example, large carnivores became extinct in medieval times, and human impacts on vegetation have been profound (Ratcliffe 1984). Of the 1423 native British vascular plant species, one in 10 suffered

Fig. 14.4 The forest/farmland boundary, Mount Oku, Cameroon. Mount Oku is part of the area covered by BirdLife International's Kilum-Ijim Forest Project, North West Province, Cameroon. Loss of montane forest is a serious problem, threatening species such as Bannerman's touraco *Tauraco bannermanni* and the banded wattle-eye *Platysteira laticincta* (Coulthard 1996). (Photo by BirdLife.)

decline of at least 20% between 1930 and 1960, and 317 higher plant species were listed as nationally rare (i.e. occurring in fifteen 10 km^2 grid squares or less) in 1983 (Perring & Farrell 1983). Many animal and plant species declining in numbers or range are acutely vulnerable to ecological change (Wynne *et al.* 1995). Since 1900, the UK has lost one mammal species, six birds, two fish, 144 invertebrates and 62 plants (Brown 1994).

Agriculture's impact on biodiversity in the UK has been increasingly negative (Sheail 1995), with ploughing of grassland and heathland for arable, conversion of meadows to grass leys, reclamation of wetlands, abandonment of land systems such as water meadows and the impact of linked ecological change such as the decline of rabbits with myxomatosis in the 1950s. Government agricultural support fuelled rapid economic change from the 1940s. Farms became more capital intensive and larger, and farm workers (and farmers) fewer, while mechanization, pesticides and inorganic fertilizers increased productivity (Moore 1969; Bowers & Cheshire 1983; Adams 1986). There has been a massive and general reduction in the extent of seminatural habitat (Fuller 1987; Peterken & Hughes 1990). Remaining areas of lowland habitat have been degraded through the withdrawal of traditional management, eutrophication and other forms of pollution. Patches of seminatural vegetation have become smaller and increasingly isolated, raising the threat of loss of diversity as local extinctions occur (Peterken & Hughes 1990).

Industrialization and urbanization have also impacted upon the biosphere through pollution. Pollution was the main stimulus for the United Nations Conference on the Human Environment in Stockholm in 1972. Although at that time pollution was seen to be a problem of industrialized economies (and wealthy countries), it has become a significant problem in Third World countries, which often lack the environmental and antipollution safeguards that have become standard in the industrialized world (Park 1986). The most serious and intractable pollution problems in Third World countries often arise in urban areas, from the disposal of wastes and by-products of industrial processes, to more mundane problems such as sewage disposal and pollution from motor vehicles. Such problems exist on a particularly grand scale and acute form in those countries in the Third World which have most successfully industrialized, such as India, Malaysia and Brazil. It is ironic that these problems are greatest in these newly industrialized countries which have been most successful in achieving the 'development' for which others strive. Pollution has been part of the price paid for 'success' in development, or at least part of the price of the right to enter the development race.

Industrial pollution in the Third World can be exacerbated because of the inappropriateness of technology, lack of technology—for example to clean up discharges—or poor governance (e.g. bureaucratic inefficiency and corruption). The lack of regulations to protect the environment or people can itself attract polluting industries, since lack of pollution controls (like cheap labour) cuts the cost of production, or rather transfers costs to the host environment and community. In a sense, therefore, pollution is a hidden subsidy for Third World industry, and there is a danger of transnational companies, in particular, moving industrial plants that are highly polluting to locations in the Third World as a direct result of environmental protection policies in industrialized countries (e.g. Suckcharoen *et al.* 1978), or developing projects with unsolved downstream environmental impacts, such as toxic mine tailings (Adams 1990).

14.5 Sustainable development

The notion that development and environmental conservation could and should be tackled in an integrated way began to grow in the 1960s. One root of such thinking was the argument that the science of ecology was relevant to development. This idea was integral in the work of the International Biological Programme (IBP), established in 1964 to consider 'the biological basis of man's welfare' (Waddington 1975), and subsequently in the United Nations Educational, Scientific and Cultural Organization (UNESCO) Man and the Biosphere Programme (Caldwell 1984). In the 1960s, international conservation and environmental organizations began to try to interpret their ideas and demands in the context of development, as a means of gaining influence on decisions made about natural resource use in the Third World (Boardman 1981; McCormick 1989). This approach bore fruit in books such as *The Careless Technology* (Farvar & Milton 1973) and a handbook for development planners, *Ecological Principles for Economic Development* (Dasmann *et al.* 1973).

A significant watershed in this proactive approach of trying to integrate development and conservation was the United Nations Conference on the Human Environment held in Stockholm in June 1972 (Adams 1990). At Stockholm, Third World countries feared that their need for development would be overwhelmed by concerns for environmental quality and industrialization, and that discussion of global resource management would mean loss of sovereignty over resources and control by industrialized countries (Biswas & Biswas 1984). It was argued that poor countries suffered the 'pollution of poverty', and the phrase 'sustainable development' was coined to suggest that wealth was possible without environmental destruction, and that the path to it could be found both by Third World countries struggling with poverty and First World countries faced by pollution (Clarke & Timberlake 1982).

Stockholm was simply one landmark in a continuing process of developing conceptual relations between conservation and development. In 1975, work began on a strategy for nature conservation. In 1977, the United Nations Environment Programme (UNEP) commissioned the International Union for the Conservation of Nature and Natural Resources (IUCN) to draft a document to provide 'a global perspective on the myriad conservation problems that beset the world and a means of identifying the most effective solutions to the priority problems' (Munro 1978). *The World Conservation Strategy* (WCS) was written by IUCN with finance provided by UNEP and the World Wildlife Fund (IUCN 1980; Boardman 1981; McCormick 1986). Thinking about sustainable development was subsequently further developed in the report of the World Commission on Environment and Development, *Our Common Future* (Brundtland 1987), the follow-up to the WCS, *Caring for the Earth* (IUCN 1991), and in the products of the United Nations Conference on Environment and Development (UNCED) at Rio in 1992, particularly the labyrinthine Agenda 21.

A remarkably consistent core of ideas flows through all of these documents, a 'mainstream' that has persisted through the two decades between Stockholm and Rio (Adams 1990, 1993). This persistence is not surprising, since all have been inspired, drafted and negotiated in a limited number of conference halls and committee rooms by a common pool of international environmental executives. The chief current in this mainstream is a vision of sustainable development that is influenced by science, by principles of wildlife conservation, by a global vision of harmonious international relations, and an emphasis on the rational management of resources to maximize human welfare.

The World Conservation Strategy highlighted three objectives. The first was the need to maintain 'life-support systems'. These systems, 'governed, supported or strongly moderated' by ecosystems, included agricultural land and soil, forests and coastal and freshwater ecosystems, variously threatened by soil erosion, pesticide-resistant pests, deforestation and pollution. The second objective was to preserve genetic diversity, both in locally adapted crops or livestock and in wild species, as an insurance (for example against crop diseases) and an investment for the future (e.g. crop breeding or pharmaceuticals). The third objective was the sustainable development of species and ecosystems, particularly fisheries, cropped wild species, timber resources and grazing land. The WCS called for rational planning and land use allocation, site-based protection of ecosystems and banks of genetic material. It proposed national conservation strategies to review development objectives in the light of the conservation objectives, to identify obstacles and propose cost-effective ways of overcoming them, and to select priority ecosystems and species for conservation.

Our Common Future emphasized economic and social development concerns, especially the goals of meeting basic needs, building environmental factors into economic decision making, and maintaining and revitalizing the world economy. It argued that growth was the only effective solution to poverty, in both industrialized and non-industrialized countries. However, it suggested that growth must be of a new kind—sustainable, environmentally aware, egalitarian, integrating economic and social development—'material- and energy-intensive and more equitable in its impact' (Brundtland 1987). The Brundtland report's route to sustainable development was through equitable trade—better access to global markets for the products of developing countries, lower interest rates, greater technology transfer, and significantly larger commercial and aid capital flows (Brundtland 1987).

Caring for the Earth turns back somewhat to the conservation focus of the WCS. It offers nine 'principles to guide the way towards sustainable societies'. These blend environmentalist and ethical ideas ('keep within the Earth's carrying capacity' and 'respect and care for the community of life') with humanitarian concerns ('improve the quality of human life'). It is strongly pragmatic, arguing for a national framework for integrating development and conservation. It is also carefully drafted, calling for 'development that is both people-centred, concentrating on improving the human condition, and conservation-based, maintaining the variety and productivity of nature' (IUCN 1991). Like its predecessors it stresses the common ground between conservation and development, saying 'we have to stop talking about conservation and development as if they were in opposition, and recognize that they are essential parts of one indispensable process'.

Debate about sustainable development has now been swamped by the proceedings of the United Nations Conference on Environment and Development (the Earth Summit) in Rio in 1992 (Holmberg *et al.* 1993). The global media storm delighted in highlighting the contradictions of contrast between privileged delegates and those living in Rio de Janeiro's *favelas*, and was critical of the conference for failing to live up to inflated expectations and usher in a new environmental world order. Despite the doubling of the size of the Global Environmental Facility, UNCED failed to stimulate the scale of financial support necessary to implement Agenda 21. Approximately US$2.5 billion was pledged, compared to some US$125 billion needed (Holmberg *et al.* 1993). However, the cumbersome and costly UNCED process did produce more than hot air, in the form of the Rio Declaration, the Forest Principles, the Convention on Biodiversity, the Climate Treaty and Agenda 21. It also saw the creation of the Commission on Sustainable Development. The Biodiversity Convention was signed by 152 countries (the USA, which at first refused, did so later) and came into force on 29 December 1993.

There was a clear and deep conflict of interest at Rio between industrialized and non-industrialized countries, epitomized by the obvious lack of enthusiasm of the USA for the whole UNCED process. The gulf between countries in the industrialized north and the underdeveloped south became obvious before the conference. They differed over which were the chief problems (poverty and its attendant problems for less industrialized countries, global atmospheric change and the biodiversity implications of tropical deforestation for industrialized countries), and who should find and pay for solutions. As at Stockholm in 1972, there was fear on the part of Third World countries that their attempts to industrialize would be stifled by restrictive international agreements on atmospheric emissions. They also feared that their freedom to use their natural resources would be constrained by environmental agreements imposed by industrialized countries that had themselves become wealthy while polluting freely and removing their forest cover.

These tensions between governments are clear in the UNCED documents. The Rio Declaration was not the strong and sharp 'Earth Charter' originally conceived by the conference chairman, Maurice Strong, but contains a long list of 27 principles—'a bland declaration that provides something for everybody' (Holmberg *et al.* 1993). Agenda 21 was much larger, with over 500 pages in 40 chapters. The text of these documents was subjected to minute scrutiny and wearying debate by diplomats to ensure that they were not committed to anything unexpected. Holmberg *et al.* (1993) point out that these represent 'a global consensus and political commitment at the highest level on development and environmental cooperation', but they are also masterpieces of vagueness, and both the texts themselves and their interpretation are often contradictory. Thus while the Rio Declaration opens with the uncontroversial principle that '... human beings are at the centre of concerns for sustainable development. They are entitled to a healthy and productive harmony with nature', the US delegation released an 'interpretive statement' arguing against Principle 3 that 'development is not a right ... on the contrary, development is a goal we all hold' (Holmberg *et al.* 1993). Similarly, Principle 7 states blandly that 'states shall cooperate in a spirit of global partnership to conserve, protect and restore the health and integrity of the world's ecosystem', but this is balanced by Principle 2, which protects national interest by recognizing the sovereign right of countries to develop as they wish (Holmberg *et al.* 1993). The USA also rejected any interpretation of Principle 7 that implied any form of international liability.

14.6 Conservation vs development?

The whole idea of sustainable development has been formulated as a means of discovering and promoting an integrated approach to development and conservation. The traditional opposition between the two is thereby, at the rhetorical level at least, overcome. However, the relationship between them is extremely complex and somewhat more paradoxical than might at first appear. Here again it is necessary to go back in time, for it is the changes wrought by development that have driven the formation and growth of the conservation movement. The origins of conservation are also complex. Grove (1987, 1990b, 1992) describes the emergence of ideas about environmental management in the British Empire, in India, the oceanic islands of the Indian and Atlantic Oceans (Mauritius and St Helena), the West Indies and Cape Colony, as a result of the interaction of imperial trade, a rising sensitivity to Romanticism and the growth of science from the mid-eighteenth century. These ideas involved in particular a concern for deforestation and the idea of links between forests and 'desiccation' among colonial scientists, and they led to policies of forest control and the creation of forest reserves. Forest reserves had been established on Tobago and St Vincent in the West Indies in the eighteenth century, and in India in the nineteenth century. In the Cape, legislation to reserve land was passed in 1946 to prevent soil erosion on open areas close to Cape Town, and forest reserves were established in 1858 (Grove 1987).

The establishment of conservation institutions in Europe and North America also began in the nineteenth century (Allen 1976; Sheail 1976; Nash 1983; Worster 1985), and the now-familiar pattern of national parks began to appear in the 1880s and 1890s in Canada, south Australia and New Zealand (Fitter 1978). In the UK, the movement was much influenced by perceptions on the part of patrician pioneers of the need to hold back forces of change, or to save particular places from destructive forces (e.g. Sheail 1976; Adams 1986, 1996; Evans 1992; Jenkins & James 1994). Change in the countryside and the encroachment of the town and its industries led to the establishment of late-Victorian preservation societies, the National Trust for Places of Historic Interest and Natural Beauty, and the Commons, Open Space and Footpaths Preservation Society. The scandal of overcollection of birds' feathers for the millinery trade led to the establishment of what became the Royal Society for the Protection of Birds. These have been followed in the twentieth century by a bewildering variety of conservation and environmental organizations.

In the post-war period, sweeping social and economic changes were brought about by new and more efficient forms of human organization, for example the rise of factory production systems (called Fordism after the US car manufacturer and his standardized Model T), suburbanization and profound shifts in material culture. These changes were opposed, both in cultural opposition to 'Americanization', and in romantic protest against fragmentation, alienation and loss of community life (Chambers 1986; Veldman 1993). Concern about landscape change and habitat loss was both a part of this opposition to development, and a focus for wider concerns.

Change in the economy, and the environmental changes it caused, contributed to the rise of environmentalism in Europe and North America from the 1960s onwards (Hays 1987; Adams 1996). As part of that environmentalism, conservation found itself both opposing mainstream patterns of development, and also a product of the changes that development had created. Indeed, some commentators have argued that concern for the conservation of species and ecosystems is something only indulged in by those who have already benefited from the development process—that conservation is an activity allowed by affluence (Cotgrove & Duff 1980). There is some support for this observation. Globally, support for conservation is most developed in industrialized economies and, within those economies, membership of conservation organizations is strongly skewed towards wealthy and middle-class households. Furthermore, it is recognized that conservationists sometimes protect the quality of their own immediate environment very fiercely, to the exclusion of wider considerations (the 'NIMBY' or 'not in my back yard' syndrome), and that 'wildlife' (in nature reserves or national parks) is enjoyed most often and most readily by those with the money to drive or fly to see it.

On the other hand, of course, conservationists would argue that they are seeking to protect the environment for wider human benefit, and in particular they are concerned precisely about the environmental needs of those who are not wealthy, notably the Third World peasant farmer or fish-catcher, for whom environmental quality is an essential prerequisite for survival (Blaikie 1985; Adams 1990). Of course, the emphasis given to particular features of the environment by a First World environmentalist and a Third World farmer might be quite different. The farmer might be much less worried by the loss of rare species or the ozone hole than the environmentalist, for example, although both are likely to worry about the loss of forest resources or a declining fishery, and see the need for unpolluted rivers.

Although it might be argued that the interests of people-orientated development and a wildlife-orientated conservation will often be shared, conventional approaches to conservation have tended to place people and conservation in opposition. They have done so primarily through the establishment of protected areas. The IUCN Commission on National Parks and Protected Areas recognizes six different categories of protected area. These embrace a wide range of objectives, including 'strict nature reserve', 'national park' and 'protected landscape'. The enthusiasm with which Third World countries have established protected areas is remarkable. In sub-Saharan Africa protected areas cover some 2.4 million hectares (about 10% of the land area). Five countries (Benin, Seychelles, Tanzania, Uganda and Zambia) have designated more than 20% of their land area. Furthermore, 33 countries in sub-Saharan Africa have signed the World Heritage Convention, 20 countries take part in the UNESCO Man and the Biosphere Programme, and 17 countries have signed the Ramsar Convention on wetlands (Olindo & Mankoto 1993).

The main conventional approach to conservation has been through the creation of protected areas that exclude people as residents, prevent consumptive use and minimize human impacts. This 'fences and fines' approach (Wells *et al.* 1992) has been widely used in the Third World, but has led to conflict with the economic interests, and rights, of local people. Many government conservation departments have their roots in agencies established to prevent hunting (or 'poaching').

Turton (1987) describes the impact of the Omo National Park (established in 1966) on the Mursi of the Omo Valley in Ethiopia. Although this area was perceived by the Wildlife Conservation Department as 'wilderness', its ecology was in fact created and maintained by the Mursi economy of cattle herding, dry season cultivation and flood-retreat farming along the Omo River, as well as (traditionally) hunting for ivory and skins for trade. Designation of the park initially had little effect on the Mursi, since little conservation activity took place. However, in the 1970s Mursi were forced by drought into the Mago area in larger numbers and in near destitution, and human use of the park intensified. A report in 1978 proposing a second park to the east, the Mago National Park, saw the Mursi as a threat to conservation and proposed exclusion and resettlement. Turton argues that if the Mursi were effectively excluded from the parks, their economy would be destroyed.

It is now widely accepted by conservationists that there can be no success for their endeavours independent of development's goal of meeting human needs. Adams & McShane (1992) point out that 'conservation will either contribute to solving the problems of the rural poor who live day to day with wild animals, or those animals will disappear'. This recognition reflects both the pragmatic realization that in the Third World, at least, protected areas will not endure if they do not have the support of local people, and also the recognition of the moral claims of the human poor.

The importance of the links between environment and development, between attempts to maintain and improve environmental quality and the conditions of life of the poor, is now recognized. 'Development' processes in one place can impoverish both the economy and environment of others, for example through pollution or drastic economic change. Furthermore, access to natural resources (e.g. cultivable land, fuelwood or fish) is uneven, and those who are poor can be so marginalized economically that their lives degrade the very resources that sustain them (Blaikie 1985). Their poverty and the deterioration of their environments are two sides of the same coin.

Blaikie & Brookfield (1987) argue that 'land degradation can undermine and frustrate economic development, while low levels of economic development can in turn have a strong causal impact on the incidence of land degradation'. Blaikie emphasizes the political dimensions of rights over resources, stressing the need for those seeking to understand environment–development problems to explore the links between environment, economy and society—what he calls 'political ecology'. Poverty and environmental degradation, driven by the development process, combine to expose the poor in particular to risk and hazard.

14.7 Conservation and community: beyond fortress conservation

Ideas about the place of people in national parks have been changing rapidly in the 1980s to consider the needs of local people: 'having gained the support of government planners and decision makers, the most important task is to win the understanding and cooperation of the local people in the vicinity of the proposed parks' (Blower 1984). Debates at the Third and Fourth World Congresses on National Parks and Protected Areas in Bali in 1982 and Caracas in 1992 mark this change (McNeely & Miller 1984; McNeely 1993). At the 1992 congress, the adverse impacts of parks on local people were widely recognized and discussed, not least by delegates from indigenous groups who came and spoke. A much higher priority was accorded to the need both to integrate parks into economic development planning frameworks and to prioritize the interests (and recognize the rights) of local people (Kemf 1993; McNeely 1993). There is now a strong consensus that local people must be involved in management decisions about protected areas (Fig. 14.5), and that they should benefit from them economically. The support of local people is seen to be essential to their long-term survival and integrity (see Cartwright 1991; Hannah 1992).

Gradually, the old confrontational approach between park managers and local people is changing to one which attempts to address the needs of local communities. Zube & Busch (1990) identify four key features of the new approach to protected area management: (i) that local people should participate in park management and operations; (ii) that services (e.g. education and health care) should be delivered to local communities by park authorities; (iii) that allowance should be made for traditional land uses to continue (e.g. hunting and gathering, agriculture, religious practices, pastoralism); and (iv) that local people should be involved in park-related tourism.

Success in changing the ways conservation is practised will require

Fig. 14.5 Local people mark the boundary of the forest, Mount Kilum, Cameroon. BirdLife International's Kilum-Ijim Forest Project involves participatory work with local communities to prevent the encroachment of farms into remaining areas of forest (Coulthard 1996). (Photo by BirdLife.)

effective responses to the obvious institutional weaknesses of both protected area systems and the government wildlife administrations which operate them (Olindo & Mankoto 1993). Legislation is often weak or confusing, and conservation or protected area policies are frequently divorced from policy development elsewhere in government (e.g. in agriculture, forestry, river basin planning or tourism). Conservation staff are often undertrained and badly paid, and government bureaucracies are frequently immobilized for lack of operating expenses or means of transport. Institutional weakness can be tackled by training and targeted external support, and institutional overlap and inertia through innovative approaches to planning.

The slogan of the Fourth World Congress on National Parks in 1992 in Caracas, Venezuela was 'Parks for Life'. It emphasized the idea of expanding partnerships for conservation, and McNeely (1996) offers some suggestions as to how this might be done (Table 14.1). First, he argues that a protected

Table 14.1 Ten principles for successful partnerships between protected area managers and local people. (From McNeely 1996.)

1 Provide benefits to local people
2 Meet local needs
3 Plan holistically
4 Plan protected areas as a system
5 Plan site management individually, with linkages to the system
6 Define objectives for management
7 Manage adaptively
8 Foster scientific research
9 Form networks of supporting institutions
10 Build public support

area must provide direct benefits to local people, that therefore its benefit/
cost ratio to local people (before, or as well as, nationally) must be positive
if it is to prosper, and that those people must be involved in its planning.
Second, he suggests that protected areas planned and managed in such a
way as to meet local needs as well as biodiversity conservation goals. Third,
the protected area must be planned in a way that is integrated with
surrounding human uses. Fourth, the protected area must be planned as a
system that addresses both national and international objectives, while,
fifth, being managed individually in collaboration with local people. Sixth,
protected areas need clear management objectives. Seventh, they need to
be managed adaptively to respond to climatic change, economic change,
population growth, warfare or other threats. Eighth, it is important that
protected area management fosters scientific research in both the natural
and social sciences. Ninth, it is important that protected areas are supported
by a network of institutions, including national, regional and local governments, universities, private landowners, non-governmental organizations
(NGOs), businesses and local cooperatives. Finally, it is vital that protected
area managers seek to build public support, through the mass media and
institutions such as universities, museums, zoos or botanic gardens
(McNeely 1996).

The importance of taking the needs, ideas and aspirations of local
people seriously in conservation planning was for too long unrecognized
by conservationists, but is now part of the language of conservation
planning. Methods such as participatory rural appraisal (PRA) (Nichols
1991; Chambers 1994) have now been integrated into the professional
conservationist's portfolio (Fig. 14.6). Caldecott (1996), for example, emphasizes the need to ensure that 'all parts of a project are both socially and
environmentally durable', and that reliance on local participation can help
ensure that projects 'put down mental, social, financial and institutional
"roots" among the people of the project area'. 'Co-management' is now
an accepted concept in national parks in North America and Australia
(e.g. Hill & Press 1994).

In some cases, a community's cultural values and practices can provide
a solid basis for an effective conservation programme (Kleymeyer 1994),
as in the case of institutions for coastal zone management in Indonesia
(Zerner 1994). However, other commentators warn against excessive
optimism about the supposed 'new paradigm' of community conservation,
pointing out that 'the troubling question of whether communities actually
can resolve resource conflicts and slow environmental degradation better
than centralized authority remains' (Western & Wright 1994). Furthermore,
invoking 'the community' as a concept may well not provide a solution
to conservation problems, particularly if the terms of local people's involvement in decisions about land and resources are set by outsiders. As
Murphree (1994) comments, 'imposed community-based conservation is
a contradiction in terms, and implies an exercise in futility'.

Fig. 14.6 The development of microprojects in the Hadejia-Nguru Wetlands has been based on extensive participatory planning with local communities (for appropriate methods see Chambers 1994; Caldecott 1996). (Photo by D.H.L. Thomas.)

14.8 Conservation with development

There is increasing interest in projects that attempt to combine both conservation and development under a single project umbrella (Fig. 14.7). Such projects are often labelled 'integrated conservation–development projects' (ICDPs) (Wells *et al.* 1992; Barrett & Arcese 1995), 'people and park projects' (Hannah 1992) or 'conservation with development projects' (Stocking & Perkin 1992). Uncertainties about 'community conservation' approaches are magnified where conservation involves significant investment in resource development.

Those interested in exploring ways to integrate conservation and development now have a range of studies to draw on (Brandon & Wells 1992; Hannah 1992; Wells & Brandon 1992). The experience at Amboseli in Kenya in particular indicates how hard it is to integrate wildlife conservation and the interests of local people. Amboseli lies in land grazed by the Maasai in the dry season. Cattle comprise up to 60% of large-animal biomass at this season (Lindsay 1987). Amboseli lies within an extensive area designated as a game reserve by the colonial government in southern Kenya in 1899, a few years after the Maasai had been hit by famine following the rinderpest pandemic (Waller 1988, 1990). Amboseli National

Reserve was created in 1952 as one of a series of smaller reserves within this area. Hunting was banned, but Maasai grazing continued. In 1961, the district council assumed control of the area, but local Maasai continued to fear loss of further grazing rights, and demanded formal ownership of the area. Although the council received some entrance fee revenue and hunting licence fees, conflict developed and large game animals began to be killed. Because of this 'poaching', Amboseli was declared a national park in 1974, under a complex agreement whereby the Maasai gave up the right to graze within the park in return for joint ownership of surrounding bushlands in group ranches. These received water supplies, compensation for lost production through wildlife grazing, the development of lodges and wildlife viewing circuits, and developments such as a school and a dispensary. Developments were funded by the World Bank (Hannah 1992). The Kajiado District Council retained control of lodges on 160 ha in the heart of the park.

The Amboseli programme has been claimed as a success, and cited as a model for the coexistence of conservation and economic development through the generation of revenues from non-consumptive use of wildlife (Western 1982). However, many reviews are more critical (Lindsay 1987; Hannah 1992; Wells & Brandon 1992). Revenues have been generated by the park from campsite fees and the sale of wood and gravel, but financial benefit to the Maasai has been small. Although a school and cattle dip have been built, the borehole and pipeline system constructed to supply water outside the park has broken down (Lindsay 1987). Killings of rhinoceros and elephant began again in the early 1980s. Lindsay (1987)

Fig. 14.7 Contouring steep fields, Mount Kupe, Cameroon. BirdLife International's Mount Kupe Forest Conservation Project (South West Province, Cameroon) seeks to conserve montane forest containing species such as endemic Mount Kupe bush-shrike *Malacanotus kupeensis*. Work includes the use of simple technologies such as this A-frame to build terraces on cultivated hillsides to minimize soil erosion, and hence enhance the sustainability of agricultural production (Coulthard 1996). (Photo by BirdLife.)

argues that the Amboseli Park Plan did not offer an acceptable and sustainable alternative for the Maasai excluded from the park. Income to the Maasai was too little and too unpredictable, and continuing cultural, social and economic change among the Maasai undermined static assumptions about the long-term acceptability and sustainability of the group ranching system.

Parkipuny (1991) explores the rights and role of Maasai people in conservation in Kenya and Tanzania. He argues that the Maasai Mara National Reserve functions perfectly well in conservation terms under the ownership of the Narok County Council, and generates sizeable economic benefits for local people. He then argues that the Maasai also have both a right to be in the Ngorogoro Conservation Area in Tanzania and do no harm. The latter point is taken up and substantially corroborated by Homewood & Rodgers (1991).

Many commentators (e.g. Stocking & Perkin 1992; Barrett & Arcese 1995) point out that ICDPs perform poorly in both developmental and conservation terms. Experience with the first generation of ICDPs is mixed (Brandon & Wells 1992; Stocking & Perkin 1992; Wells & Brandon 1992). Wells & Brandon (1992) note that 'linking conservation and development objectives is in fact extremely difficult, even at a conceptual level'. Conservationists may have been naive in assuming that a commitment on paper to sustainability and participation or 'bottom-up' planning would yield successful projects where more conventional development projects have a poor record.

Barrett & Arcese (1995) query the biological sustainability of ICDPs, and particularly the problems of setting a 'sustainable harvest' in practice. They also criticize their socioeconomic rationale. They analyse who harvests illegally in protected areas, and the monetary and non-monetary reasons why they do so. They suggest that ICDPs do not decouple rural livelihoods from unsustainable patterns of wildlife harvest, and that positive impacts of the project on the local economy are transient and dependent on the maintenance of flows of project revenues. They conclude that ICDPs are 'no more than short-term palliatives'.

Stocking & Perkin (1992) provide a case study of ICDPs in action in the East Usambaras Agricultural Development and Environmental Conservation Project in Tanzania. The East Usambaras reach an altitude of 1500 m and support submontane forests with a very high level of endemism. The IUCN project began in 1987 with three aims: to improve the living standards of the people, to protect the functions of the forest (particularly its role as a catchment for downstream water supply) and to preserve biological diversity. Traditional conservation objectives were deliberately de-emphasized to stress revenue generation and development. After 4 years, achievements were modest. A vast range of project activities had been begun, from agricultural extension to attempts to control illegal pit-sawing, most with limited success. Problems included lack of funds, leading in turn to a lack of breadth in technical expertise, and the way in which capital and energy were dissipated in too wide a range of activities. Behind many of these problems lay the lack of a proper feasibility study at the planning stage (Caldecott 1996).

Stocking & Perkin (1992) conclude that ICDPs are hard to transfer from paper to reality. As projects, they are inherently highly complex and

(a)

Fig. 14.8 (a) Traditional cantilever bridge; (b) traditional suspension bridge. The Himalayan Jungle Project in the Palas Valley, Kohistan District, Pakistan, is run by BirdLife International with Worldwide Fund for Nature (WWF) Pakistan, World Pheasant Association (WPA) Pakistan and the North West Province Forest and Wildlife Department, under an agreement with the people of Bar Palas. The conservation interests of this area include the western tragopan *Tragopan melanocephalus*. Work has included a great diversity of activities, including the construction of suspension foot bridges as part of a wider disaster rehabilitation plan. (Photos by N. Bean, BirdLife.)

(b)

demand high levels of skill on the part of project staff. They also demand substantial funds and a realistic (i.e. slow) timescale. Their chances of success depend on the local perception of the project, and this is vulnerable to the public failure of particular components. ICDPs resemble other 'integrated' development projects fashionable in the 1970s and 1980s, both in their complexity and in the vulnerability to poor performance that this complexity brings. Clear and precise objectives, careful evaluation of the costs and benefits of project components at the level of the individual household, long-term commitment to funding and strong local participatory linkages are essential. It is obvious that projects of this sort will not be cheap to implement, and will not yield results quickly.

Experience with development 'microprojects' in support of conservation in the Hadejia-Nguru Wetlands in Nigeria leads to a similar solution (Adams & Thomas 1993). The scale of implementation here was relatively small, but a series of technical innovations were developed in collaboration with local communities, including a fish pond, improved bee hives for honey production, donkey ploughs, tree nurseries and embankments to control flood depth and timing to allow transplanting of rice (Fig. 14.9). Several of these initiatives were successful at a technical level, but their development was slow and careful, and there were significant constraints on their

Fig. 14.9 Improved bee hive design was one of a series of development microprojects initiated by the Hadejia-Nguru Wetland Conservation Project. It was technically successful, but widespread adoption was constrained by sociopolitical and cultural factors (Adams & Thomas 1995). (Photo by D.H.L. Thomas.)

Fig. 14.10 The fish pond was the first of a series of development microprojects through which the Hadejia-Nguru Wetlands Conservation Project has sought to identify strategies that can improve livelihoods in an area affected by drought and upstream dam construction in a sustainable way (Adams & Thomas 1995). (Photo by D.H.L. Thomas.)

implementation on a large scale. Their direct contribution to conservation priorities in the wetlands was slight (Adams & Thomas 1996).

There can be dangers from a conservation perspective if emphasis on development leads to such a profound de-emphasis of conservation goals that they are no longer seriously addressed. Oates (1995) argues that precisely this has happened in the case of the Okumu Forest Reserve in southwest Nigeria. She is very critical of the sustainable development rhetoric in *Caring for the Earth,* and blames this for new conservation programmes that have accelerated forest loss to small farmers. There may be a need to distinguish between the merits of development interventions of this kind as a contribution to local livelihoods and their contribution to conservation. While at a rhetorical level it may be desirable to argue that conservation and development can go hand in hand through a joint programme, development expenditure for conservation purposes may not give results that are cost effective in either livelihood or conservation terms.

14.9 Conservation for development

Ideas of sustainable development developed in *The World Conservation Strategy* (IUCN 1980) and its successors have had a profound effect on conservation strategies as well as on development thinking. Two shifts in conservation thinking stand out. The first involves a shift away from a traditional concern for protected areas and species to an awareness of ecosystem function, and particularly for the common interests between the development needs of the poor and the imperatives of conservation. The second shift is based on the so-called 'sustainable use' debate, the idea that sustainable harvesting of wild species can both secure economic revenues and sustain populations of target species.

An example of a renewed focus on ecosystems is the work of the IUCN Wetlands Programme. There is now growing international concern about wetlands because of their enormous economic importance, their sensitivity to ecological change and their attractiveness for other land users such as intensive agriculture and industrialization (Maltby 1988; Turner 1991). Wetland environments began to receive widespread attention following signing of the Ramsar Convention on Wetlands of International Importance, Especially as Waterfowl Habitat in 1971. This defined wetlands very broadly to include the widest possible variety of freshwater, brackish and shallow marine ecosystems (Dugan 1990). The IUCN Wetlands Programme's approach builds upon the principles of 'wise use' of wetlands adopted by the Contracting Parties to the Ramsar Convention in 1987 (Hollis 1990; Maltby 1988).

The IUCN Wetlands Programme approach is to seek routes to sustainable development that build upon their ecological and economic productivity and maintain the ecological and physical processes that sustain them (Dugan 1990). Thus, in a range of African countries (including Uganda, Tanzania and Zimbabwe), the IUCN has been involved with governments in developing national wetland conservation and management programmes, as well as working with national organizations and local people to explore the sustainable routes to the utilization of specific wetlands. The IUCN approach is seeking to replace the idea that wetlands need protection from people in the hope of protecting their wildlife, with an approach that relates the dynamics of the wetland ecosystems to the resources they can provide to local people.

In Nigeria, the IUCN is involved as a partner in the Hadejia-Nguru Wetlands Conservation Project, an attempt to promote integrated and sustainable use of the extensive flood plains of the Hadejia and Jama'are Rivers against pressures of upstream water abstraction, drought and demands for canalization downstream. The Hadejia-Nguru Wetlands consist of a complex of seasonally and permanently flooded land mixed with dry farmland, fed by the waters of the Hadejia and Jama'are Rivers as they drain northeastwards towards Lake Chad. The annual rainfall of the area is less than 500 mm, falling in a single wet season (roughly May to September), and is highly variable; river flow is highly seasonal, with almost 80% of runoff in August and September (Hollis *et al.* 1993). Up to 1 million people live in the wetlands, with a sizeable economy based on rainy season and dry season agriculture, fishing and grazing (Barbier *et al.* 1991). The area sustains internationally important populations of migratory birds, and has recently been incorporated into the Chad Basin National Park (Adams & Thomas 1996).

The Hadejia-Nguru Wetlands Conservation Project, established in 1987, aims to promote conservation and sustainable development of the

Hadejia-Jama'are Wetlands. This has taken place on two scales. The first involved the creation of community-level development 'microprojects' (mentioned above). The second addressed a much larger-scale critique of management of the water resources of the whole basin, and in particular the *unsustainable* development represented by dam construction and irrigation upstream of the wetlands.

The economy and ecology of the Hadejia-Jama'are flood plain are threatened by dams and water abstraction for irrigation upstream within the Hadejia and Jama'are river basins, particularly in low rainfall years (Hollis *et al.* 1993). Lack of floodwater presents a threat to both wildlife conservation interests and economic activities such as fishing and agriculture that transcends any conflicts that might exist between them (Adams & Thomas 1996). Both wildlife and the local economy demanded a focus on sustainability on a large scale as well.

The project has attempted to demonstrate the nature and value of economic activities in the wetlands (Barbier *et al.* 1991). In the past, decision makers have assumed that the area is relatively unimportant economically. Demonstration of the economic importance of the Hadejia-Jama'are flood plain provides a potentially strong argument for the maintenance of flood flows, which might convince government planners. There is a high degree of common ground between the interests of conservation (focused on the maintenance of wetland flooding for migratory birds) and those of peasant farmers and fishermen interested in the maintenance of floods for their livelihood. In this example, wildlife conservation has sought to engage directly with decisions that affect the environmental (and developmental) needs of local people.

A second significant shift in conservation thinking and practice is in the area of the utilization of wildlife, or in making wildlife 'pay its way' (Eltringham 1994). This can be done through non-consumptive use, through wildlife tourism. However, there are environmental costs to the tourist industry (not least resource consumption and pollution from air flights), and opportunities may be limited, particularly in those protected areas that do not share the attractiveness of the grassland savannas of East Africa or the gorillas of Zaire. Eltringham (1994) also points out that tourism is a fickle industry, 'subject to the vagaries of the political climate'.

More controversial, perhaps, is the consumptive use of wildlife. This may take the form of hunting by local people (e.g. for bushmeat), killing in return for a licence fee by big-game hunters, or through the collection of marketable or consumable natural products (e.g. rainforest rattans or turtle eggs). The key transformation of thinking here involves viewing wildlife simply as an economic resource that should be exploited in an effective and sustainable way. In 1990, the IUCN established a specialist group on sustainable use, and began to develop guidelines for the utilization of wild species. The resulting debate saw polarization within the IUCN, with members from Third World countries speaking out against the traditional protectionist approaches to conservation advocated by First World organizations, and calling for new approaches to conservation that explicitly recognized the dependence of rural communities on wild species as resources (Allen & Edwards 1995).

There are now a number of examples of use-based conservation projects and programmes. In Zambia, controlled hunting is an element of the Lwangwa Integrated Resource Development Project, begun in 1987. Revenue from safari hunting in the Lupande Game Reserve is used for development projects in the local area, as well as to finance the cost of game guards in the South Lwangwa National Park. Game guards are locally recruited, and both safari hunting and community game harvesting (particularly of hippopotamus) take place legally (Wells & Brandon 1992). This approach was obviously judged a success, since some of its principles were extended in 1987 to 10 other game management areas in Zambia under the ADMADE programme. Revenue from safari and other hunting fees is used to meet wildlife management costs (40% to these activities within the game management area itself plus 15% to the national parks system and 10% to the Zambian Tourist Bureau) and to generate revenue for local community projects (35%).

Wildlife utilization has been developed further in Zimbabwe. Wildlife ranching has become important economically as well as in conservation terms since populations of species such as the black rhinoceros and cheetah within Zimbabwe are mostly on private land. There has also been development of communal wildlife utilization projects under the Communal Areas Management Programme for Indigenous Resources (CAMPFIRE). This is similar to Zambia's ADMADE programme, but places greater emphasis on communal initiation and control of the programme and hunting activity. About 12.7% of Zimbabwe lies within national parks and equivalent areas, and a further 2.4% lies within forest reserves (Child 1995). Wildlife policy in Zimbabwe was reviewed in the late 1960s, and the Parks and Wildlife Act 1975 allowed authority over wildlife resources to be devolved to the district level (Metcalfe 1994). The revolutionary step of CAMPFIRE was the *de facto* granting of authority to local authorities, such that they can make decisions about hunting and conservation, and gain from revenues so generated. Outside the protected area, wildlife is considered a renewable resource to be managed to create maximum human welfare, and the CAMPFIRE model is seen to offer a form of conservation that is both popular and affordable (Child 1995).

The most controversial dimension of the debate about the consumptive use of wildlife is probably the harvesting of elephants. It has been widely argued that the high value of ivory means that it is economically desirable to treat elephants as a resource, and that it is scientifically possible to set sustainable harvest levels (e.g. Barbier *et al.* 1990). This approach has been pursued strongly by the countries of southern Africa such as Zimbabwe (Hill 1995), although opponents of elephant hunting (chiefly motivated by ethical considerations, or fears that institutional failure in measures to control hunting would allow indiscriminate and illegal killing to continue, particularly in East and West Africa) won their case for a total ban on international trade in ivory at the meeting of Contracting Parties to the Convention on International Trade in Endangered Species (CITES) in 1989. The relative success of CAMPFIRE and other community-based consumptive use projects in Africa is reviewed by Barbier (1992). This may be an effective way forward for conservation, at least for those areas of Africa with extensive savannas, good tourist infrastructure and a ready supply of wealthy safari hunters. However, as Allen & Edwards (1995) point out, the debate about consumptive use cuts deeply into established patterns of conservation thinking. While to many observers in the Third World it looks like pragmatic common sense, to others (including many conservationists in western urban societies) sustainable use is 'a very threatening concept because it challenges our perceptions of what conservation is about'.

14.10 Conclusion

There is increasing confidence among conservationists that economic development and conservation go hand in hand to the potential benefit of local people, and the availability of funds through the Global Environmental Facility since Rio has brought considerable investment in projects aimed at working in precisely this way. Enthusiasm for local people as conservationists is matched, in the view of the World Bank and bilateral donors such as USAID, by the reflection that this approach to conservation avoids the high costs of old-style interventionist conservation bureaucracies.

There are, however, serious questions about how far conservation goals can be met if people are allowed to enter national parks (the 'core' areas of zoned protected areas) to pursue economic activities, and how much 'governance' is needed to make conservation work—or how effective conservation is if organized locally. It may be that there is no ideal standard solution to the relationship between conservation and development. Experiments such as CAMPFIRE obviously offer exciting models for the conservation of some kinds of wildlife outside protected areas that may be more widely applicable. Similarly, non-consumptive use of wildlife clearly has application in many different circumstances, whether gorilla tourism in Zaire, turtle tourism in Costa Rica, or 'wilderness' tourism in the Canadian Rockies National Parks.

However, there are obvious limits to the extent that particular targets for the conservation of biodiversity can be made to coincide with the desires of local people. Wildlife may not always pay its way directly. There may, therefore, also be a place for strict exclusive conservation as it may be the only way to protect some environments from conflicting human demands. This may be easy to pay for and gain legitimacy for in rich countries, but may have to be paid for by international cross-subsidy in poor countries. The challenge for conservationists is to recognize where traditional strict conservation has a place, and to create the institutions that will pay for it and the value systems that will give it legitimacy (Fig. 14.11).

If 'conservation by local people' becomes a dogma, it is unlikely to be any more universally applicable than the dogma of 'no human use' that preceded it. If it fails locally and repeatedly (as it may well do), then there is likely to be a backlash, and this could be very damaging for the emerging relationship between conservation and development. Prins (1992) is among those who are unconvinced by the argument that conservation and human use (in his case pastoralism) are compatible: 'Nature reserves have little to do with "wise use of natural resources": we are closing our eyes if we think that allowing people to invade protected areas can result in a harmonious relation between them, their livestock and wildlife'.

Similar arguments could be made about the more general relationship between conservation and development. The rhetoric of sustainable development has offered an unprecedented opportunity for environ-mentalists and developers to sit down together and talk, to great mutual advantage. Much has been learned, but the comfortable assumption that this solves problems of conflict between the demands of human economies on the environment, and conflict between the environmental 'haves' (the First World and the Third World élite) and 'have nots', is quite unreal. Sustainable development is a phrase that defines the terms for a debate about the shape of human society and economy, and the terms of human relations with the rest of the biosphere. That debate has barely begun.

Fig. 14.11 Successful conservation initiatives demand originality and surprising skills, as this photograph of the Mount Kupe Forest Project Football Club (in South West Province, Cameroon) about to play the first match of the 1992 season demonstrates (Coulthard 1996). (Photo by R. Stone, BirdLife.)

References

Adams, J.S. & McShane, T.O. (1992) *The Myth of Wild Africa: Conservation Without Illusion*. New York: W.W. Norton & Company.

Adams, W.M. (1986) *Nature's Place: Conservation Sites and Countryside Change*. Hemel Hempstead: Allen & Unwin.

Adams, W.M. (1990) *Green Development: Environment and Sustainability in the World*. London: Routledge.

Adams, W.M. (1993) Sustainable development and the greening of development theory. In: Schuurman, F. (ed.) *Beyond the Impasse: New Directions in Development Theory*. London: Zed Books, pp. 207–22.

Adams W.M. (1995) *Future Nature: a Vision for Conservation*. London: Earthscan.

Adams, W.M. (1996) *Future Nature: a Vision for Conservation*. London: Earthscan.

Adams, W.M. & Thomas, D.H.L. (1995) Mainstream sustainable development: the challenge of putting theory into practice. *Journal of International Development* **5**, 591–604.

Adams, W.M. & Thomas, D.H.L. (1996) Conservation and sustainable resource use in the Hadejia-Jama'are Valley, Nigeria. *Oryx* **30**, 131–42.

Adams, J.M. & Woodward, F.I. (1989) Patterns in tree species richness as a test of the glacial extinction hypothesis. *Nature* **339**, 699–701.

Aizen, M.A. & Feinsinger, P. (1994) Forest fragmentation, pollination, and plant reproduction in a Chaco dry forest, Argentina. *Ecology* **75**, 330–51.

Albert, T.F. (1962) The effect of DDT on the sperm production of the domestic fowl. *Auk* **79**, 104–7.

Allen, C.M. & Edwards, S.R. (1995) The sustainable-use debate: observation from IUCN. *Oryx* **29**, 92–8.

Allen, D.E. (1976) *The Naturalist in Britain*. London: Penguin.

Anderson, D.M. & Grove, R.H. (1987) The scramble for Eden: past, present and future in African conservation. In: Anderson, D.M. & Grove, R.H. (eds) *Conservation in Africa*. Cambridge: Cambridge University Press, pp. 1–12.

Anderson, R.M. & May, R.M. (1991) *Infectious Diseases of Humans, Dynamics and Control*. Oxford: Oxford University Press.

Anderson, S. (1994) Area and endemism. *Quarterly Review of Biology* **69**, 451–71.

Anderson, S. & Marcus, L.F. (1993) Effect of quadrat size on measurement of species density. *Journal of Biogeography* **20**, 421–8.

Andow, D.A., Kareiva, P.M., Levin, S.A. & Okubo, A. (1990) Spread of invading organisms. *Landscape Ecology* **4**, 177–88.

André, H.M., Noti, M-I. & Lebrun, P. (1994) The soil fauna: the other last biotic frontier. *Biodiversity and Conservation* **3**, 45–56.

Andreasen, J.K. (1985) Insecticide resistance in mosquito fish of the Lower Rio Grande Valley of Texas—an ecological hazard. *Archives of Environmental Contamination and Toxicology* **14**, 573–7.

Andren, H. & Angelstam, P. (1988) Elevated predation rates as a consequence of edge effects in habitat islands: experimental evidence. *Ecology* **69**, 544–7.

Angel, M.V. (1994) Spatial distribution of marine organisms: patterns and processes. In: Edwards, P.J., May, R.M. & Webb, N.R. (eds) *Large-scale Ecology and Conservation Biology*. Oxford: Blackwell Scientific Publications, pp. 59–109.

Angermeier, P.L. (1994) Does biodiversity include artificial diversity? *Conservation Biology* **8**, 600–2.

Angermeier, P.L. & Karr, J.R. (1994) Biological integrity vs biological diversity as policy directives. *BioScience* **44**, 690–7.

Anon. (1992) *Global Biodiversity Strategy: Guidelines for Action to Save, Study and Use Earth's Biotic Wealth Sustainably and Equitably*. Washington, DC: World

Resources. Institute (WRI), The World
Conservation Union (IUCN) and United Nations
Environment Programme (UNEP).

Anon. (1997) *A Moving Story*. Godalming, Surrey:
World Wide Fund for Nature.

Archie, M., Mann, L. & Smith, W. (1993) *Partners in
Action: Environmental Social Marketing and
Environmental Education*. Washington, DC:
Academy for Educational Development.

Ashton, P.A. & Abbott, R.J. (1992) Multiple origins
and genetic diversity in the newly arisen
allopolyploid species, *Senecio cambrensis Rosser*
(Compositae). *Heredity* **68**, 25–32.

Austad, I. (1988) Tree pollarding in western Norway.
In: Birks, H.H., Birks, H.J.B., Kaland, P.E. & Moe,
D. (eds) *The Cultural Landscape: Past, Present and
Future*. Cambridge: Cambridge University Press,
pp. 11–29.

Austin, M.P., Cunningham, R.B. & Fleming, P.M.
(1984) New approaches to direct gradient analysis
using environmental scalars and statistical curve-
fitting procedures. *Vegetation* **55**, 11–27.

Austin, M.P. & Margules, C.R. (1986) Assessing
representativeness. In: Usher, M.B. (ed.) *Wildlife
Conservation Evaluation*. London: Chapman & Hall,
pp. 46–67.

Baker, C.S. & Palumbi, S.R. (1994) Which whales
are hunted? A molecular genetic approach to
monitoring whaling. *Science* **265**, 1538–9.

Balick, M.J. & Beitel, J.M. (1989) *Lycopodium* spores
used in condom manufacture: associated health
hazards. *Economic Botany* **43**, 373–7.

Balle, L. (1994) An index of evenness and its
associated diversity measure. *Oikos* **70**, 167–71.

Balmford, A. (1996) Extinction filters and current
resilience: the significance of past selection
pressures for conservation biology. *Tree* **11**, 193–6.

Balmford, A. & Long, A. (1995) Across-country
analyses of biodiversity congruence and current
conservation effort in the tropics. *Conservation
Biology* **9**, 1539–47.

Barbier, E.B. (1992) Community-based development
in Africa. In: Swanson, T.M. & Barbier, E.B. (eds)
*Economics for the Wilds: Wildilife, Wildlands, Diversity
and Development*. London: Earthscan, pp. 103–35.

Barbier, E.B., Adams, W.M. & Kimmage K. (1991)
*Economic Valuation of Wetland Benefits: the
Hadejia-Jama'are Floodplain, Nigeria*. LEEC Paper
DP-91–02. London: International Institute for
Environment and Development.

Barbier, E.B., Burgess, J. & Folke, C. (1994) *Paradise
Lost? The Ecological Economics of Biodiversity*. London:
Earthscan.

Barbier, E.B., Burgess, J.C., Swanson, T.M. & Pearce,
D.W. (1990) *Elephants, Economics amd Ivory*.
London: Earthscan.

Barbour, M., Pavlik, B., Drysdale, F. & Lindstrom S.
(1993) *California's Changing Landscapes: Diversity
and Conservation of California Vegetation*.
Sacramento: California Native Plant Society.

Barker, R.J. (1958) Notes on some ecological effects
of DDT sprayed on elms. *Journal of Wildlife
Management* **22**, 269–74.

Barkham, J. (1994) Climate change and British
wildlife. *British Wildlife* **5**, 169–80.

Barrett, C.B. & Arcese, P. (1995) Are integrated
conservation–development projects (ICDPs)
sustainable? On the conservation of large
mammals in subSaharan Africa. *World Development*
23, 1073–84.

Barrett, G.W. (1968) The effects of an acute
insecticide stress on a semi-enclosed grassland
ecosystem. *Ecology* **49**, 1019–35.

Barrett, S. (1994) Self-enforcing international
agreements. *Oxford Economic Papers* **46**, 878–94.

Baur, B. & Schmidt, B. (1996) Spatial and temporal
patterns of genetic diversity within species.
In: Gaston, K.J. (ed.) *Biodiversity: a Biology of
Numbers and Difference*. Oxford: Blackwell Science,
pp. 169–201.

Beard, J.S. (1969) *A Descriptive Catalogue of West
Australian Plants*. Perth: SGAP.

Beck, B.B., Kleiman, D.G., Dietz, J.M. *et al.* (1991)
Losses and reproduction in reintroduced golden
lion tamarins *Leontopithecus rosalia*. *Dodo* **27**, 50–61.

Beck, B.B., Rapaport, L.G. & Wilson, A.C. (1994)
Reintroduction of captive-born animals. In: Olney,
P.J.S., Mace, G.M. & Feistner, A.T.C. (eds) *Creative
Conservation: Interactive Management of Wild and
Captive Animals*. London: Chapman & Hall,
pp. 265–86.

Beddington, J.R. & Basson, M. (1994) The limits to
exploitation on land and sea. *Philosophical
Transactions of the Royal Society London Series B* **343**,
87–92.

Beddington, J.R. & May, R.M. (1977) Harvesting
natural populations in a randomly fluctuating
environment. *Science* **197**, 463–5.

Begon, M., Harper, J.L. & Townsend, C.R. (1996)
Ecology: Individuals, Populations and Communities.
Oxford: Blackwell Scientific Publications.

Behnke, R.H. & Scoones, I. (1991) *Rethinking Range
Ecology: Implications for Range Management in Africa*.
London: Overseas Development Institute/
International Institute for Environment and
Development.

Belbin, L. (1995) A multivariate approach to the selection of biological reserves. *Biodiversity and Conservation* **4**, 951–63.

Bellrose, F.C. (1959) Lead poisoning as a mortality factor in waterfowl populations. *Bulletin of the Illinois Natural History Survey* **27**, 235–88.

Berkefeld, K. (1993) Eine Nachweismöglichlkeit für Kondombenutzung bei Sexualdelikten. *Archiv für Kriminologie* **192**, 37–41.

Berkes, F. (ed.) (1989) *Common Property Resources: Ecology and Community-based Sustainable Development*. London: Belhaven Press.

Bernard, R.F. (1966) DDT residues in avian tissues. *Journal of Applied Ecology* **3** (Suppl.), 193–8.

Beshkarev, A.B., Swenson, J.E., Angelstam, P., Andren, H. & Blagovidov, A.B. (1994) Long-term dynamics of hazel grouse populations in source- and sink-dominated pristine taiga landscapes. *Oikos* **71**, 375–80.

Beverton, R.J.H. & Holt, S.J. (1957) *On the Dynamics of Exploited Fish Populations*. London: H.M. Stationery Office.

Bibby, C.J. & Etheridge, B. (1993) Status of the hen harrier *Circus cyaneus* in Scotland 1988–89. *Bird Study* **40**, 1–11.

Bierregaard Jr, R.O., Lovejoy, T.E., Kapos, V., dos Santos, A.A. & Hutchings, R.W. (1992) The biological dynamics of tropical rainforest fragments. *BioScience* **42**, 859–66.

Birkhead, M. & Perrins, C. (1985) The breeding biology of the mute swan *Cygnus olor* on the River Thames with special reference to lead poisoning. *Biological Conservation* **32**: 1–11.

Birks, H.J.B. (1986) Late-Quaternary biotic changes in terrestrial and lacustrine environments, with particular reference to north-west Europe. In: Berglund, B.E. (ed.) *Handbook of Holocene Palaeoecology and Palaeohydrology*. Chichester: Wiley, pp. 3–65.

Biswas, M.R. & Biswas, A.K. (1984) Complementarity between environment and development processes. *Environmental Conservation* **11**, 35–44.

Blackburn, T.M. & Gaston, K.J. (1995) What determines the probability of discovering a species?: a study of South American oscine passerine birds. *Journal of Biogeography* **22**, 7–14.

Blackburn, T.M. & Gaston, K.J. (1996) Spatial patterns in the species richness of birds in the New World. *Ecography*, **19**, 369–76.

Blaikie, P. (1985) *The Political Economy of Soil Erosion*. London: Longman.

Blaikie, P. & Brookfield, H. (1987) *Land Degradation and Society*. London: Methuen.

Blanchard, K.A. (1995) Reversing population declines in seabirds on the north shore of the Gulf of St Lawrence, Canada. In: Jacobson, S.K. (ed.) *Conserving Wildlife: International Education and Communication Approaches*. New York: Columbia University Press, pp. 51–63.

Blower, J. (1984) National Parks for developing countries. In: McNeely, A. & Miller, K.R. (eds) *National Parks, Conservation and Development: the Role of Protected Areas in Sustaining Society*. Washington, DC: Smithsonian Institute Press, pp. 722–7.

Blum, A. (1987) Students' knowledge and beliefs concerning environmental issues in four countries. *Journal of Environmental Education* **18**(3), 7–13.

Boardman, R. (1981) *International Organizations and the Conservation of Nature*. Bloomington: Indiana University Press.

Bobbink, R. & Willems, J.H. (1987) Increasing dominance of *Brachypodium pinnatum* (L.) *Beauv.* in chalk grassland: a threat to a species-rich ecosystem. *Biological Conservation* **40**, 301–14.

Boorman, L.A. & Fuller, R.M. (1981) The changing status of reedswamp in the Norfolk Broads. *Journal of Applied Ecology* **18**, 241–69.

Borg, K., Wanntorp, H., Erne, K. & Hanko, E. (1969) Alkyl mercury poisoning in terrestrial Swedish wildlife. *Viltrevy* **6**, 301–79.

Bormann, F.H. & Likens, G.E. (1979) Catastrophic disturbance and the steady state in northern hardwood forests. *American Scientist* **67**, 660–9.

Boucher, G. & Lambshead, P.J.D. (1995) Ecological biodiversity of marine nematodes in samples from temperate, tropical and deep-sea regions. *Conservation Biology* **9**, 1594–604.

Bourn, D. (1978) Cattle, rainfall and tsetse in Africa. *Journal of Arid Environments* **1**, 9–61.

Bowers, J.K. & Cheshire, P. (1983) *Agriculture, the Countryside and Land Use*. London: Methuen.

Brander, K. (1981) Disappearance of the common skate *Raia batis* from Irish Sea. *Nature* **290**, 48–9.

Brandon, K.E. & Wells, M. (1992) Planning for people and parks: design dilemmas. *World Development* **20**, 557–70.

Breeze, D.J. (1992) The great myth of Caledon. *Scottish Forestry* **46**, 331–5.

Brener, A.G.F. & Ruggiero, A. (1994) Leaf-cutting ants (*Atta* and *Acromyrmex*) inhabiting Argentina: patterns in species richness and geographical range sizes. *Journal of Biogeography* **21**, 391–9.

Brenton, T. (1994) *The Greening of Machiavelli*. London: Earthscan and Royal Institute of International Affairs.

Brewer, A. & Williamson, M. (1994) A new relationship for rarefaction. *Biodiversity and Conservation* **3**, 373–9.

Briggs, J.C. (1994) Species diversity: land and sea compared. *Systematic Biology* **43**, 130–5.

Brooks, T.M., Pimm, S.L. & Collar, N.J. (1997) Deforestation predicts the number of threatened birds in insular South-East Asia. *Conservation Biology* **11**, 382–4.

Brothers, N. (1991) Albatross mortality and associated bait loss in the Japanese longline fishery in the Southern Ocean. *Biological Conservation* **55**, 255–68.

Browder, J. (1988) Public policy and deforestation in Brazilian Amazon. In: Repetto, R. & Gillis, M. (eds) *Public Policies and Misuse of Forest Resources*. New York: Cambridge University Press.

Brown, A.W.A. (1978) *Ecology of Pesticides*. New York: Wiley.

Brown, J.H. (1984) On the relationship between abundance and distribution of species. *American Naturalist* **124**, 255–79.

Brown, J.H. (1995) *Macroecology*. Chicago: University of Chicago Press.

Brown, J.H. & Kurzius, M.A. (1987) Composition of desert rodent faunas: combinations of coexisting species. *Annales Zoologici Fennici* **24**, 227–37.

Brown, K. (1994) Biodiversity. In: Pearce, D. (ed.) *Blueprint 3*. London: Earthscan, pp. 98–114.

Browne, J. (1983) *The Secular Ark: Studies in the History of Biogeography*. New Haven: Yale University Press.

Bruenig, E.F. (1996) *Conservation and Management of Tropical Rainforests*. Wallingford: CAB International.

Brundtland, H. (1987) *Our Common Future*. Oxford: Oxford University Press for the World Commission on Environment and Development.

BSP\CI\TNC\WCS\WRI\WWF (1995) *A Regional Analysis of Geographic Priorities for Biodiversity Conservation in Latin America and the Caribbean*. Washington, DC: Biodiversity Support Programme.

Buck, N.A., Estesen, B.J. & Ware, G.W. (1983) DDT moratorium in Arizona: residues in soil and alfalfa after 12 years. *Bulletin of Environmental Contamination and Toxicology* **31**, 66–72.

Budiansky, S. (1993) The doomsday myths. *UN News and World Report* **13 December**, 81–3.

Budiansky, S. (1994) Extinction or miscalculation? *Nature* **370**, 104–5.

Budiansky, S. (1995) *Nature's Keepers*. New York: Free Press.

Buechner, M. (1987) Conservation in insular parks: simulation models of factors affecting the movement of animals across park boundaries. *Biological Conservation* **41**, 57–76.

Burbidge A.A. & McKenzie, N.L. (1989) Patterns in the modern decline of Western Australia's vertebrate fauna: causes and conservation implications. *Biological Conservation* **50**,143–98.

Butler, D. & Merton, D. (1992) *The Black Robin: Saving the World's Most Endangered Bird*. Oxford: Oxford University Press.

Butterfield, B.R., Csuti, B. & Scott, J.M. (1994) Modelling vertebrate distribution for gap analysis. In: Miller, R.I. (ed.) *Mapping the Diversity of Nature*. London: Chapman & Hall, pp. 53–68.

Buzas, M.A. & Culver, S.J. (1994) Species pool and dynamics of marine palaeocommunities. *Science* **264**, 1439–41.

Byrne, R.W. (1993) Complex leaf-gathering skills of mountain gorillas (*Gorilla g. beringei*). Variability and standardisation. *American Journal of Primatology* **31**, 241–61.

Cade, T.J. & Temple, S.A. (1995) Management of threatened bird species: evaluation of the hands-on approach. *Ibis* **137** (Suppl. 1), 161–72.

Cade, T.J., Enderson, J.H., Thelander, C.G. & White, C.M. (1988) *Peregrine Falcon Populations: their Management and Recovery*. Boise: The Peregrine Fund Inc.

Cairncross, F. (1991) *Costing the Earth*. London: Business Books in association with The Economist Books.

Cairncross, F. (1995) *Green, Inc.: Guide to Business and the Environment*. London: Earthscan.

Caldecott, J. (1996) *Designing Conservation Projects* Cambridge: Cambridge University Press.

Caldwell, L.K. (1984) Political aspects of ecologically sustainable development. *Environmental Conservation* **11**, 299–308.

Campbell, M.M. (1976) Colonisation of *Aphytis melinus Debach* (Hymenoptera, Aphelinidae) and *Aonidiella aurantii* (Mask.) (Hemiptera, Coccidae) on *Citrus* in South Australia. *Bulletin of Entomological Research* **65**, 659–68.

Carlsson, H., Carlsson, L., Wallin, C. & Wallin, N-E. (1991) Great tits incubating empty nest cups. *Ornis Svecica* **1**, 51–2.

Carlton, J.T. & Geller, J. (1993) Ecological roulette: the global transport and invasion of non-indigenous marine organisms. *Science* **261**, 78–82.

Carson, R. (1962) *Silent Spring*. London: Houghton Mifflin.

Cartwright, J. (1991) Is there hope for conservation

in Africa? *Journal of Modern African Studies* **29**, 355–71.

Caswell, H. (1989) *Matrix Population Models.* Sunderland, MA: Sinauer Associates.

Chambers, I. (1986) *Popular Culture: the Metropolitan Experience.* London: Routledge.

Chambers, R. (1983) *Rural Development: Putting the Last First.* London: Longman.

Chambers, R. (1994) The origins and practice of participatory rural appraisal. *World Development* **22**(7), 953–69.

Chase, A. (1987) *Playing God in Yellowstone: the Destruction of America's First National Park*, 2nd edn. San Diego: Jovanovich.

Child, G. (1995) Managing wildlife in Zimbabwe. *Oryx* **29**, 171–7.

Christensen, N.L., Agee, J.K., Brussard, P.F. *et al.* (1989) Interpreting the Yellowstone fires of 1988. *Bioscience* **39**, 678–85.

Clarke, C.W. (1973) The economics of overexploitation. *Science* **181**, 630–4.

Clarke, C.W. (1981) Bioeconomics. In: May, R.M. (ed.) *Theoretical Ecology*, 2nd edn. Oxford: Blackwell Scientific Publications, pp. 387–418.

Clarke, C.W. (1989) Bioeconomics. In: Roughgarden, J., May, R.M. & Levin, S.A. (eds) *Perspectives in Ecological Theory*. Princeton: Princeton University Press, pp. 275–86.

Clarke, R. & Timberlake, L. (1982) *Stockholm Plus Ten: Promises Promises? The Decade Since the 1972 UN Environment Conference.* London: Earthscan.

Clements, F.E. (1916) Plant succession: an analysis of the development of vegetation. *Carnegie Institute of Washington Publication* **242**, 1–512.

Coase, R. (1960) The problem of social cost. *Journal of Law and Economics* **3**, 1–44.

Cohen, J.E. (1995) *How Many People Can the Earth Support?* New York: W.W. Norton & Company, Inc.

Collar, N.J., Crosby, M.J. & Stattersfield, A.J. (1994) *Birds to Watch*, Vol. 2: *The World List of Threatened Birds*. BirdLife Conservation Series No. 4. Cambridge, UK: BirdLife International.

Collins, N.M., Sayer, J.A. & Whitmore, T.C. (eds) (1991) *The Conservation Atlas of Tropical Forests: Asia and the Pacific.* London: Macmillian Press for the International Union for Conservation of Nature and Natural Resources.

Colman, D. (1991) Land purchase as a means of providing public goods from agriculture. In: Hanley, N. (ed.) *Farming and the Countryside: an Economic Analysis.* Oxford: CAB International.

Commission of the European Communities vs Kingdom of Spain (1993) Case c-355/90.

European Community Reports I-2 (in French).

Common, M. (1988) *Environmental and Resource Economics: An Introduction.* London: Longman.

Conant, R. & Collins, J.T. (1991) *A Field Guide to Reptiles and Amphibians of Eastern and Central North America.* Boston: Houghton Mifflin Company.

Connell, J.H. (1978) Diversity in tropical rainforests and coral reefs. *Science* **199,** 1302–10.

Connell, J.H. & Slatyer, R.O. (1977) Mechanisms of succession in natural communities and their role in community stability and organization. *American Naturalist* **111**, 1119–44.

Connor, E.F. & McCoy, L.F. (1979) The statistics and biology of the species–area relationship. *American Naturalist* **113**, 791–833.

Conroy, M.J., Cohen, Y., James, F.C., Matsinos, Y.G. & Maurer, B.A. (1995) Parameter estimation, reliability, and model improvement for spatially explicit models of animal populations. *Ecological Applications* **5**, 17–19.

Cook, R.E. (1969) Variation in species density of North American birds. *Systematic Zoology* **18**, 63–84.

Cook, R.M., Sinclair, A. & Stefánsson, G. (1997) Potential collapse of North Sea cod stocks. *Nature* **385**, 521–2.

Cooke, A.S. (1973) Shell-thinning in avian eggs by environmental pollutants. *Environmental Pollution* **4**, 85–152.

Cooke, B.K. & Stringer, A. (1982) Distribution and breakdown of DDT in orchard soil. *Pesticides Science* **13**, 545–51.

Corbridge, S.E. (1992) *Debt and Development.* London: Edward Arnold.

Corbridge, S.E. (1995) *Development Studies: a Reader.* London: Edward Arnold.

Cory, J.S., Hirst, M.L., Williams, T. *et al.* (1994) First field trial of a genetically improved baculovirus insecticide. *Nature* **370**, 138–40.

Cotgreave, P. (1993) The relationship between body size and abundance in animals. *Trends in Ecology and Evolution* **8**, 244–8.

Cotgrove, S. & Duff, A. (1980) Environmentalism, middle-class radicalism and politics. *Sociological Review* **28**(2), 333–51.

Coulthard, N. (1996) Conservation in the community. *World Birdwatch* **18**, 12–15.

Council of the European Community (1992) Council Directive 92/43/EEC on the conservation of natural habitats and of wild fauna and flora. *Official Journal of the European Communities* **L206**, 7–50.

Cowan, M. & Shenton, R. (1995) The invention of

development. In: Crush, J. (ed.) *Power of Development*. London: Routledge, pp. 27–43.

Cowling, R.M. (ed.) (1992) *The Ecology of Fynbos: Nutrients, Fire and Diversity*. Cape Town: Oxford University Press.

Coyne, A. & Adamowicz, W. (1992) Modelling choice of site for hunting bighorn sheep. *Wildlife Society Bulletin* **20**, 26–33.

Crawley, M.J. (1987) What makes a community invasible? In: Gray, A.J., Crawley, M.J. & Edwards, P.J. (eds) *Colonization, Succession and Stability*. Oxford: Blackwell Scientific Publications, pp. 429–53.

Crawley, M.J. (1990) The population dynamics of plants. *Philosophical Transactions of the Royal Society of London Series B* **330**, 125–40.

Crawley, M.J. (1997a) Plant herbivore interactions. In: Crawley, M.J. (ed.) *Plant Ecology*, 2nd edn. Oxford: Blackwell Science, pp. 401–74.

Crawley, M.J. (1997b) *Aliens. The Population Biology of Non-indigenous Plants*. Oxford: Oxford University Press.

Crawley, M.J., Hails, R.S., Rees, M., Kohn, D. & Buxton, J. (1993) Ecology of transgenic oilseed rape in natural habitats. *Nature* **363**, 620–3.

Crawley, M.J., Harvey, P.H. & Purvis, A. (1996) Comparative ecology of the native and alien floras of the British Isles. *Philosophical Transactions of the Royal Society of London, Series B*, **351**, 1251–9.

Crick, H.A.P. & Ratcliffe, D.A. (1995) The peregrine, *Falco peregrinus* breeding population of the United Kingdom in 1991. *Bird Study* **42**, 1–19.

Croxall, J.P., Rothery, P., Pickering, S.P.C. & Prince, P.A. (1990) Reproductive performance, recruitment and survival of wandering albatrosses *Diomedea exulans* at Bird Island, South Georgia. *Journal of Animal Ecology* **59**, 775–96.

Crush, J. (ed.) (1995) *Power of Development*. London: Routledge.

Curio, E. (1993) Proximate and developmental aspects of antipredator behaviour. *Advances in the Study of Behaviour* **22**, 135–238.

Curio, E., Ernst, U. & Vieth, W. (1978) Cultural transmission of enemy recognition: one function of mobbing. *Science* **202**, 899–901.

Curnutt, J., Lockwood, J., Luh, H-K., Nott, P. & Russell, G. (1994) Hotspots and species diversity. *Nature* **367**, 326–7.

Currie, D.J. (1991) Energy and large-scale patterns of animal- and plant-species richness. *American Naturalist* **137**, 27–49.

Curtis, J.T. & Partch, M.L. (1948) Effect of fire on the competition between blue grass and certain prairie plants. *American Midland Naturalist* **39**, 437–43.

Cushing, D.H. (1988) *The Provident Sea*. Cambridge: Cambridge University Press.

Dahl, A.L. (1986) *Review of the Protected Areas System in Oceania*. Gland, Switzerland: International Union for Conservation of Nature and Natural Resources and United Nations Environment Programme.

Daly H.E. (1992) *Steady State Economics*, 2nd edn. London: Earthscan.

D'Antonio, C.M. & Vitousek, P.M. (1992) Biological invasions by exotic grasses, the grass/fire cycle, and global change. *Annual Review of Ecology and Systematics* **23,** 63–87.

Darwin, C. (1888) *The Origin of Species by Means of Natural Selection or the Preservation of Favoured Races in the Struggle for Life*, 6th edn. London: John Murray.

Dasmann, R.F. (1972) Towards a system for classifying natural regions of the world and their representation by national parks and reserves. *Biological Conservation* **4**, 247–55.

Dasmann, R.F. (1973) *Biotic Provinces of the World*. Occasional Paper No. 9. Gland, Switzerland: International Union for the Conservation of Nature and Natural Resources.

Dasmann, R.F., Milton, J.P. & Freeman, P.H. (1973) *Ecological Principles for Economic Development*. Chichester: Wiley.

Davis, G.J. & Howe, R.W. (1992) Juvenile dispersal, limited breeding sites, and the dynamics of metapopulations. *Theoretical Population Biology* **41**, 184–207.

Davis, M.B. (1981) Quaternary history and the stability of plant communities. In: Wert, D., Shugart, H.H. & Botkin, D.B. (eds) *Forest Succession: Concepts and Applications*. Berlin: Springer-Verlag, pp. 132–53.

Davis, M.B. (1994) Ecology and paleoecology begin to merge. *Trends in Ecology and Evolution* **9**, 357–8.

Davis, S.D., Heywood, V.H. & Hamilton, A.C. (eds) (1994) *Centres of Plant Diversity*, Vol. 1: *Europe, Africa, South West Asia and the Middle East*. Cambridge: IUCN Publications.

Davis, S.D., Heywood, V.H. & Hamilton, A.C. (eds) (1995) *Centres of Plant Diversity*, Vol. 2: *Asia, Australasia and the Pacific*. Cambridge: IUCN Publications.

De Klemm, C. & Shine, C. (1993) *Biological Diversity Conservation and the Law*. Gland, Switzerland: International Union for the Conservation of Nature and Natural Resources.

DeAngelis, D.L. & Gross, L.J. (eds) (1992). *Individual-based Models and Approaches in Ecology*. New York: Chapman & Hall.

Della Sala, D.A., Olson, D.M., Barth, S.E., Crane, S.L. & Primm, S.A. (1995) Forest health: moving beyond rhetoric to restore healthy landscapes in the inland Northwest. *Wildlife Society Bulletin* **23**, 346–56.

Diamond, J.M. (1972) Biogeographic kinetics: estimation of relaxation times for avifaunas of Southwest Pacific Islands. *Proceedings of the National Academy of Sciences of the USA* **69**, 3199–203.

Diamond, J.M. (1975a) The island dilemma: lessons of modern biogeographic studies for the design of natural preserves. *Biological Conservation* **7**, 129–46.

Diamond, J.M. (1975b) Assembly of species communities. In: Cody, M.L. & Diamond, J.M. (eds) *Ecology and Evolution of Communities*. Massachussetts: Belknap, pp. 342–444.

Diamond, J.M. (1976) Island biogeography and conservation: strategy and limitations. *Science* **193**, 1027–9.

Diamond, J.M. (1984) 'Normal' extinctions of isolated populations. In: Nitecki, M.H. (ed.) *Extinctions*. Chicago: University of Chicago Press, pp. 191–246.

Diamond, J.M. (1989) Overview of recent extinctions. In: Western, D. and Pearl, M. (eds) *Conservation for the Twenty-first Century*. New York: Oxford University Press, pp. 37–41.

Diamond, J.M. & May, R.M. (1976) Island biogeography and the design of natural reserves. In: May, R.M. (ed.) *Theoretical Ecology: Principles and Applications*. Philadelphia: W.B. Saunders, pp. 228–52.

Dickson, J. (1993) Scottish woodlands: their ancient past and precarious present. *Scottish Forestry* **47**, 73–8.

Dinerstein, E. & Wikramanayake, E.D. (1993) Beyond 'hotspots': how to prioritise investments to conserve biodiversity in the Indo-Pacific region. *Conservation Biology* **7**, 53–65.

Dinerstein, E., Olson, D.M., Graham, D.J. *et al.* (1995) *A Conservation Assesssment of the Terrestrial Ecoregions of Latin America and the Caribbean*. Washington, DC: World Bank.

Doak, D.F. (1995) Source–sink models and the problem of habitat degradation: general models and applications to the Yellowstone grizzly. *Conservation Biology* **9**, 1370–9.

Doak, D.F. & Mills, L.S. (1994) A useful role for theory in conservation. *Ecology* **75**, 615–26.

Dolman, P.M. & Sutherland, W.J. (1991) Historical clues to conservation. *New Scientist* **1751**, 40–3.

Dolman, P.M. & Sutherland, W.J. (1992) The ecological changes of Breckland grass heaths and the consequences of management. *Journal of Applied Ecology* **29**, 402–13.

Donovan, T.M., Lamberson, R.H., Kimber, A., Thompson III, F.R. & Faaborg, J. (1995a) Modeling the effects of habitat fragmentation on source and sink demography of neotropical migrant songbirds. *Conservation Biology* **9**, 1396–407.

Donovan, T.M., Thompson III, F.R., Faaborg, J. & Probst, J.R. (1995b) Reproductive success of migratory birds in habitat sources and sinks. *Conservation Biology* **9,** 1380–95.

Drake, J.A., Mooney, H.A., di Castri, F. *et al.* (eds) (1989) *Biological Invasions, a Global Perspective*. Chichester: John Wiley.

Dugan, P.J. (ed.) (1990) *Wetland Conservation: a Review of Current Issues and Required Action*. Gland, Switzerland: International Union for the Conservation of Nature and Natural Resources.

Dung, V.V., Giao, P.M., Chinh, N.N., Tuoc, D., Arctander, P. & McKinnon, J. (1993) A new species of living bovid from Vietnam. *Nature* **363**, 443–5.

Dunn, E. (1995) *Global Impacts of Fisheries on Seabirds*. London: Royal Society for the Protection of Birds.

Dunning Jr, J.B., Stewart, D.J., Danielson, B.J. *et al.* (1995) Spatially explicit population models: current forms and future uses. *Ecological Applications* **5**, 3–11.

Dynesius, M. & Nilsson, C. (1994) Fragmentation and flow regulation of river systems in the northern third of the world. *Science* **266**, 753–62.

Easterbrook, G. (1995) *A Moment on the Earth*. New York: Viking.

Eggleton, P. (1994) Termites live in a pear-shaped world: a response to Platnick. *Journal of Natural History* **28**, 1209–12.

Eggleton, P., Williams, P.H. & Gaston, K.J. (1994) Explaining global termite diversity: productivity or history? *Biodiversity and Conservation* **3**, 318–30.

Ehrlich, P.R. (1994) Energy use and biodiversity loss. *Philosophical Transactions of the Royal Society London Series B* **344**, 99–104.

Ehrlich, P.R. & Daily, G.C. (1993) Population extinction and saving biodiversity. *Ambio* **22**, 64–8.

Ehrlich, P.R. & Ehrlich, A.H. (1981) *Extinction: the Causes and Consequences of the Disappearance of Species*. New York: Random House.

Ehrlich, P.R. & Wilson, E.O. (1991) Biodiversity studies: science and policy. *Science* **253**, 758–62.

Elliott, J.M. (1994) *Quantitative Ecology and the Brown Trout*. Oxford: Oxford University Press.

Ellis, D.H., Dobrott, S.J. & Goodwin, J.G. (1977) Reintroduction techniques for masked bobwhites. In: Temple, S.A. (ed.) *Endangered Birds: Management Techniques for Preserving Threatened Species*. London: Croom Helm, pp. 345–54.

Elmes, G.W. & Thomas, J.A. (1992) Complexity of species conservation in managed habitats—interaction between Maculinea butterflies and their ant hosts. *Biodiversity and Conservation* **1**, 155–69.

Elton, C.S. (1958) *The Ecology of Invasions by Animals and Plants*. London: Methuen.

Eltringham, S.K. (1994) Can wildlife pay its way? *Oryx* **28**, 163–8.

Erwin, T.L. (1982) Tropical forests: their richness in Coleoptera and other arthropod species. *Coleopterists Bulletin* **36**, 74–5.

Escobar, A. (1995) Imagining a post-development era. In: Crush, J. (ed.) *Power of Development*. London: Routledge, pp. 211–27.

Essen, L. (1991) A note on the lesser white-fronted goose *Anser erythropus* in Sweden and the result of a re-introduction scheme. *Ardea* **79**, 305–6.

Etheridge, B., Summers, R. & Green, R.E. (1997) The effects of illegal killing and destruction of nests by humans on the population dynamics of the hen harrier *Circus cyaneus* in Scotland. *Journal of Applied Ecology* **34**, 1081–1105.

Etnier D.A. & Starnes, W.C. (1993) *The Fishes of Tennessee*. Knoxville: University of Tennessee Press.

European Environment Agency (1996) *Environmental Taxes: Implementation and Environmental Effectiveness*. Copenhagen: European Environment Agency.

Evans, D. (1992) *A History of Nature Conservation in Great Britain*. London: Routledge.

Evans, M.I. (1994) *Important Bird Areas in the Middle East*. Cambridge, UK: BirdLife International.

Evelyn, J. (1664) *Sylva, or a Discourse of Forest-Trees*. London.

Fahrig, L. & Merriam, G. (1985) Habitat patch connectivity and population survival. *Ecology* **66**, 762–8.

FAO (1992) *The Forest Resources of the Tropical Zone by Main Ecological Regions*. Rome: Food & Agricultural Organization.

Farvar, M.T. & Milton, J.P. (eds) (1973) *The Careless Technology: Ecology and International Development*. London: Stacey.

Fernald, M.L. (1970) *Gray's Manual of Botany*. New York: Van Nostrand.

Fielder, P.L. & Jain, S.K. (eds) (1992) *Conservation Biology: the Theory and Practice of Nature Conservation*. London: Chapman & Hall.

Fisher, A.C. (1980) *Resource and Environmental Economics*. Cambridge: Cambridge University Press.

FitzGibbon, C.D., Mogaka, H. & Fanshawe, J.H. (1995) Subsistence hunting in Arabuko-Sokoke Forest, Kenya, and its effects on mammal populations. *Conservation Biology* **9**, 1116–26.

Fjeldså, J. & Rahbek, C. (1997) Species richness and endemism in South American birds: implications for the design of networks of nature reserves. In: Laurence, W.F., Bierregaard, R. & Mortiz, C. (eds) *Tropical Forest Remnants: Ecology, Management and Conservation of Fragmental Communities*. Chicago: University of Chicago Press, pp. 466–82.

Fleming, T.H. (1973) Numbers of mammal species in North and Central American forest communities. *Ecology* **54**, 555–63.

Forey, P.L., Humphries, C.J. & Vane-Wright, R.I. (eds) (1994) *Systematics and Conservation Evaluation*. Oxford: Oxford University Press.

Frankel, O.H., Brown, H.D. & Burdon, J.J. (1995) *The Conservation of Plant Biodiversity*. Cambridge: Cambridge University Press.

Frankham, R. (1994) Genetic management of captive population for reintroductions. In: Serena, M. (ed.) *Reintroduction Biology of Australian and New Zealand Fauna*. Chipping Norton: Surrey Beatty & Sons, pp. 31–4.

Frankham, R. (1995) Conservation genetics. *Annual Review of Genetics* **29**, 305–27.

Frankham, R. & Loebel, D.A. (1992) Modelling problems in conservation genetics using captive *Drosophila* populations: rapid genetic adaptation to captivity. *Zoo Biology* **11**, 333–42

Freeberg, W.H. & Taylor, L.E. (1961) *Philosophy of Outdoor Education*. Minneapolis: Burgess Publishing.

Freedman, B. (1995) *Environmental Ecology*, 2nd edn. London: Academic Press.

Fuller, R.M. (1987) The changing extent and conservation interest of lowland grasslands in England and Wales: a review of grassland surveys 1930–84. *Biological Conservation* **40**, 281–300.

Garrod, G. & Willis, K. (1995) Valuing the benefits of the South Downs Environmentally Sensitive Area. *Journal of Agricultural Economics* **46**(2), 160–73.

Gaston, K.J. (1991a) The magnitude of global insect species richness. *Conservation Biology* **5**, 283–96.

Gaston, K.J. (1991b) Estimates of the near-imponderable: a reply to Erwin. *Conservation Biology* **5**, 566–8.

Gaston, K.J. (1991c) Body size and probability of description: the beetle fauna of Britain. *Ecological Entomology* **16**, 505–8.

Gaston, K.J. (1993) Spatial patterns in the description and richness of the Hymenoptera. In: LaSalle, J. & Gauld, I.D. (eds) *Hymenoptera and Biodiversity.* Wallingford: CAB International, pp. 277–93.

Gaston, K.J. (1994a) *Rarity.* London: Chapman & Hall.

Gaston, K.J. (1994b) Spatial patterns of species description: how is our knowledge of the global insect fauna growing? *Biological Conservation* **67**, 37–40.

Gaston, K.J. (1996a) What is biodiversity? In: Gaston, K.J. (ed.) *Biodiversity: a Biology of Numbers and Difference.* Oxford: Blackwell Science, pp. 1–9.

Gaston, K.J. (ed.) (1996b) *Biodiversity: a Biology of Numbers and Difference.* Oxford: Blackwell Science.

Gaston, K.J. (1996c) Species richness: measure and measurement. In: Gaston, K.J. (ed.) *Biodiversity: a Biology of Numbers and Difference.* Oxford: Blackwell Science, pp. 77–113.

Gaston, K.J. (1996d) Biodiversity—congruence. *Progress in Physical Geography* **20**, 105–12.

Gaston, K.J. & Blackburn, T.M. (1994) Are newly discovered bird species small-bodied? *Biodiversity Letters* **2**, 16–20.

Gaston, K.J. & Blackburn, T.M. (1995) Mapping biodiversity using surrogates for species richness: macro-scales and New World birds. *Proceedings of the Royal Society of London Series B* **262**, 335–41.

Gaston, K.J. & Blackburn, T.M. (1997) How many birds are there? *Biodiversity and Conservation* **6**, 615–25.

Gaston, K.J. & Hudson, E. (1994) Regional patterns of diversity and estimates of global insect species richness. *Biodiversity and Conservation* **3**, 493–500.

Gaston, K.J. & Mound, L.A. (1993) Taxonomy, hypothesis testing and the biodiversity crisis. *Proceedings of the Royal Society of London Series B* **251**, 139–42.

Gaston, K.J. & Williams, P.H. (1993) Mapping the world's species—the higher taxon approach. *Biodiversity Letters* **1**, 2–8.

Gaston, K.J. & Williams, P.H. (1996) Spatial patterns in taxonomic diversity. In: Gaston, K.J. (ed.) *Biodiversity: a Biology of Numbers and Difference.* Oxford: Blackwell Science, pp. 202–29.

Gaston, K.J., Blackburn, T.M. & Loder, N. (1995a) Which species are described first?: the case of North American butterfly species. *Biodiversity and Conservation* **4**, 119–27.

Gaston, K.J., Gauld, I.D. & Hanson, P. (1997) The size and composition of the hymenopteran fauna of Costa Rica. *Journal of Biogeography* **23**, 105–13.

Gaston, K.J., Scoble, M.J. & Crook, A. (1995b) Patterns in species description: a case study using the Geometridae (Lepidoptera). *Biological Journal of the Linnean Society* **55**, 225–37.

Gaston, K.J., Williams, P.H., Eggleton, P. & Humphries, C.J. (1995c) Large scale patterns of biodiversity: spatial variation in family richness. *Proceedings of the Royal Society of London Series B* **260**, 149–54.

Gavin, T.A. & Sherman, P.W. (1995) Proposition 80. *Conservation Biology* **9**, 1343–4.

Gentry, A.H. (ed.) (1990) *Four Neotropical Forests.* New Haven: Yale University Press.

George, J.L. & Frear, D.E.H. (1966) Pesticides in the Antarctic. *Journal of Applied Ecology* **3** (Suppl.), 155–67.

Getz, W.M. & Haight, R.G. (1989) *Population Harvesting: Demographic Models of Fish, Forest and Animal Resources.* Princeton: Princeton University Press.

Gibbons, D.W., Reid, J.B. & Chapman, R.A. (1993) *The New Atlas of Breeding Birds in Britain and Ireland: 1988–1991.* London: Poyser.

Gillison, A.N. & Brewer, K.R.W. (1985) The use of gradient directed transects or gradsects in natural resource survey. *Environmental Management* **20**, 103–27.

Gleason, H.A. (1926) The individualistic concept of the plant association. *Bulletin of the Torrey Botanical Club* **543**, 7–26.

Godfray, H.C.J. (1995) Field experiments with genetically manipulated insect viruses: ecological issues. *Trends in Ecology and Evolution* **10**, 465–9.

Godfray, H.C.J., Agassiz, D.J.L., Nash, D.R. & Lawton, J.H. (1995) The recruitment of parasitoid species to two invading herbivores. *Journal of Animal Ecology* **64**, 393–402.

Godwin, H. (1975) *History of British Flora,* 2nd edn. Cambridge: Cambridge University Press.

Goldschmidt, T., Witte, F. & Wanink, J. (1993) Cascading effects of the introduced nile perch on the detritivorous/phytoplanktivorous species in the sublittoral areas of Lake Victoria. *Conservation Biology* **7**, 686–99.

Goodman, D. (1987) The demography of chance extinction. In: Soulé, M.E. (ed.) *Viable Populations for Conservation.* Cambridge: Cambridge University Press, pp. 11–34.

Gordon, H.S. (1954) The economic theory of a common-property resource: the fishery. *Journal of Political Economics* **62**, 124–42.

Gore A. (1992) *Earth in the Balance: Forging a New Common Purpose.* London: Earthscan.

Gosling, L.M. & Baker, S.J. (1987) Planning and monitoring an attempt to eradicate coypu from Britain. *Symposia of the Zoological Society of London* **58**, 99–113.

Goulet, D. (1992) Development: creator and destroyer of values. *World Development* **20**, 467–75.

Grassle, J.F. & Maciolek, N.J. (1992) Deep-sea species richness: regional and local diversity estimates from quantitative bottom samples. *American Naturalist* **139**, 313–41.

Graveland, J. (1990) Effects of acid precipitation on reproduction in birds. *Experimentia* **46**, 960–70.

Graveland, J., van Derwal, R., van Balen, H. & van Noordwijk, A. (1994) Poor reproduction in forest passerines from decline in snail abundance. *Nature* **368**, 446–8.

Gray, A.J., Marshall, D.F. & Raybould, A.F. (1991) A century of evolution in *Spartina anglica*. *Advances in Ecological Research* **21**, 1–62.

Greathead, D.J. (1995) Benefits and risks of classical biological control. In: Hokkanen, H.M.T. & Lynch, J.M. (eds) *Biological Control: Benefits and Risks*. Cambridge: Cambridge University Press, pp. 53–63.

Griffith, B., Scott, J.M., Carpenter, J.W. & Reed, C. (1989) Translocation as a species conservation tool: status and strategy. *Science* **245**, 477–80.

Grimmett, R.F.A. & Jones, T.A. (1989) *Important Bird Areas in Europe*. Cambridge: International Council for Bird Preservation.

Groombridge, B. (ed.) (1992) *Global Biodiversity: Status of the Earth's Living Resources*. London: Chapman & Hall.

Grove, A.T. & Rackham, O. (in press) *Towards a Historical Ecology of Southern Europe*. New Haven: Yale University Press.

Grove, R.H. (1987) Early themes in African conservation: the Cape in the Nineteenth Century. In: Anderson, D.M. & Grove, R.H. (eds) *Conservation in Africa: People, Policies and Practice*. Cambridge: Cambridge University Press, pp. 21–40.

Grove, R.H. (1990) The origins of environmentalism. *Nature* **345**: 11–14.

Grove, R.H. (1992) origins of western environmentalism. *Scientific American* **267**: 42–7.

Grubb, M., Koch, M., Munson, A., Sullivan, F. & Thomson, K. (1993) *The Earth Summit Agreements: a Guide and Assessment*. London: Earthscan and Royal Institute of International Affairs.

Gulland, J.A. (1983) *Fish Stock Assessment: A Manual of Basic Methods*. Chichester: Wiley-Interscience.

Haas, C.A. (1995) Dispersal and use of corridors by birds in wooded patches on an agricultural landscape. *Conservation Biology* **9**, 845–54.

Hadfield, M.G. (1986) Extinction in Hawaiian achatinelline snails. *Malacologia* **27**, 67–81.

Hadfield, P. (1994) If you can't save them, freeze them. *New Scientist* **141**, 1910, 10.

Haila, Y. & Kouki, J. (1994) The phenomenon of biodiversity in conservation biology. *Annales Zoologici Fennici* **31**, 5–18.

Hall, M.A. (1996) On by-catches. *Review of Fish Biology and Fisheries* **6**, 319–52.

Hambler, C. & Speight, M.R. (1995) Biodiversity conservation in Britain: science replacing tradition. *British Wildlife* **6**, 137–47 (see also replies on pp. 337–8 and 405).

Hammond, P.M. (1992) Species inventory. In: Groombridge, B. (ed.) *Global Biodiversity: Status of the Earth's Living Resources*. London: Chapman & Hall, pp. 17–39.

Hammond, P.M. (1994) Practical approaches to the estimation of the extent of biodiversity in speciose groups. *Philosophical Transactions of the Royal Society of London Series B* **345**, 119–36.

Hammond, P.M. (1995) Described and estimated species numbers: an objective assessment of current knowledge. In: Allsopp, D., Hawksworth, D.L. & Colwell, R.R. (eds) *Microbial Diversity and Ecosystem Function*. Wallingford: CAB International, pp. 29–70.

Hanley, N. (1993) Controlling water pollution using market mechanisms: results from empirical studies. In: Turner, R.K. (ed.) *Sustainable Environmental Economics and Management*. London: Belhaven Press.

Hanley, N. & Craig, S. (1991) Wilderness development decisions and the Krutilla–Fisher model: the case of Scotland's 'Flow Country'. *Ecological Economics* **4**, 145–64.

Hanley, N. & Spash, C. (1994) *Cost–Benefit Analysis and the Environment*. Cheltenham: Edward Elgar.

Hanley, N. & Sumner, C. (1995) Applying the Coase theorem to red deer in the Scottish Highlands. *Journal of Environmental Management* **43**, 87–95.

Hanley, N., Shogren, J. & White, B. (1996) *Environmental Economics in Theory and Practice*. Basingstoke: MacMillan.

Hanley, N., Spash, C. & Walker, L. (1995) Problems in valuing the benefits of biodiversity protection. *Environmental and Resource Economics* **5**, 249–72.

Hannah, L. (1992) *African People, African Parks: an Evaluation of Development Initiatives as a Means of Improving Protected Area Conservation in Africa*. Washington, DC: USAID.

Hannah, L., Lohse, D., Hutchinson, C., Carr, J.L. & Lankerani, A. (1994) A preliminary inventory of human disturbance of world ecosystems. *Ambio* **23**, 246–50.

Hanski, I. & Gilpin, M. (1991) Metapopulation dynamics—brief history and conceptual domain. *Biological Journal of the Linnean Society* **42,** 3–16.

Hanski, I. & Simberloff, D. (1997) The metapopulation approach, its history, conceptual domain and application to conservation. In: Hanski, I. & Gilpin, M.E. (eds) *Metapopulation Biology, Ecology, Genetics & Evolution*. San Diego: Academic Press, pp. 5–62.

Hanski, I. & Thomas, C.D. (1994) Metapopulation dynamics and conservation: a spatially explicit model applied to butterflies. *Biological Conservation* **68**, 167–80.

Hanski, I., Poyry, J., Pakkala, T. & Kuussaari, M. (1995) Multiple equilibria in metapopulation dynamics. *Nature* **377**, 618–21.

Hardin, G. (1968) The tragedy of the commons. *Science* **162**, 1243–8.

Harper, J.L. & Hawksworth, D.L. (1994) Biodiversity: measurement and estimation (Preface). *Philosophical Transactions of the Royal Society London Series B* **345**, 5–12.

Harrison, S. (1994) Metapopulations and conservation. In: Edwards, P.J., Webb, N.R. & May, R.M. (eds) *Large-scale Ecology and Conservation Biology*. Oxford: Blackwell Science, pp. 111–28.

Harrison, S. & Taylor, A.D. (1997) Empirical evidence for metapopulation dynamics: a critical review. In: Hanski, I. & Gilpin, M.E. (eds) *Metapopulation Dynamics: Ecology, Genetics and Evolution*. San Diego: Academic Press, pp. 27–42.

Harrison, S., Stahl, A. & Doak, D. (1993) Spatial models and spotted owls: exploring some biological issues behind recent events. *Conservation Biology* **7**, 950–3.

Harvey, P.H. & Pagel, M.D. (1991) *The Comparative Method in Evolutionary Biology*. Oxford: Oxford University Press.

Hastings, A. (1994) Conservation and spatial structure: theoretical approaches. *Lecture Notes in Biomathematics* **100,** 494–504.

Hastings, A. & Harrison, S. (1994) Metapopulation dynamics and genetics. *Annual Review of Ecology and Systematics* **25**, 167–88.

Hawkins, C.P. & MacMahon, J.A. (1989) Guilds: the multiple meanings of a concept. *Annual Review of Entomology* **34**, 423–51.

Hawksworth, D.L. (1991) The fungal dimension of biodiversity: magnitude, significance, and conservation. *Mycological Research* **95**, 441–56.

Hawksworth, D.L. (ed.) (1995) *Biodiversity: Measurement and Estimation*. London: Chapman & Hall.

Hays, S.P. (1987) *Beauty, Health and Permanence: Environmental Politics in the United States, 1955–85*. Cambridge: Cambridge University Press.

Hayward, I.M. & Druce, G.C. (1919) *The Adventive Flora of Tweedside*. Arbroath: T. Buncle & Co.

Hedrick, P.W., Lacy, R.C., Allendorf, F.W. & Soulé, M.E. (1996) Directions in conservation biology: comments on Caughley. *Conservation Biology* **10**, 1312–20.

Heessen, H.J.L. & Daan, N. (1996) *Long-term Changes in Ten Non-target Fish Species in the North Sea. ICES Journal of Marine Sciences* **53**, 1063–78.

Heil, G.W. & Diemont, W.H. (1983) Raised nutrient levels change heathland into grassland. *Vegetatio* **53**, 113–20.

Henderson, C. (1984) Publicity strategies and techniques for Minnesota's nongame wildlife checkoff. *Transactions North American Wildlife and Natural Resource Conference* **49**, 181–9.

Henderson, N. & Sutherland, W.J. (1996) Two truths about discounting and their environmental consequences. *Trends in Ecology and Evolution* **11**, 527–8.

Hengeveld, R. (1989) *Dynamics of Biological Invasions*. London: Chapman & Hall.

Hengeveld, R. (1990) *Dynamic Biogeography*. Cambridge: Cambridge University Press.

Hengeveld, R. (1994) Small-step invasion research. *Trends in Ecology and Evolution* **9**, 339–42.

Heywood, V.H. (ed.) (1995) *Global Biodiversity Assessment*. Cambridge: Cambridge University Press.

Hilborn, R. & Walters, C.J. (1992) Quantitative stock assessment: choice, dynamics and uncertainty. London: Chapman & Hall.

Hill, D.A. & Robertson, P. (1988) *The Pheasant*. Oxford: Blackwell Scientific Publications.

Hill, E.F., Heath, R.G., Spann, J.W. & Williams, J.D. (1975) *Lethal Dietary Toxicities of Environmental Pollutants to Birds*. Special Scientific Report—Wildlife No. 191. Washington, DC: US Fish and Wildlife Service.

Hill, K.A. (1995) Conflicts over development and environmental values: the international ivory trade in Zimbabwe's historical context. *Environment and History* **1**, 335–49.

Hill, M.A. & Press, A.J. (1994) Kakadu National Park: an Australian experience in comanagement.

In: Western, D., White, R.M. and Strum, S.C. (eds) *Natural Connections: Perspectives in Community-based Conservation*. Washington, DC: Island Press, pp. 135–57.

Hines, J.M., Hungerford, H.R. & Tomera, A.N. (1986/87) Analysis and synthesis of research on responsible environmental behavior: a meta-analysis. *Journal of Environmental Education* **18**, 1–8.

Hirons, G., Goldsmith, B. & Thomas, G. (1995) Site management planning. In: Sutherland, W.J. & Hill, D.A. (eds) *Managing Habitats for Conservation*. Cambridge: Cambridge University Press, pp. 22–41.

Hobbs, R.J. (1991) Disturbance as a precursor to weed invasion in native vegetation. *Plant Protection Quarterly* **6**, 99–104.

Hobbs, R.J. (1992) The role of corridors in conservation: solution or bandwagon. *Trends in Ecology and Evolution* **7**, 389–92.

Hobbs, R.J. & Huenneke, L.F. (1992) Disturbance, diversity, and invasion: implications for conservation. *Conservation Biology* **6**, 324–37.

Hodge, I. (1995) *Environmental Economics*. Basingstoke: MacMillan.

Hodkinson, I.D. & Casson, D. (1990) A lesser predilection for bugs: Hemiptera (Insecta) diversity in tropical rain forests. *Biological Journal of the Linnean Society of London* **43**, 101–9.

Holdgate, M. (1996) *From Care to Action: Making a Sustainable World*. London: Earthscan and International Union for the Conservation of Nature and Natural Resources.

Hollis, G.E. (1990) Environmental impacts of development on wetlands in arid and semiarid lands. *Hydrological Sciences Journal* **35**, 411–28.

Hollis, G.E., Adams, W.M. & Kano, A. (eds) (1993) *Hydrology and Sustainable Resource Development of a Sahelian Floodplain Wetland*. Gland, Switzerland: International Union for the Conservation of Nature and Natural Resources, Wetlands Programme.

Holm, L.G., Plunkett, D.L., Pancho, J.V. & Herberger, J.P. (1977) *The World's Worst Weeds: Distribution and Biology*. Honolulu: University of Hawaii Press.

Holmberg, J. & Sandbrook, R. (1992) Sustainable development: what is to be done? In: Holmberg, J. (ed.) *Policies for a Small Planet*. London: Earthscan.

Holmberg, J., Thomson, K. & Timberlake, L. (1993) *Facing the Future: Beyond the Earth Summit*. London: Earthscan and International Institute for Environment and Development.

Holt, R.D. & Lawton, J.H. (1994) The ecological consequences of shared natural enemies. *Annual Review of Ecology and Systematics* **25**, 495–520.

Homewood, K. & Rodgers, W.A. (1987) Pastoralism, conservation and the overgrazing controversy. In: Anderson, D.M. & Grove, R.H. (eds) *Conservation in Africa: People, Policies and Practice*. Cambridge: Cambridge University Press, pp. 111–28.

Homewood, K. & Rodgers, W.A. (1991) *Masailand Ecology*. Cambridge: Cambridge University Press.

Horowitz, M.M. & Little, P.D. (1987) African pastoralism and poverty: some implications for drought and famine. In: Glantz, M.H. (ed.) *Drought and Hunger in Africa*. Cambridge: Cambridge University Press, pp. 59–82.

Horwich, R. & Lyon, J. (1993) *A Belizean Rain Forest: the Community Baboon Sanctuary*. Wisconsin: Howlers Forever Inc.

Horwich, R.H. & Lyon, J. (1995) Multilevel conservation and education at the community baboon sanctuary, Belize. In: Jacobson, S.K. (ed.) *Conserving Wildlife: International Education and Communication Approaches*. New York: Columbia University Press, pp. 235–53.

Howe, R.W., Davis, G.J. & Mosca, V. (1991) The demographic significance of 'sink' populations. *Biological Conservation* **57**, 239–55.

Howell, E.A. & Jordan III, W.R. (1991) Tallgrass prairie restoration in the North American midwest. In: Spellerberg, F., Goldsmith, F.B. and Morris, M.G. (eds) *The Scientific Management of Temperate Communities for Conservation*. Oxford: Blackwell Science, pp. 395–414.

Hudson, R.H., Tucker, R.K. & Haegele, M.A. (1984) *Handbook of Toxicity of Pesticides to Wildlife*, 2nd edn. US Department of International Fish and Wildlife Service, Resource Publication No. 153. Washington, DC: Department of International Fish and Wildlife Service.

Humphries, C.J., Williams, P.H. & Vane-Wright, R.I. (1995) Measuring biodiversity value for conservation. *Annual Review of Ecology and Systematics* **26**, 93–111.

Hunter Jr, M.L. (1996) *Fundamentals of Conservation Biology*. Oxford: Blackwell Science.

Huntley, B. (1991) Historical lessons for the future. In: Spellerberg, I.F., Goldsmith, F.B. & Morris, M.G. (eds) *The Scientific Management of Temperate Communities for Conservation*. Oxford: Blackwell Science, pp. 473–503.

Huston, M.A. (1994) *Biological Diversity: the Coexistence of Species on Changing Landscapes*. Cambridge: Cambridge University Press.

Hutchings, J.A. & Myers, R.A. (1994) What can be learnt from the collapse of a renewable resource—

Atlantic cod *Gadus morhua* of Newfoundland and Labrador. *Canadian Journal of Fisheries and Aquatic Sciences* **51**, 2126–46.

Hyde, W. (1989) Marginal costs of managing endangered species: the red cockaded woodpecker. *Journal of Agricultural Economics Research* **41**(2), 12–19.

ICBP (1992) *Putting Biodiversity on the Map: Priority Areas for Global Conservation.* Cambridge: International Council for Bird Preservation.

ICES (1995) *Report of the Advisory Committee on Fishery Management. Stocks in the North Sea (sub-area IV).* Copenhagen: International Council for the Exploration of the Sea.

Irish, J. (1989) Biospeleology of Dragon's Breath Cave, South West Africa/Namibia. In: *Proceedings of the 7th Entomological Congress, Entomological Society of Southern Africa, Pietermaritzburg.* p. 73.

IUCN (1980) *The World Conservation Strategy.* Geneva: International Union for Conservation of Nature and Natural Resources, United Nations Environment Programme and World Wildlife Fund.

IUCN (1991) *Caring for the Earth: a Strategy for Sustainable Living.* Gland, Switzerland: International Union for Conservation of Nature and Natural Resources.

IUCN Commission on National Parks and Protected Areas (1994) *Parks for Life: Action for Protected Areas in Europe.* Gland, Switzerland: International Union for Conservation of Nature and Natural Resources.

IUCN-SSC (Species Survival Commission) (1994) *IUCN Red List Categories.* Gland, Switzerland: International Union for Conservation of Nature and Natural Resources.

Jackmann, H., Berry, P.S.M. & Imae, H. (1995) Tusklessness in African elephants—a future trend. *African Journal of Ecology* **33**, 230–5.

Jacobson, S.K. (1991) Evaluation model for developing, implementing and assessing conservation education programs: examples from Belize and Costa Rica. *Environmental Management* **15**(2), 143–50.

Jacobson, S.K. (ed.) (1995) *Conserving Wildlife: International Education and Communication Approaches.* New York: Columbia University Press.

Jacobson, S.K. & Padua, S.M. (1995) A systems model for conservation education in parks: examples from Malaysia and Brazil. In: Jacobson, S.K. (ed.) *Conserving Wildlife: International Education and Communication Approaches.* New York: Columbia University Press, pp. 3–15.

Jaksic, F.M. & Medel, R.G. (1990) Objective

recognition of guilds: testing for statistically significant species clusters. *Oecologia* **82**, 87–92.

James, H.F. & Olson, S.L. (1991) Descriptions of thirty-two new species of birds from the Hawaiian islands: Part II. Passeriformes. *Ornithological Monographs* **46**. Washington, DC: American Ornithologists' Union.

Janzen, D.H. (1986) Biogeography of an unexceptional place: what determines the saturniid and sphingid moth fauna of Santa Rosa National Park, Costa Rica, and what does it mean to conservation biology? *Brenesia* **25/26**, 51–87.

Jenkins, J. & James, P. (1994) *From Acorn to Oak Tree: the Growth of the National Trust.* London: MacMillan.

Jenks, S.M. & Wayne, R.K. (1992) Problems and policy for species threatened by hybridization: the red wolf as a case study. In: MCullough, D.R. & Barrett, R.H. (eds) *Wildlife 2001: Populations.* London: Elsevier Applied Science.

Jennersten, O. (1988) Pollination in *Dianthus deltoides* (Caryophyllaceae): effects of habitat fragmentation on visitation and seed set. *Conservation Biology* **4**, 359–66.

Jensen, S., Johnels, A.G., Olsson, M. & Westermark, T. (1972) The avifauna of Sweden as indicators of environmental contamination with mercury and chlorinated hydrocarbons. *Proceedings of the International Ornithological Congress* **15**, 455–65.

Jewell, P.A. (1980) Ecology and management of game animals and domestic livestock in African savannas. In: Harris, D.R. (ed.) *The Human Ecology of Savanna Environments.* London: Academic Press, pp. 353–81.

Johnson, G., Walker, C.H. & Dawson, A. (1994) Interactive effects of prochloraz and malathion in pigeon, starling and hybrid red-legged partridge. *Environmental Toxicology and Chemistry* **13**, 115–20.

Johnson, N.C. (1995) *Biodiversity in the Balance: Approaches to Setting Geographic Conservation Priorities.* Washington, DC: Biodiversity Support Programme.

Johnson, S.P. (1993) *The Earth Summit: the United Nations Conference on Environment and Development (UNCED).* London: Graham and Trotman.

Joint Committee for the Conservation of British Insects (1986) Insect re-establishment—a code of conservation practice. *Antennae* **10**, 13–18.

Jones, C.G., Lawton, J.H. & Shachak, M. (1994) Organisms as ecosystem engineers. *Oikos* **69**, 373–86.

Jordan III, W.R., Gilpin, M.E. & Aber, J.D. (1987) Restoration ecology: ecological restoration as a technique for basic research. In: Jordan III, W.R., Gilpin, M.E. & Aber, J.D. (eds) *Restoration Ecology.* Cambridge: Cambridge University Press, pp. 3–21.

Kadmon, R. & Shmida, A. (1990) Spatiotemporal demographic processes in plant populations: an approach and a case study. *American Naturalist* **135**, 382–97.

Kareiva, P. (1996) Developing a predictive ecology for non-indigenous species and ecological invasions. *Ecology* **77**, 1651–2.

Kareiva, P. & Stark, J. (1994) Environmental risks in agricultural biotechnology. *Chemistry and Industry* **2**, 52–5.

Kareiva, P., Parker, I.M. & Pascual, M. (1996) Can we use experiments and models in predicting the invasiveness of genetically engineered organisms? *Ecology* **77**, 1670–5.

Karr, J.R. (1982) Avian extinction on Barro Colorado Island, Panama: a reassessment. *American Naturalist* **119**, 220–39.

Kaufman, D.M. (1995) Diversity of New World mammals: universality of the latitudinal gradients of species and bauplans. *Journal of Mammalogy* **76**, 322–34.

Keddy, P.A. (1981) Experimental demography of the sand-dune annual, *Cakile edentula*, growing along an environmental gradient in Nova Scotia. *Journal of Ecology* **69**, 615–30.

Keddy, P.A. (1982) Population ecology on an environmental gradient: *Cakile edentula* on a sand dune. *Oecologia* **52**, 348–55.

Kelleher, G., Bleakley, C. & Wells, S. (eds) (1995) *A Global Representative System of Marine Protected Areas,* Vols 1–4. Washington, DC: Great Barrier Reef Marine Park Authority, World Bank and International Union for Nature Conservation and Natural Resources.

Kelleher, G. & Kenchington, R. (1992) *Guidelines for Establishing Marine Protected Areas.* Gland, Switzerland: International Union for Conservation of Nature and Natural Resources.

Kemf, E. (ed.) (1993) *The Law of the Mother: Protecting Indigenous Peoples in Protected Areas.* San Francisco: Sierra Club Books.

Kemp, N., Dilger, M., Burgess, N. & Van Dung, C. (1997) The saola *Pseudoryx nghetinhensis* in Vietnam—new information on distribution and habitat preferences, and conservation needs. *Oryx* **31**, 37–44.

Kettlewell, H.B.D. (1973) *The Evolution of Melanism. The Study of a Recurring Necessity. With Special Reference to Industrial Melanism in the Lepidoptera.* Oxford: Clarendon Press.

Kindvall, O. & Ahlen, I. (1992) Geometrical factors and metapopulation dynamics of the bush cricket, *Metrioptera bicolor Philippi* (Orthoptera: Tettigoniidae). *Conservation Biology* **6**, 520–9.

King, M. (1995). *Fisheries Biology, Assessment and Management.* Oxford: Fishing News Books.

Kirch, P.V. (1984) *The Evolution of the Polynesian Chiefdoms.* Cambridge: Cambridge University Press.

Kirkwood, G.P., Beddington, J.R. & Rossouw, J.A (1994) Harvesting species of different lifespans. In: Edwards, P.J., May, R.M. & Webb, N.R. (eds) *Large-scale Ecology and Conservation Biology.* Oxford: Blackwell Science, pp. 199–227.

Kitching, I. (1996) Identifying complementary areas for conservation in Thailand: an example using owls, hawkmoths and tiger beetles. *Biodiversity and Conservation* **5**, 841–58.

Klein, B.C. (1989) Effects of forest fragmentation on dung and carrion beetle communities in central Amazonia. *Ecology* **70**, 1715–25.

Kleymeyer, C.D. (1994) Cultural traditions and community-based conservation. In: Western, D., White, R.M. & Strum, S.C. (eds) *Natural Connections: Perspectives in Community-based Conservation.* Washington, DC: Island Press, pp. 323–46.

Kline, V.M. & Howell, E.A. (1987) Prairies. In: Jordan III, W.R., Gilpin, M.E. & Aber, J.D. (eds) *Restoration Ecology: a Synthetic Approach to Ecological Research.* Cambridge: Cambridge University Press.

Knick, S.T. & Rotenberry, J.T. (1995) Landscape characteristics of fragmented shrubsteppe habitats and breeding passerine birds. *Conservation Biology* **9**, 1059–71.

Kornberg, H. & Williamson, M.H. (eds) (1987) *Quantitative Aspects of the Ecology of Biological Invasions.* London: The Royal Society.

Kouki, J., Niemelä, P. & Viitasaari, M. (1994) Reversed latitudinal gradient in species richness of sawflies (Hymenoptera, Symphyta). *Annales Zoologici Fennici* **31**, 83–8.

Krebs, C.J. (1972) *Ecology.* New York: Harper & Row.

Kruess, A. & Tscharntke, T. (1994) Habitat fragmentation, species loss and biological control. *Science* **264**, 1581–4.

Kunin, W.E. & Lawton, J.H. (1996) Does biodiversity matter? Evaluating the case for conserving species. In: Gaston, K.J. (ed.) *Biodiversity: a Biology of Numbers and Difference.* Oxford: Blackwell Science, pp. 283–308.

Kvalseth, T.O. (1991) Note on biological diversity, evenness, and homogeneity measures. *Oikos* **62**, 123–7.

Laan, R. & Verboom, B. (1990) Effect of pool size and isolation on amphibian communities. *Biological Conservation* **54**, 251–62.

Lachance, S. & Mangan, P. (1990) Performance of domestic, hybrid and wild strains of brook trout, *Salvelinus fortinalis*, after stocking: the impact of intra- and interspecific competition. *Canadian Journal of Fisheries and Aquatic Sciences* **47**, 2278–84.

Lack, D. (1976) *Island Biology: Illustrated by the Landbirds of Jamaica*. Oxford: Blackwell Science.

Laevastu, T. (1996) *Exploitable Marine Ecosystems: their Behaviour and Management*. Oxford: Fishing News Books.

Lake, R. (1996) *New and Additional? Financial Resources for Biodiversity Conservation in Developing Countries, 1987–94*. Cambridge, UK: BirdLife International and Royal Society for the Protection of Birds.

Lamberson, R.H., McKelvey, K.S., Noon, B.R. & Voss, C. (1992) A dynamic analysis of northern spotted owl viability in a fragmented forest landscape. *Conservation Biology* **6,** 505–12.

Lambshead, P.J.D. (1993) Recent developments in marine benthic biodiversity research. *Oceanis* **19**, 5–24.

Lande, R. (1993) Risks of population extinction from demographic and environmental stochasticity and random catastrophies. *American Naturalist* **142**, 911–27.

Lande, R., Engen, S. & Saether, B. (1994) Optimal harvesting, economic discounting and extinction risk in fluctuating populations. *Nature* **372**, 88–90.

Lande, R., Engen, S. & Saether, B. (1995) Optimal harvesting of fluctuating populations with a risk of extinction. *American Naturalist* **145**, 728–45.

Law, R. & Grey, D.R. (1989) Evolution of yields from populations with age-specific cropping. *Evolutionary Ecology* **3**, 343–59.

Lawton, J.H. (1992) There are not 10 million kinds of population dynamics. *Oikos* **63**, 337–8.

Lawton, J.H. & Brown, V.K. (1993) Functional redundancy. In: Schulze, E-D. & Mooney, H.A. (eds) *Biodiversity and Ecosystem Function*. Berlin: Springer-Verlag, pp. 255–70.

Leader-Williams, N., Harrison, J. & Green, M.J.B. (1990) Designing protected areas to conserve natural resources. *Science Progress* **74**, 189–204.

Leemans, R. (1996) Biodiversity and global change. In: Gaston, K.J. (ed.) *Biodiversity: a Biology of Numbers and Difference*. Oxford: Blackwell Science, pp. 367–87.

Leigh, E.G., Wright, S.J., Herre, E.A. & Putz, F.E. (1993) The decline of tree diversity on newly isolated tropical islands: a test of a null hypothesis and some implications. *Evolutionary Ecology* **7**, 76–102.

Lélé, S.M. (1991) Sustainable development: a critical review. *World Development* **19**: 607–21.

Levins, R.A. (1970) Extinction. *Lectures on Mathematics in the Life Sciences* **2**, 75–107.

Ligon, J.D., Stacey, P.B., Conner, R.N., Bock, C.E. & Adkisson, C.S. (1986) Report of the American Ornithologists' Union Committee for the conservation of the red-cockaded woodpecker. *Auk* **103**, 848–55.

Lindsay, W.K. (1987) Integrating Parks and Pastoralists: Some Lessons from Amboseli. In: Anderson, D.M. & Grove, R.H. (eds) *Conservation in Africa: People, Policies and Practive*. Cambridge University Press, pp. 149–68.

Lines, M. (1995) Dynamics and uncertainty. In: Folner, H., Landisgabel, H. & Opschoor, H. (eds). *Principles of Environmental and Resource Economics*. Cheltenham: Edward Elgar, pp. 67–105.

Lipton, M. & Longhurst, R. (1989) *New Seeds and Poor People*. Baltimore: Johns Hopkins University Press.

Liu, J., Dunning, J.B. & Pulliam, H.R. (1995) Potential effects of a forest management plan on Bachman's sparrows (*Aimophila aestivalis*): linking a spatially explicit model with GIS. *Conservation Biology* **9**, 62–75.

Lombard, A.T. (1995) The problems with multispecies conservation: do hotspots, ideal reserves and existing reserves coincide? *South African Journal of Zoology* **30**, 145–63.

Longcore, J.R. & Stendel, R.C. (1977) Shell thinning and reproductive impairment in black ducks after cessation of DDE dosage. *Archives of Environmental Contamination and Toxicology* **6**, 293–304.

Looyen, R.C. & Bakker, J.P. (1987) Utilization of different salt-marsh communities by cattle and geese. In: Huiskes, A.H.L., Blom, C.W.P. & Rozema, J. (ed.) *Vegetation between Land and Sea*. Dordrecht: Junk, pp. 54–64.

Lovejoy, T.E. (1980) A projection of species extinctions. In: Barney, G.O. (ed.) *The Global 2000 Report to the President*, Vol. II. Washington, DC: Council on Environmental Quality, pp. 328–9.

Low, D.A. & Lonsdale, J.A. (1976) Towards a new order 1945–1963. In: Low, D.A. & Lonsdale, J.M. (eds) *History of East Africa*, volume III. Oxford: Clarendon Press.

Lubina, J.A. & Levin, S.A. (1988) The spread of a reinvading species: range expansion in the California sea otter. *American Naturalist* **131**, 526–43.

Lynch, M., Conery, J. & Burger, R. (1995) Mutation accumulation and the extinction of small populations. *American Naturalist* **146**, 489–518.

McAllister, D.E. (1991) What is biodiversity? *Canadian Biodiversity* **1**, 4–6.

McAllister, D.E., Schueler, F.W., Roberts, C.M. & Hawkins, J.P. (1994) Mapping and GIS analysis of the global distribution of coral reef fishes on an equal-area grid. In: Miller, R.I. (ed.) *Mapping the Diversity of Nature*. London: Chapman & Hall, pp. 155–75.

MacArthur, R.H. & Wilson, E.O. (1963) An equilibrium theory of insular zoogeography. *Evolution* **17**, 373–87.

MacArthur, R.H. & Wilson, E. (1967) *The Theory of Island Biogeography*. Princeton: Princeton University Press.

McCloskey, J.M. & Spalding, H. (1989) A reconnaissance-level inventory of the amount of wilderness remaining in the world. *Ambio* **18**, 221–7.

McCormick, J.S. (1989) *Reclaiming Paradise: the Global Environmental Movement*. Bloomington: Indiana University Press.

McCullagh, P. & Nelder, J.A. (1983) *Generalized Linear Modelling*. London: Chapman & Hall.

McGee, T.G. (1995) Eurocentrism and geography: reflections on an Asian urbanization. In: Crush, J. (ed.) *Power of Development*. London: Routledge, pp. 192–207.

Macilwain, C. (1996) Bollworms chew hole in gene-engineered cotton (news item). *Nature* **382**, 289.

McIntosh, R.P. (1995) H.A. Gleason's 'individualistic concept' and theory of animal communities: a continuing controversy. *Biological Reviews* **70**, 317–57.

Mack, R.N. (1981) Invasion of *Bromus tectorum* L. into western North America: an ecological chronicle. *Agro-ecosystems* **7**, 145–65.

McKelvey, K., Noon, B.R. & Lamberson, R.H. (1993) Conservation planning for species occupying fragmented landscapes: the case of the northern spotted owl. In: Kareiva, P.M., Kingsolver, J.G. & Huey, R.B. (eds) *Biotic Interactions and Global Change*. Sunderland: Sinauer Associates, pp. 424–50.

MacKenzie, D. (1995). The cod that disappeared. *New Scientist* **147**, 24–9.

McKinney, M.L. (1990) Trends in body-size evolution. In: McNamara, K. (ed.) *Evolutionary Trends*. London: Belhaven Press, pp. 75–118.

McKinney, M.L., Lockwood, J.L. & Frederick, D.R. (1996) Does ecosystem and evolutionary stability

include rare species? *Palaeogeography, Palaeoclimatology and Palaeoecology*.

MacKinnon, J. & De Wulf, R. (1994) Designing protected areas for giant pandas in China. In: Miller, R.I. (ed.) *Mapping the Diversity of Nature*. London: Chapman & Hall, pp. 127–42.

MacKinnon, J. & MacKinnon, C. (1986a) *Review of the Protected Areas System in the Afrotropical Realm*. Gland, Switzerland: International Union for Conservation of Nature and Natural Resources.

MacKinnon, J. & MacKinnon, K. (1986b) *Review of the Protected Areas System in the Indo-Malayan Realm*. Gland, Switzerland: International Union for Conservation of Nature and Natural Resources.

McLeon, I.G., Lundie-Jenkins, G. & Jarman, P.J. (1994) Training captive rufous hare-wallabies to recognise predators. In: Serena, M. (ed.) *Reintroduction Biology of Australian and New Zealand Fauna*. Chipping Norton: Surrey Beatty & Sons, pp. 177–82.

MacMillan, D., Hanley, N. & Buckland, S. (1996) A contingent valuation study of uncertain environmental gains. *Scottish Journal of Political Economy* **43**, 519–33.

McNeely, J.A. (1993a) Foreword. In: de Klemm, C. & Shine, C. (eds) *Biological Diversity Conservation and the Law*. Gland, Switzerland: International Union for the Conservation of Nature and Natural Resources.

McNeely, J.A. (1993b) Economic incentives for conserving biodiversity: lessons for Africa. *Ambio* **22**, 144–50.

McNeely J.A. (1995) Human influences on biodiversity. In: Hegwood, V.H. (ed.) *Global Biodiversity Assessment*. Cambridge: Cambridge University Press, pp. 733–83.

McNeely, J.A. (1996) Partnerships for conservation: an introduction. In: McNeely, J.A. (ed.) *Expanding Partnerships in Conservation*. Washington DC: Island Press, pp. 1–10.

McNeely, J.A. & Miller, K.R. (eds)(1984) *National Parks, Conservation and Development: the Role of Protected Ares in Sustaining Society*. Washington, DC: Smithsonian Institute Press.

McNeely, J.A., Miller, K.R., Reid, W.V., Mittermeier, R.A. & Werner, T.B. (1990) *Conserving the World's Biological Diversity*. Gland, Switzerland: International Union for the Conservation of Nature and Natural Resources.

Maehr, D.S. & Cox, J.A. (1995) Landscape features and panthers in Florida. *Conservation Biology* **9**, 1008–19.

Magsalay, P., Brooks, T., Dutson, G. & Timmins, R.

(1995) Extinction and conservation on Cebu. *Nature* **373**, 294.

Magurran, A.E. (1988) *Ecological Diversity and its Measurement*. London: Croom Helm.

Major, J. (1988) Endemism: a botanical perspective. In: Myers, A.A. & Giller, P.S. (eds) *Analytical Biogeography: an Integrated Approach to the Study of Animal and Plant Distributions*. London: Chapman & Hall, pp. 117–46.

Maler K-G. (1989) The acid rain game. In: Folmer, H. & van Ierland, E. (eds) *Valuation and Policy Making in Environmental Economics*. Amsterdam: Elsevier.

Mallet, J. (1996) The genetics of biological diversity: from varieties to species. In: Gaston, K.J. (ed.) *Biodiversity: a Biology of Numbers and Difference*. Oxford: Blackwell Science, pp. 13–53.

Maltby, E. (1988) *Waterlogged Wealth: Why Waste the World's Wet Places?* London: Earthscan.

Malthus, T.R. (1798) *An Essay on the Principle of Population*. Homewood: Richard D. Irwin, Inc. edition (1963).

Mann, C.C. & Plummer, M.L. (1995) *Noah's Choice*. New York: Alfred A Knopf.

Marchant, J.H., Hudson, R., Carter, S.P. & Whittington, P. (1990) *Population Trends in British Breeding Birds*. Thetford: British Trust for Ornithology.

Margules, C.M. (1986) Conservation evaluation in practice. In: Usher, C.M.B. (ed.) *Wildlife Conservation*. London: Chapman & Hall, pp. 297–314.

Margules, C.R. & Austin, M.P. (1991) *Nature Conservation: Cost Effective Biological Surveys and Data Analysis*. Melbourne: Commonwealth Scientific and Industrial Research Organisation.

Margules, C.R. & Stein, J.L. (1989) Problems in the distribution of species and the selection of nature reserves: an example from *Eucalyptus* forests in south-eastern New South Wales. *Biological Conservation* **50**, 219–38.

Margules, C.R., Creswell, I.D. & Nicholls, A.O. (1994) A scientific basis for establishing networks of protected areas. In: Forey, P.L., Humphries, C.J. & Vane-Wright, R.I. (eds) *Systematics and Conservation Evaluation*. Oxford: Clarendon Press, pp. 327–50.

Margules, C.R., Nicholls, A.O. & Pressey, R.L. (1988) Selecting networks of reserves to maximise biological diversity. *Biological Conservation* **46**, 63–76.

Martinez, N.D. (1996) Defining and measuring functional aspects of biodiversity. In: Gaston, K.J.

(ed.) *Biodiversity: a Biology of Numbers and Difference*. Oxford: Blackwell Science, pp. 114–48.

Marzolf, R. (1988) Konza prairie research natural area of Kansas State University. *Transactions of Kansas Academic Sciences* **91**, 24–9.

May, R.M. (1973) *Stability and Complexity in Model Ecosystems*. Princeton: Princeton University Press.

May, R.M. (1990) How many species? *Philosophical Transactions of the Royal Society of London Series B* **330**, 293–304.

May, R.M. (1992) Biodiversity: bottoms up for the oceans. *Nature* **357**, 278–9.

May, R.M. (1994a) Biological diversity: differences between land and sea. *Philosophical Transactions of the Royal Society of London Series B* **343**, 105–11.

May, R.M. (1994b) Conceptual aspects of the quantification of the extent of biological diversity. *Philosophical Transactions of the Royal Society of London Series B* **345**, 13–20.

May, R.M. & Nee, S. (1995) The species alias problem. *Nature* **378**, 447–8.

May, R.M., Lawton, J.H. & Stork, N.E. (1995) Assessing extinction rates. In: Lawton, J.H. & May, R.M. (eds) *Extinction Rates*. Oxford: Oxford University Press, pp. 1–24.

Meffe, G.K. & Carroll, C.R. (1994) *Principles of Conservation Biology*. Sunderland: Sinauer Associates.

Mendelssohn, H. (1994). Experimental releases of waldrapp ibis *Geronticus eremita*: an unsuccessful trial. *International Zoo Yearbook* **33**, 79–85.

Menges, E.S. (1990) Population viability analysis for an endangered plant. *Conservation Biology* **4**, 52–62.

Metcalfe, S. (1994) The Zimbabwe Communal Areas Management Programme for Indigenous Resources (CAMPFIRE). In: Western, D., White, R.M. & Strum, S.C. (eds) *Natural Connections: Perspectives in Community-based Conservation*. Washington, DC: Island Press, pp. 161–92.

Miller, K.R. & Lanou, S.M. (1995) *National Biodiversity Planning: Guidelines Based on Early Experiences around the World*. Washington, DC; Nairobi; Gland, Switzerland: World Resources Institute, United Nations Environment Programme and the World Conservation Union.

Miller, R.R., Williams, J.D. & Williams, J.E. (1989) Extinctions of North American fishes during the past century. *Fisheries* **14**, 22–38.

Milner-Gulland, E.J. (1994) A population model for the management of the saiga antelope. *Journal of Applied Ecology* **31**, 25–39.

Minnich, R.A., Barbour, M.G., Burk, J.H. & Fernaud, R.F. (1995) Sixty years of change of California coniferous forests of the San Bernardino mountains. *Conservation Biology* **9**, 902–14.

Missfeldt, F. (1995) *Game Theoretic Modelling of Transboundary Pollution Control.* Discussion Papers in Ecological Economics No. 95/3. Stirling: University of Stirling.

Mitchell, M.K. & Stapp, W.B. (1992) *Field Manual for Water-quality Monitoring: an Environmental Education Program for Schools*, 6th edn. Dexter: Thomson-Shore Printers.

Mittermeier, R.A. & Werner, T.B. (1990) Wealth of plants and animals unites 'megadiversity' countries. *Tropicus* **4**, 1–5.

Mladenoff, D.J., Sickley, T.A., Haight, R.G. & Wydeven, A.P. (1995) Regional landscape analysis and prediction of favorable gray wolf habitat in the northern Great Lakes region. *Conservation Biology* **9**, 279–94.

Mollison, D. (1977) Spatial contact model for ecological and epidemic spread. *Journal of the Royal Statistical Society Series B* **39**, 283–326.

Mooney, H.A. & Berbardi, G. (eds) (1990) *Introduction of Genetically Modified Organisms into the Environment.* New York: John Wiley.

Moore, N.W. (1969) Experience with pesticides and the theory of conservation. *Biological Conservation* **1**, 201–7.

Moore N.W. (1987) *The Bird of Time: the Science and Politics of Nature Conservation—a Personal Account.* Cambridge: Cambridge University Press.

Mosquin, T. (1996) A conceptual framework for the ecological functions of biodiversity. *Global Biodiversity* **4**, 2–16.

Moyle, P.B. & Leidy, R.A. (1992) Loss of biodiversity in aquatic ecosystems: evidence from fish faunas. In: Fiedler, P.L. & Jain, S.K. (eds) *Conservation Biology: the Theory and Practice of Nature Conservation, Preservation, and Management.* New York: Chapman & Hall, pp. 127–67.

Moyle, P.B. & Light, T. (1996) Fish invasions in California: do abiotic factors determine success? *Ecology* **77**, 1660–9.

Munasinghe, M. & McNeely, J. (eds) (1994) *Protected Area Economics and Policy: Linking Conservation and Sustainable Development.* Washington, DC: World Bank and International Union for the Conservation of Nature and Natural Resources.

Munro, D.A. (1978) The 30 years of IUCN. *Nature and Resources* **14**(2), 14–18.

Munson, A. (1993) The UN Convention on Biological Diversity. In: Grubb, M., Koch, M., Munson, A., Sullivan, F. & Thomson, K. (eds) *The Earth Summit Agreements.* London: Earthscan.

Murphree, M.W. (1994) The role of institutions in community-based conservation. In: Western, D., White, R.M. & Strum, S.C. (eds) *Natural Connections: Perspectives in Community-based Conservation.* Washington, DC: Island Press, pp. 403–27.

Murton, R.K. Theale, R.J.P. & Thompson, J. (1972) Ecological studies of the feral pigeon (*Columba livia*) var. I. Population, breeding biology and methods of control. *Journal of Applied Ecology* **9**, 875–89.

Myers, A.A. & Giller, P.S. (eds) (1988) *Analytical Biogeography: an Integrated Approach to the Study of Animal and Plant Distributions.* London: Chapman & Hall.

Myers, N. (1988) Threatened biotas: 'hotspots' in tropical forests. *Environmentalist* **8**, 1–20.

Myers, N. (1990) The biodiversity challenge: expanded hot-spots analysis. *Environmentalist* **10**, 243–56.

Myers, N. (1994) Tropical deforestation: rates and patterns. In: Brown, K. & Pearce, D.W. (eds) *The Causes of Tropical Deforestation.* London: University College Press, pp. 27–40.

Myers, N. & Simon, J.L. (1994) *Scarcity or Abundance?* New York: W.W. Norton & Company.

Myers, S.A., Millam, J.R., Roudybush, T.E. & Grav, G.R. (1988) Reproductive success of hand-reared vs parent-reared cockatiels (*Nymphicus hollandicus*). *Auk* **105**, 536–42.

Nabhan, G.P. (1995) Cultural parallax in viewing North American habitats. In: Feidler, P.L. & Jain, S.K. (eds) *Conservation Biology: the Theory and Practice of Nature Conservation, Preservation and Management.* London: Chapman & Hall, pp. 87–101.

Naeem, S., Thompson, L.J., Lawler, S.P., Lawton, J.H. & Woodfin, R.M. (1994) Declining biodiversity can alter the performance of ecosystems. *Nature* **368**, 734–7.

Naiman, R.J., Johnston, C.A. & Kelley, J.C. (1988) Alteration of North American streams by beavers. *BioScience* **38**, 750–2.

Nash, R. (1973) *Wilderness and the American Mind.* New Haven: Yale University Press.

Nash, D.R., Agassiz, D.J.L., Godfray, H.C.J. & Lawton, J.H. (1995) The pattern of spread of invading species: two leaf-mining moths colonising Great Britain. *Journal of Animal Ecology* **64**, 225–33.

National Fish and Wildlife Foundation (1995) Review and assessment of information on the federally endangered Karner blue butterfly (*Lycaeides melissa samulelis*).

Washington, DC: National Fish and Wildlife Foundation.

Navrud, S. (1988) Estimating social benefits from environmental improvements from reduced acid deposition. In: Folmer, H. & van Ierland, E. (eds) *Valuation and Policy Making in Environmental Economics*. Amsterdam: Elsevier.

Navrud, S. & Veisten, K. (1996) Validity of non-use values in contingent valuation. Paper to the European Association of Environmental and Resource Economists Conference, Lisbon.

Nee, S., Holmes, E.C., May, R.M. & Harvey, P.H. (1994) Estimating extinction from molecular phylogenies. *Philosophical Transactions of the Royal Society of London Series B* **344**, 77–82.

Nee, S., Mooers, A.O. & Harvey, P.H. (1992) The tempo and mode of evolution revealed from molecular phylogenies. *Proceedings of the National Academy of Sciences of the USA* **89**, 8322–6.

Neubert, M, Kot, M. & Lewis, M.A. (1995) Dispersal and pattern formation in a discrete-time predator–prey model. *Theoretical Population Biology* **48**, 7–43.

Newton, I. (1972) *Finches*. London: Collins.

Newton, I. (1979) *Population Ecology of Raptors*. Berkamsted: T. and A.D. Poyser.

Newton, I. (1986) *The Sparrowhawk*. Calton: T. and A.D. Poyser.

Newton, I. (1988) Determination of critical pollutant levels in wild populations, with examples from organochlorine insecticides in birds of prey. *Environmental Pollution* **55**, 29–40.

Newton, I. & Wyllie, I. (1992) Recovery of a sparrowhawk population in relation to declining pesticide contamination. *Journal of Applied Ecology* **20**, 476–84.

Nicholls, A.O. & Margules, C.R. (1993) An upgraded reserve selection algorithm. *Biological Conservation* **64**, 165–9.

Nichols, P. (1991) *Social Survey Methods*. London: Oxfam Books.

Nilsson, C. (1986) Methods of selecting lake shorelines as nature reserves. *Biological Conservation* **35**, 269–91.

Nilsson, S.I. & Duinker, P. (1987) The extent of forest decline in Europe. *Environment* **29**, 4–31.

Norris, K.S. & Jacobson, S.K. (in press) Content analysis of tropical conservation education programs: elements of success. *Journal of Environmental Education*.

Norse, E.A. (ed.) (1993) *Global Marine Biological Diversity: a Strategy for Building Conservation into Decision Making*. Washington, DC: Island Press.

North, D. (1990) *Institutions, Institutional Change and Economic Performance*. Cambridge: Cambridge University Press.

Norton, B.G. (1987) *Why Preserve Natural Variety?* Princeton: Princeton University Press.

Norton, B.G. (1994) On what we should save: the role of culture in determining conservation targets. In: Forey, P.L. Humphries, C.J. & Vane-Wright, R.I. (eds) *Systematics and Conservation Evaluation*. Oxford: Clarendon Press, pp. 23–9.

Noss, R.F. (1987) From plant communities to landscapes in conservation inventories: a look at the Nature Conservancy (USA). *Biological Conservation* **41**, 11–37.

Noss, R.F. (1990) Indicators for monitoring biodiversity: a hierarchical approach. *Conservation Biology* **4**, 355–64.

Nott, M.P. & Pimm, S.L. (1997) The evaluation of biodiversity as a target for conservation. In: Ostfeld, R.S. & Pickett, S.T.A. (eds) *Principles of the New Ecology*. London: Chapman & Hall.

Nott, M.P., Rogers, E., & Pimm, S.L. (1995) Modern extinctions in the kilo-death range. *Current Biology* **5**, 14–17.

NRC (1996) *The Economic and Non-economic Value of Biodiversity*. Washington, DC: National Research Council.

Oates, J.F. (1995) The dangers of conservation by rural development—a case study from the forests of Nigeria. *Oryx* **29**, 115–22.

O'Connor, R.J. & Mead, C.J. (1984) The stock dove in Britain, 1930–80. *British Birds* **77**, 181–201.

Odum, W.E. (1982) Environmental degradation and the tyranny of small decisions. *Bioscience* **32**, 728–9.

OECD (1994) *Managing the Environment: the Role of Economic Instruments*. Paris: Organization for Economic Cooperation and Development.

Øien, I.J. (1997) Secrets of the lesser white-fronted goose. *World Birdwatch* **19**, 9–12.

Okubo, A. (1980) *Diffusion and Ecological Problems: Mathematical Models*. Berlin: Springer-Verlag.

Olindo, P. & Mankoto M.M. (1993) *Regional Review of Protected Areas in Sub-Saharan Africa*. Cambridge: Protected Areas Data Unit, World Conservation Monitoring Centre.

Olson, S.L. & James, H.F. (1982) Fossil birds from the Hawaiian islands: evidence for wholesale extinction by man before western contact. *Science* **217**, 633–5.

Olson, S.L. & James, H.F. (1991) Descriptions of thirty-two new species of birds from the Hawaiian islands: Part I. Non-passeriformes. *Ornithological Monographs* **45**. Washington, DC: American Ornithologists' Union.

Olwig, K.R. (1995) Reinventing common nature: Yosemite and Mount Rushmore. In: Cronon, W. (ed.) *Uncommon Ground: Towards Reinventing Nature.* New York: Norton, pp. 378–408.

Opdam, P. (1990) Metapopulation theory and habitat fragmentation: a review of holarctic breeding bird studies. *Landscape Ecology* **5**, 93–106.

Ostrom, E. (1990) *Governing the Commons: the Evolution of Institutions for Collective Action.* Cambridge: Cambridge University Press.

OTA (US Congress Office of Technology Assessment) (1987) *Technologies to Maintain Biological Diversity.* Washington, DC: US Government Printing Office.

Pain, D.J. (1991) Lead poisoning of waterfowl: a review. In: Pain, D.J. (ed.) *Lead poisoning in Waterfowl.* International Waterfowl and Wetlands Research Bureau Special Publication No. 16. Slimbridge, UK: International Waterfowl and Wetlands Research Bureau, pp. 7–13.

Palmer, M.W. & White, P.S. (1994) Scale dependence and the species–area relationship. *American Naturalist* **144**, 717–40.

Papastavrou, V. (1996) Sustainable use of whales: whaling or whale watching? In: Taylor, V.J. & Dunstone, N. (eds) *The Exploitation of Mammal Populations.* London: Chapman & Hall, pp. 102–16.

Park, C.C. (ed.) (1986) *Environmental Policies: an International Review.* London: Croom Helm.

Parker, G.L., Petrie, M.J. & Sears, D.T. (1992) Waterfowl distribution relative to wetland acidity. *Journal of Wildlife Management* **56**, 268–74.

Parker, I.M. & Kareiva, P. (1996) Assessing the risk of invasion for genetically engineered crops: acceptable evidence and reasonable doubt. *Biological Conservation* **78**, 193–203.

Parkipuny, M.S. (1991) Pastoralism, conservation and development in the greater Serengeti region. International Institute for Environment and Development *Dryland Network Paper* **26**.

Parslow, J. (1973) *Breeding Birds of Britain and Ireland: a Historical Survey.* Berkhamsted: T. and A.D. Poyser.

Patterson, B.D. (1994) Accumulating knowledge on the dimensions of biodiversity: systematic perspectives on neotropical mammals. *Biodiversity Letters* **2**, 79–86.

Peakall, D.B. & Kiff, L.F. (1988) DDE contamination in peregrines and American kestrels and its effects on reproduction. In: Cade, T.J., Enderson, J.H., Thelander, C.G. & White, C.M. (eds) *Peregrine Falcon Populations. Their Management and Recovery.* Boise, OK: The Peregrine Fund Inc., pp. 337–50.

Pearce, D. & Moran, D. (1994) *The Economic Value of Biodiversity.* London: Earthscan.

Pearce, D., Markyanda, A. & Barbier, E. (1988) *Blueprint for a Green Economy.* London: Earthscan.

Peet, R.K. (1974) The measurement of species diversity. *Annual Review of Ecology and Systematics* **5**, 285–307.

Pegoraro, K. & Thaler, E. (1994) Introduction of waldrapp ibis *Geronticus eremita* on the basis of family bonding: a successful pilot study. *International Zoo Yearbook* **33**, 74–9.

Peres, C.A. (in press) Evaluating the impact and sustainability of subsistence hunting at multiple Amazonian forest sites. In: Robinson, J.G. & Bennett, E.L. (eds) *Evaluating the Sustainability of Hunting in Tropical Forests.* New York: Columbia University Press.

Perring, F.H. & Farrell, L. (1983) *British Red Data Book 1. Vascular Plants.* Lincoln: Royal Society for Nature Conservation.

Perrings, C., Mäler, K-G., Folke, C., Holling, C.S. & Jansson, B-O. (eds) (1995a) *Biodiversity Loss: Economic and Ecological Issues.* Cambridge: Cambridge University Press.

Perrings, C., Mäler, K-G., Folke, C., Holling, C.S. & Jansson, B-O. (1995b) Introduction: framing the problem of biodiversity loss. In: Perrings, C. Mäler, K-G., Folke, C., Holling, C.S. & Jansson, B-O. (eds) *Biodiversity Loss: Economic and Ecological Issues.* Cambridge: Cambridge University Press, pp. 1–17.

Petanidou, T., Ellis, W.E. & Ellis-Adam, A.C. (1995) Ecogeographical patterns in the incidence of brood parasitism in bees. *Biological Journal of the Linnean Society* **55**, 261–72.

Peterken, G.F. (1996) *Natural Woodland: Ecology and Conservation in Northern Temperate Regions.* Cambridge: Cambridge University Press.

Peterken, G.F. & Hughes, F.M.R. (1990) The changing lowlands. In: Bayliss-Smith, T.P. & Owens, S.E. (eds) *Britain's Changing Environment from the Air.* Cambridge: Cambridge University Press, pp. 48–76.

Peters, C.M., Gentry, A.H. & Mendelsohn, R.O. (1989) Valuation of an Amazonian rainforest. *Nature* **339**, 655–6.

Philip, C. & Lord, T. (1996) *The Plant Finder.* London: Royal Horticultural Society.

Pianka, E.R. (1966) Latitudinal gradients in species diversity: a review of concepts. *American Naturalist* **100**, 33–46.

Piatt, J.F., Lensink, C.J., Butler, W., Kendziorek, M. & Nysewander, D.R. (1990) Immediate impact of the 'Exxon Valdez' oil spill on marine birds. *Auk* **107**, 387–97.

Pickett, S.T.A. & Thompson, J.N. (1978) Patch

dynamics and the design of nature reserves. *Biological Conservation* **13**, 27–37.

Pickett, S.T.A., Parker, V.T. & Fiedler, P.L. (1992) The new paradigm in ecology: implications for conservation biology above the species level. In: Fiedler, P.L. & Jain, S.K. (eds) *Conservation Biology: the Theory and Practice of Nature Conservation, Preservation, and Management*. New York: Chapman & Hall, pp. 65–88.

Pimentel, D. (1995) Biotechnology: environmental impacts of introducing crops and biological agents in North American agriculture. In: Hokkanen, H.M.T. & Lynch, J.M. (eds) *Biological Control: Benefits and Risks*. Cambridge: Cambridge University Press, pp. 13–29.

Pimm, S.L. (1987) The snake that ate Guam. *Trends in Ecology and Evolution* **2**, 293–5.

Pimm, S.L. (1991) *The Balance of Nature? Ecological Issues in the Conservation of Species and Communities*. Chicago: University of Chicago Press.

Pimm, S.L. & Askins, R.A. (1995) Forest losses predict bird extinction in eastern North America. *Proceedings of the National Academy of Sciences of the USA* **92**, 9343–7.

Pimm, S.L., Diamond, J.M., Reed, T.M., Russell, G.J. & Verner, J.M. (1993) Times to extinction for small populations of large birds. *Proceedings of the National Academy of Sciences of the USA* **90**, 10871–5.

Pimm, S.L., Jones, H.L. & Diamond, J.M. (1988) On the risk of extinction. *American Naturalist* **132**, 757–85.

Pimm, S.L., Moulton, M.P. & Justice, L.J. (1994) Bird extinctions in the central Pacific. *Philosophical Transactions of the Royal Society of London Series B* **344**, 27–33.

Pimm, S.L., Russell, J., Gittleman, J.L. & Brooks, T.M. (1995) The future of biodiversity. *Science* **269**, 347–50.

Pine, R.H. (1994) New mammals not so seldom. *Nature* **368**, 593.

Pitcher, T.J. & Hart, P.J.B. (1982) *Fisheries Ecology*. London: Croom Helm.

Pope, J.G. (1972) An investigation of the accuracy of virtual population analysis. *ICNAF Research Bulletin* **9**, 65–74.

Potts, G.R. (1986) *The Partridge: Pesticides, Predation and Conservation*. London: Collins.

Potts, G.R. & Aebischer, N.J. (1995) Population dynamics of the grey partridge *Perdix perdix* 1793–1993: monitoring, modelling and management. *Ibis* **137** (Suppl.), 29–37.

Poulin, R. (1995) Phylogeny, ecology, and the richness of parasite communities in vertebrates.

Ecological Monographs **65**, 283–302.

Prendergast, J.R., Quinn, R.M., Lawton, J.H., Eversham, B.C. & Gibbons, D.W. (1994) Rare species, the coincidence of diversity hotspots and conservation strategies. *Nature* **365**, 335–7.

Press, F. & Atiyah, M. (1992) *Joint Statement on Biodiversity*. Washington, DC; London: National Academy of Sciences and British Royal Society.

Pressey, R.L. (1994) *Ad hoc* reservations: forward or backwards steps in developing representative reserve systems. *Biological Conservation* **8**, 662–8.

Pressey, R.L. (1995) Conservation reserves in NSW: crown jewels or leftovers. *Search* **26**(2), 47–51.

Pressey, R.L. & Nicholls, A.O. (1989) Efficiency in conservation evaluation: scoring vs iterative approaches. *Biological Conservation* **50**, 199–218.

Pressey, R.L. & Tully, S.L. (1994) The cost of *ad hoc* reservation: a case study in western New South Wales. *Australian Journal of Ecology* **19**, 375–84.

Pressey, R.L., Humphries, C.J., Margules, C.R., Vane-Wright, R.I. & Williams, P.H. (1993) Beyond opportunism: key principles for systematic reserve selection. *Tree* **8**, 124–8.

Pressey, R.L., Johnson, I.R. & Wilson, P.D. (1994) Shades or irreplaceability: towards a measure of the contribution of sites to a reservation goal. *Biodiversity and Conservation* **3**, 242–62.

Preston, F.W. (1960) Time and space and the variation of species. *Ecology* **41**, 611–27.

Preston, F.W. (1962) The canonical distribution of commonness and rarity. *Ecology* **43**, 185–215, 410–32.

Price, S.M. (1989) *Animal Re-introductions: the Arabian Oryx in Oman*. Cambridge: Cambridge University Press.

Prins, H.H.T. (1992) The pastoral road to extinction: competition between wildlife and traditional pastoralism in East Africa. *Environmental Conservation* **19**(2): 117–23.

Pulliam, H.R. (1988) Sources, sinks, and population regulation. *American Naturalist* **132**, 652–61.

Pulliam, H.R. & Danielson, B.J. (1991) Sources, sinks, and habitat selection: a landscape perspective on population dynamics. *American Naturalist* **137**, S50–S66.

Pulliam, H.R. & Dunning Jr, J.B. (1994) Demographic processes: population dynamics on heterogeneous landscapes. In: Meffe, G.K. & Carroll, C.R. (eds) *Principles of Conservation Biology*. Sunderland: Sinauer Associates, pp. 179–205.

Pulliam, H.R., Dunning Jr, J.B & Liu, J. (1992) Population dynamics in a complex landscape: a case study. *Ecological Applications* **2**, 165–77.

Pyne, S.J. (1982) *Fire in America: a Cultural History of Wildland and Rural Fire*. Princeton: Princeton University Press.

Pysek, P., Prach, K., Rejmanek, M. & Wade, M. (1995) *Plant Invasions: General Aspects and Special Problems*. Amsterdam: SPB Academic Publishing.

Quicke, D.L.J. & Kruft, R.A. (1995) Latitudinal gradients in North American braconid wasp species richness and biology. *Journal of Hymenoptera Research* **4**, 194–203.

Rabinovich, J.E. & Rapoport, E.H. (1975) Geographical variation of diversity in Argentine passerine birds. *Journal of Biogeography* **2**, 141–57.

Rabinowitz, D., Cairns, S. & Dillon, T. (1986) Seven forms of rarity and their frequency in the flora of the British Isles. In: Soulé, M.E. (ed.) *Conservation Biology: the Science of Scarcity and Diversity*. Sunderland: Sinauer Associates, pp. 182–204.

Rackham, O. (1976) *Trees and Woodlands in the British Landscape*. London: Dent.

Rackham, O. (1978) Archaeology and land-use history. In: Corke, D. (ed.) *Epping Forest: the Natural Aspect?* Essex Naturalist New Series No. 2, London: Essex Field Club, pp. 16–75.

Rackham, O. (1980) *Ancient Woodland: its History, Vegetation and Uses in England*. London: Edward Arnold.

Rackham, O. (1986) *The History of the Countryside*. London: Dent.

Rackham, O. (1987) The countryside: history and pseudo-history. *Historian* **14**, 13–17.

Rackham, O. (1989) *The Last Forest: the History of Hatfield Forest*. London: Dent.

Rackham, O. (1990) *Trees and Woodland in the British Landscape*, 2nd edn. London: Dent.

Rackham, O. (1992) Gamlingay Wood. *Nature in Cambridgeshire* **34**, 3–14.

Rackham, O. (1996a) Ecology and pseudo-ecology: the example of Ancient Greece. In: Shipley, G. & Salmon, J. (eds) *Human Landscapes in Classical Antiquity: Environment and Culture*. London: Routledge, pp. 16–43.

Rackham, O. (1996b) Forest history of countries without much forest: questions of conservation and savanna. In: Cavaciocchi, S. (ed.) *L'uomo e la foresta secc. XIII–XVIII*. Firenze: Le Monnier, pp. 296–326.

Rackham, O. (1997) Cressing Temple: trees, woods, and timber-framed buildings. *The Local Historian* **27**, 66–77.

Rackham, O. & Coombe, D.E. (1996) Madingley Wood. *Nature in Cambridgeshire* **38**, 27–54.

Rackham, O. & Moody, J.A. (1997) *The Making of the Cretan Landscape*. Manchester: Manchester University Press.

Rahbek, C. (1995) The elevational gradient of species richness: a uniform pattern? *Ecography* **18**, 200–5.

Ramsar Convention Bureau (1990) *Proceedings of the Fourth Meeting of the Conference of the Contracting Parties, Montreux, Switzerland, 1990*. Gland, Switzerland: Ramsar Convention Bureau.

Rands, M.R.W. (1985) Pesticide use on cereals and the survival of partridge chicks: a field experiment. *Journal of Applied Ecology* **22**, 49–54.

Rapoport, E.H. (1982) *Areography: Geographical Strategies of Species*. Oxford: Pergamon Press.

Ratcliffe, D.A. (1970) Changes attributable to pesticides in egg breakage frequency and eggshell thickness in some British birds. *Journal of Applied Ecology* **7**, 67–107.

Ratcliffe, D.A. (1980) *The Peregrine Falcon*. Calton: T. and A.D. Poyser.

Ratcliffe, D.A. (1984) Post-medieval and recent changes in British vegetation: the culmination of human influence. *New Phytologist* **98**, 73–100.

Rathcke, B.J. & Jules, E.S. (1993) Habitat fragmentation and plant–pollinator interactions. *Current Science* **65**, 273–7.

Raven, P.H. (1987) The scope of the plant conservation problem worldwide. In: Bramwell, D., Hamann, O., Heywood, V. & Synge, H. (eds) *Botanic Gardens and the World Conservation Strategy*. London: Academic Press, pp. 19–29.

Raven, P.H. (1988) Our diminishing tropical forests. In: Wilson, E.O. (ed.) *Biodiversity*. Washington, DC: National Academy Press, pp. 119–122.

Ray, C., Gilpin, M.E. & Smith, A.T. (1991) The effect of conspecific attraction on metapopulation dynamics. *Biological Journal of the Linnean Society* **42**, 123–34.

Read, H.J. (ed.) (1992) *Pollard and Veteran Tree Management*. London: City of London Corporation.

Rebelo, A.G. & Siegfried, W.R. (1992) Where should nature reserves be located in the Cape Floristic Region, South Africa?—models for spatial configuration of a reserve network aimed at maximising the protection of floral diversity. *Conservation Biology* **6**, 243–52.

Redclift, M. (1987) *Sustainable Development: Exploring the Contradictions*. London: Methuen.

Rees, M. (1993) Trade-offs among dispersal strategies in British plants. *Nature* **366**, 150–2.

Rees, M. (1996) Evolutionary ecology of seed

dormancy and seed size. *Philosophical Transactions of the Royal Society London Series B* **351**, 1299–1308.

Reid, D. (1995) *Sustainable Development: an Introductory Guide.* London: Earthscan.

Reid, W.V. (1992) How many species will there be? In: Whitmore, T.C. & Sayer, J.A. (eds) *Tropical Deforestation and Species Extinction.* London: Chapman & Hall, pp. 55–73.

Reid, W.V. & Miller, K.R. (1989) *Keeping Options Alive: the Scientific Basis for Conserving Biodiversity.* Washington, DC: World Resources Institute.

Reining, C. & Heinzman, R. (1992) Nontimber forest products in the Petén, Guatemala: why extractive reserves are critical for both conservation and development. In: Plotkin, M & Famolare, L. (eds) *Sustainable Harvest and Marketing of Rainforest Products.* Washington, DC: Island Press, pp. 110–17.

Rejmánek, M. & Richardson, D.M. (1996) What attributes make some plant species more invasive. *Ecology* **77**, 1655–6.

Repetto, R. (1988) *The Forest for the Trees: Government Policies and the Misuse of Forest Resources.* Washington, DC: World Resources Institute.

Repetto, R. & Gillis, M. (eds) (1988) Public policies and Misuse of Forest Resources. New York: Cambridge University Press.

Rex, M.A. (1981) Community structure in the deep-sea benthos. *Annual Review of Ecology and Systematics* **12**, 331–54.

Rhymer, J.M. & Simberloff, D. (1996) Extinction by hybridization and introgression. *Annual Review of Ecology and Systematics* **27**, 83–109.

Ribbink, A.J. (1994) Biodiversity and speciation of freshwater fishes with particular reference to African cichlids. In: Giller, P.S., Hildrew, A.G. & Raffaelli, D.G. (eds) *Aquatic Ecology: Scale, Pattern and Process.* Oxford: Blackwell Science, pp. 261–88.

Ricker, W.E. (1981) Changes in the average size and average age of salmon. *Canadian Journal of Fisheries and Aquatic Sciences* **38**, 1636–56.

Ricklefs, R.E. (1990) *Ecology.* New York: W.H. Freeman.

Ricklefs, R.E. & Schluter, D. (eds) (1993) *Species Diversity in Ecological Communities: Historical and Geographical Perspectives.* Chicago: University of Chicago Press.

Risebrough, R.W. (1986) Pesticides and bird populations. *Current Ornithology* **3**, 397–427.

Risser, P.G. (1995) Biodiversity and ecosystem function. *Conservation Biology* **9**, 742–6.

Roberts, C.M. (1997) Ecological advice for the global fisheries crisis. *Trends in Ecology and Evolution* **12**, 35–8.

Robinson, J.G. & Redford, K.H. (1991) Sustainable harvest of neotropical forest mammals. In: Robinson, J.G. & Redford, K.H. (eds) *Neotropical Wildlife Use and Conservation.* Chicago: University of Chicago Press, pp. 415–29.

Rodwell, J.S. (ed.) (1991) *British Plant Communities: Woodlands and Scrub.* Cambridge: Cambridge University Press.

Rohde, K. (1992) Latitudinal gradients in species diversity: the search for the primary cause. *Oikos* **65**, 514–27.

Roland, J. (1993) Large-scale forest fragmentation increases the duration of tent caterpillar outbreak. *Oecologia* **93**, 25–30.

Rosenzweig, M.L. (1992) Species diversity gradients: we know more and less than we thought. *Journal of Mammalogy* **73**, 715–30.

Rosenzweig, M.L. (1995) *Species Diversity in Space and Time.* Cambridge: Cambridge University Press.

Rosenzweig, M.L. & Abramsky, Z. (1993) How are diversity and productivity related? In: Ricklefs, R.E. & Schluter, D. (eds) *Species Diversity in Ecological Communities: Historical and Geographical Perspectives.* Chicago: University of Chicago Press, pp. 52–65.

Ruckelshaus, M., Hartway, C. & Kareiva, P. (in press) Assessing the data requirements of spatially explicit population models. *Conservation Biology.*

Said, E. (1978) *Orientalism.* New York: Pantheon.

Salisbury, E. (1961) *Weeds and Aliens.* London: Collins.

Sanderson, G.C. & Bellrose, F.C. (1986) A review of the problem of lead poisoning in waterfowl. *Illinois Natural History Survey Special Publication* **4**. Champaign: Illinois Natural History Survey.

Sandlund, O.T., Hindar, K. & Brown, A.H.D. (eds) (1992) *Conservation of Biodiversity for Sustainable Development.* Oslo: Scandinavian University Press.

Sas, H. (1989) *Lake Restoration by Nutrient Control. Expectations, Experiences, Extrapolations.* Sant Augustin: Academia Verlag Richarz.

Savidge, J.A. (1987) Extinction of an island forest avifauna by an introduced snake. *Ecology* **68**, 660–8.

Schaefer, M.B. (1954) Some aspects of the dynamics of populations important to the management of commercial marine fisheries. *Inter-American Tropical Tuna Commission Bulletin* **1**, 27–56.

Schall, J.J. & Pianka, E.R. (1978) Geographical trends in numbers of species. *Science* **201**, 679.

Scheiner, S.M. & Rey-Benayas, J.M. (1994) Global patterns of plant diversity. *Evolutionary Ecology* **8**, 331–47.

Schmid, A. (1989) *Benefit–Cost Analysis*. Boulder: Westview.

Schmidheiny, S. (1992) *Changing Course: a Global Perspective on Development and the Environment*. Cambridge: MIT Press.

Schneider, S.H., Mearns, L. & Gleick, P.H. (1992) Climate change scenarios for impact assessment. In: Peters, R.L. & Lovejoy, T.E. (eds) *Global Warming and Biological Diversity*. New Haven: Yale University Press, pp. 18–55.

Schönrogge, K., Stone, G.N. & Crawley, M.J. (1995) Spatial and temporal variation in guild structure: parasitoids and inquilines of *Andricus quercuscalicis* (Hymenoptera: Cynipidae) in its native and alien ranges. *Oikos* **72**, 51–60.

Schönrogge, K., Stone, G.N. & Crawley, M.J. (1996) Alien herbivores and native parasitoids: rapid developments and structure of the parasitoid and inquiline complex in an invading gall wasp *Andricus quercuscalicis* (Hymenoptera: Cynipidae). *Ecological Entomology* **21**, 71–80.

Schuurman, F.J. (1993) *Beyond the Impasse: New Directions in Development Theory*. London: Zed Press.

Schwartz, M.W. (1994) Conflicting goals for conserving biodiversity: issues of scale and value. *Natural Areas Journal* **14**, 213–16.

Scoones, I. (1994) *Living with Uncertainty: New Directions in Pastoral Development in Africa*. London: IT Publications.

Scott, D.A. (1989) *A Directory of Asian Wetlands*. Gland, Switzerland: International Union for Conservation of Nature and Natural Resources.

Scott, D.A. (ed.) (1993) *A Directory of Wetlands in Oceania*. Slimbridge, UK; Kuala Lumpur, Malaysia: International Waterfowl and Wetlands Research Bureau and Asian Wetland Bureau.

Scott, D.A. & Carbonell, M. (1986) *A Directory of Neotropical Wetlands*. Cambridge; Slimbridge: International Union for the Conservation of Nature and Natural Resources and International Waterfowl Research Bureau.

Scott, J. (1985) *Weapons of the Weak: Everyday Forms of Peasant Resistance*. New Haven: Yale University Press.

Scott, J.M., Davis, F., Csuti, B. *et al.* (1993) Gap analysis: a geographic approach to protection of biological diversity. *Wildlife Monographs* **123**, 1–41.

Shaffer, M.L. (1990) Population viability analysis. *Conservation Biology* **4**, 39–40.

Sharpe, G.W. (1982) *Interpreting Our Environment*. New York: Macmillan.

Sheail, J. (1976) *Nature in Trust: the History of Nature Conservation in Britain*. Glasgow: Blackie.

Sheail, J. (1995) Nature protection, ecologists and the farming context: a UK historical perspective. *Journal of Rural Studies* **11**, 79–88.

Shelford, V.E. (1963) *The Ecology of North America*. Urbana: University of Illinois Press.

Shetler, S.G. (1991) Biological diversity: are we asking the right questions? In: Dudley, E.C. (ed.) *The Unity of Evolutionary Biology: Proceedings of the Fourth International Congress of Systematic and Evolutionary Biology*, 2 vols. Portland: Dioscorides Press, pp. 37–43.

Shigesada, N., Kawasaki, K. & Teramoto, E. (1986) Traveling periodic waves in heterogeneous environments. *Theoretical Population Biology* **30**, 143–60.

Shigesada, N., Kawasaki, K. & Takeda, Y. (1995) Modeling stratified diffusion in biological invasions. *American Naturalist* **146**, 229–51.

Shiva, V. (1988) *Staying Alive: Women, Ecology and Development*. London: Zed Books.

Short, J. & Smith, A. (1994) Mammal decline and recovery in Australia. *Journal of Mammalogy* **75**, 288–97.

Sibley, C.G. & Ahlquist, J.E. (1990) *Phylogeny and Classification of the Birds*. New Haven: Yale University Press.

Sibley, C.G. & Monroe Jr, B.L. (1990) *Distribution and Taxonomy of Birds of the World*. Newhaven: Yale University Press.

Sibley, C.G. & Monroe Jr, B.L. (1993) *A Supplement to Distribution and Taxonomy of Birds of the World*. Newhaven: Yale University Press.

Simberloff, D. (1972) Properties of the rarefaction diversity measurements. *American Naturalist* **106**, 414–18.

Simberloff, D. (1986) The proximate causes of extinction. In: Raup, D.M. & Jablonski, D. (eds.) *Patterns and Processes in the History of Life*. Berlin: Springer-Verlag, pp. 259–76.

Simberloff, D. (1988) The contributions of population and community biology to conservation science. *Annual Review of Ecology and Systematics* **19**, 473–511.

Simberloff, D. (1989) Which insect introductions succeed and which fail? In: Drake, J.A., Mooney, H.A., di Castri, F. *et al.* (eds) *Biological Invasions: a Global Perspective*. Chichester: Wiley, pp. 61–75.

Simberloff, D. & Dayan, T. (1991) The guild concept and the structure of ecological communities. *Annual Review of Ecology and Systematics* **22**, 115–43.

Simberloff, D., Farr, J.A., Cox, J. & Mehlman, D.W. (1992) Movement corridors: conservation bargains or poor investments. *Conservation Biology* **6**, 493–504.

Simon, J.L. & Wildavsky, A. (1984) On species loss, the absence of data, and risks to humanity. In: Simon, J.L. & Kahn, H. (eds) *The Resourceful Earth*. New York: Basil Blackwell, pp. 171–83.

Simon, J.L. & Wildavsky, A. (1993) Facts, not species, are periled. *New York Times* **142**, A23.

Sinclair, A.R.E. (1989) Population regulation in animals. In: Cherrett, J.M. (ed.) *Ecological Concepts*. Oxford: Blackwell Scientific Publications, pp. 197–241.

Sjogren-Gulve, P. (1994) Distribution and extinction patterns within a northern metapopulation of the pool frog, *Rana lessonae*. *Ecology* **75**, 1357–67.

Skellam, J.G. (1951) Random dispersal in theoretical populations. *Biometrika* **38**, 196–218.

Skole, D. & Tucker, C. (1993) Tropical deforestation and habitat fragmentation in the Amazon: satellite data from 1978 to 1988. *Science* **260**, 1905–10.

Small, M.F. & Hunter, M.L. (1988) Forest fragmentation and avian nest predation in forested landscapes. *Oecologia* **90**, 489–99.

Smith, P.G.R. & Theberge, J.B. (1986) A review of criteria for evaluating natural areas. *Environmental Management* **10**, 715–34.

Smith, V.K. & Desvouges, W. (1986) *Measuring Water Quality Benefits*. Boston: Kluwer-Nijhoff.

Snyder, N.F.R. & Synder, H.A. (1989) Biology and conservation of the Californian condor. *Current Ornithology* **5**, 175–267.

Soderquist, T.R. (1994) The importance of hypothesis testing in reintroduction biology: examples from the reintroduction of the carnivorous marsupial *Phascogale tapoatafa*. In: Serena, M. (ed.) *Reintroduction Biology of Australian and New Zealand Fauna*. Chipping Norton: Surrey Beatty & Sons, pp. 159–64.

Sohmer, S. (1994) In: Peng, C-I. & Chou, C.H. (eds) *Biodiversity and Terrestrial Ecosystems*. Monograph Series No. 14. Taipei, Taiwan: Institute of Botany, Academia Sinica, pp. 43–51.

Solbrig, O.T. (ed.) (1991) *From Genes to Ecosystems: a Research Agenda for Biodiversity*. Cambridge: International Union of Biological Sciences.

Solow, A., Mound, L.A. & Gaston, K.J. (1995) Estimating the rate of synonymy. *Systematic Biology* **44**, 93–6.

Soulé, M.E. (1983) What do we really know about extinction? In: Schonewald-Cox, C.M., Chambers, S.M., MacBryde, B. & Thomas, L. (eds) *Genetics and Conservation*. Menlo Park: Benjamin/Cummings, pp. 111–24.

Sousa, W.P. (1984) The role of disturbance in natural communities. *Annual Review of Ecology and Systematics* **15**, 353–91.

Spash, C. & Simpson, I. (1994) Utilitarian and rights based approaches for protecting sites of special scientific interest. *Journal of Agricultural Economics* **45**, 15–26.

Spotted Owl Subgroup of the Wildlife Committee of the National Forest Products Association and American Forest Council (1991) *A Multi-resource Strategy for Conservation of the Northern Spotted Owl*. Washington, DC: American Forest Resource Alliance.

Stanley, P.I. & Bunyan, P.J. (1979) Hazards to wintering geese and other wildlife from the use of dieldrin, chlorfenvinphos and carbophenothion as wheat seed treatments. *Proceedings of the Royal Society of London Series B* **205**, 31–45.

Stapp, W.B., Cromwell, M.M. & Wals, A. (1995) The global rivers environmental education network. In: Jacobson, S.K. (ed.) *Conserving Wildlife: International Education and Communication Approaches*. New York: Columbia University Press, pp. 177–97.

Steadman, D.W. (1995) Prehistoric extinctions of Pacific Island birds: biodiversity meets zooarchaeology. *Science* **267**, 1123–31.

Stehli, F.G. (1968) Taxonomic diversity gradients in pole location: the recent model. In: Drake, E.T. (ed.) *Evolution and Environment*. New Haven: Yale University Press, pp. 163–228.

Stehli, F.G., McAlester, A.L. & Helsley, C.E. (1967) Taxonomic diversity of recent bivalves, some implications for geology. *Geological Society of America Bulletin* **78**, 455–66.

Stevens, T., Echevarria, J., Glass, R., Hager, T. & More, T. (1991) Measuring the existence value of wildlife. *Land Economics* **67**(4), 390–400.

Stevens, G.C. (1989) The latitudinal gradient in geographical range: how so many species coexist in the tropics. *American Naturalist* **133**, 240–56.

Stevens, G.C. (1992) The elevational gradient in altitudinal range: an extension of Rapoport's latitudinal rule to altitude. *American Naturalist* **140**, 893–911.

Stewart, A., Pearman, D.A. & Preston, C.D. (1994) *Scarce Plants on Britain*. Peterborough: Joint Nature Conservancy Committee.

Stewart, I. (1990) Risky business. *New Scientist* **126**, insert between pp. 52 & 53.

Stickel, L.F. (1975) The costs and effects of chronic exposure to low-level pollutants in the environment. In: *Hearings before the Sub-committee on the Environment and the Atmosphere*. Washington Committee on Science and Technology, US House of Representatives, pp. 716–28.

Stocking, M. & Perkin, S. (1992) Conservation-with-development: an application of the concept in the Usumbara Mountains, Tanzania. *Transactions of the Institute of British Geographers New Series* **17**, 337–49.

Stokes,T.K., Mcglade, J.M. & Law, R. (eds) (1991) *The Exploitation of Evolving Resources*. Lecture Notes in Biomathematics. Berlin: Springer-Verlag.

Stone, C.P., Smith, C.W. & Tunison, J.T. (1992) *Alien Plant Invasions in Native Ecosystems of Hawaii: Management and Research*. Honolulu: University of Hawaii Press.

Stork, N.E. (1988) Insect diversity: facts, fiction and speculation. *Biological Journal of the Linnean Society of London* **35**, 321–37.

Strong, D.R., Southwood, T.R.E. & Lawton, J.H. (1984) *Insects on Plants*. Oxford: Blackwell Scientific Publications.

Suckcharoen, S., Nuorteva, P. & Hasanen, E. (1978) Alarming signs of mercury pollution in a freshwater area of Thailand. *Ambio* **7**: 113–16.

Sutherland, W.J. (1995) Introduction and principles of habitat management. In: Sutherland, W.J. & Hill, D.A. (eds) *Managing Habitats for Conservation*. Cambridge: Cambridge University Press, pp. 1–21.

Swanson, T. (1993) Regulating endangered species. *Economic Policy* **11**, 187.

Swanson T. (1994) The economics of extinction revisited. *Oxford Economic Papers* **46**, 800–21.

Swanson, T.M. & Barbier, E.B. (1992) Economics for the Wilds: Wildlife, Wildlands, Diversity and Development. London: Earthscan.

Swennen, C. (1972) Chlorinated hydrocarbons attacked the eider population in the Netherlands. *TNO nieuws* **27**, 556–60.

Taylor, J.D. & Taylor, C.N. (1977) Latitudinal distribution of predatory gastropods on the eastern Atlantic shelf. *Journal of Biogeography* **4**, 73–81.

Templeton, A.R. & Read, B. (1983) The elimination of inbreeding depression in a captive herd of Speke's gazelle. In: Schonewald-Cox, C.M., Chambers, S.M., MacBryde, B. & Thomas, W.L. (eds) *Genetics and Conservation*. Menlo Park, CA: Benjamin/Cummings, pp. 241–61.

Therivel, R., Wilson, E., Thompson, S., Heaney, D. & Pritchard, D. (1992) *Strategic Environmental Assessment*. London: Earthscan and Royal Society for the Protection of Birds.

Thirgood, S.J. & Heath, M.F. (1994) Global patterns of endemism and the conservation of biodiversity.

In: Forey, P.L., Humphries, C.J. & Vane-Wright, R.I. (eds) *Systematics and Conservation Evaluation*. Oxford: Clarendon Press, pp. 207–27.

Thomas, C.D. & Hanski, I.A. (1997) Butterfly metapopulations. In: Hanski, I.A. & Gilpin, M.E. (eds) *Metapopulation Biology*. San Diego: Academic Press, pp. 359–86.

Thomas, C.D. & Jones, T.M. (1993) Partial recovery of a skipper butterfly (*Hesperia comma*) from population refuges: lessons for conservation in a fragmented habitat. *Journal of Animal Ecology* **62**, 472–81.

Thomas, C.D., Thomas, J.A. & Warren, M.S. (1992) Distributions of occupied and vacant butterfly habitats in fragmented landscapes. *Oecologia* **92**, 563–7.

Thomas, J.A. (1983) The ecology and conservation of *Lysandra bellargus* in Britain. *Journal of Applied Ecology* **20**, 59–83.

Thomas, J.A. (1991) Rare species conservation: case studies of European butterflies. In: Spellerberg, I.F., Goldsmith, F.B. & Morris, M.G. (eds) *The Scientific Management of Temperate Communities for Conservation*. Oxford: Blackwell Scientific Publications, pp. 144–97.

Thomas, J.W., Forsman, E.D., Lint, J.B., Meslow, E.C., Noon, B.R. & Verner, J. (1990) *A Conservation Strategy for the Northern Spotted Owl. Report to the Interagency Scientific Committee to Address the Conservation of the Northern Spotted Owl*. Portland: US Forest Service, US Department of the Interior.

Thompson III, F.R. (1993) Simulated responses of a forest interior bird population to forest management options in central hardwood forests of the United States. *Conservation Biology* **7**, 325–33.

Tiedje, J.M., Colwell, R.K., Grossman, Y.L. *et al.* (1989) The planned introduction of genetically engineered organisms: ecological considerations and recommendations. *Ecology* **70**, 298–315.

Tietenberg, T. (1994) *Environmental Economics and Policy*. London: Harper Collins.

Tilman, D. (1988) *Plant Strategies and the Dynamics and Structure of Plant Communities*. Princeton: Princeton University Press.

Tilman, D. & Downing, J.A. (1994) Biodiversity and stability in grasslands. *Nature* **367**, 363–5.

Tindale, S. & Holtham, G. (1996) *Green Tax Reform*. London: Institute of Public Policy Research.

TNC (1996) *Priorities for Conservation: 1996 Annual Report Card for US Plant and Animal Species*. Arlington: The Nature Conservancy.

Tucker, R.K. & Crabtree, D.G. (1970) *Handbook of Toxicity of Pesticides to Wildlife*. Resource Publication

No. 84. Washington, DC: Government Printing Office.

Tuljapurkar, S. (1990) *Population Dynamics in Variable Environments*. Berlin: Springer-Verlag.

Turner, K. (1991) Economics and wetland management. *Ambio* **20**, 59–63.

Turner, M.G., Arthaud, G.J., Engstrom, R.T. *et al.* (1995) Usefulness of spatially explicit population models in land management. *Ecological Applications* **5**, 12–16.

Turner, R.K. (1993) Sustainability: principles and practice. In: Turner, R.K. (ed.) *Sustainable Environmental Economics and Management*. London: Belhaven Press pp. 1–36.

Turton, D. (1987) The Mursi and National Park development in the lower Omo Valley. In: Anderson, D.M. & Grove R.H. (eds) *Conservation in Africa: People, Policies and Practice*. Cambridge: Cambridge Univeristy Press, pp. 169–86.

Udvardy, M.D.F. (1975) A classification of the biogeographical provinces of the world. *IUCN Occasional Paper* **18**. Gland: International Union for the Conservation of Nature and Natural Resources.

Underhill, L.G. (1994) Optimal and suboptimal reserve selection algorithms. *Biological Conservation* **70**, 85–7.

UNDP (1992) *Human Development Report*. Oxford: Oxford University Press for United Nations Development Programme.

UNESCO (1996) *The World Network of Biosphere Reserves*. Paris: United Nations Educational, Scientific and Cultural Organization.

US Forest Service (1994) *Western Forest Health Initiative*. Washington, DC: US Department of Agriculture Forest Service, State and Private Forest Program.

US Forest Service and US Bureau of Land Management (1994) *Interior Columbia Basin Ecosystem Management Project. Preliminary Issues for the Development of Alternatives. Eastside Environmental Impact Statement Team for Eastern Oregon and Washington*. Walla Walla: US Forest Service and US Bureau of Land Management.

Usher, M.B. (1986) Wildlife conservation evolution: attributes, criteria and values. In: Usher, M.B. (ed.) *Wildlife Conservation Evaluation*. London: Chapman & Hall, pp. 3–44.

Usher, M.B. (1994) Biodiversity: which communities are hiding it? In: Leather, S.R., Watt, A.D., Mills, N.J. & Walters, K.F.A. (eds) *Individuals, Populations and Patterns in Ecology*. Andover: Intercept, pp. 265–73.

Van den Bosch, F., Hengeveld, R. & Metz, J.A.J. (1992) Analyzing the velocity of animal range expansion. *Journal of Biogeography* **19**, 135–50.

Van den Bosch, F., Metz, J.A.J. & Diekmann, O. (1990) The velocity of spatial population expansion. *Journal of Mathematical Biology* **28**, 529–65.

Van Wieren, S.E. (1991) The management of populations of large mammals. In: Spellerberg, I.F., Goldsmith, F.B. & Morris, M.G. (eds) *The Scientific Management of Temperate Communities for Conservation*. Oxford: Blackwell Scientific Publications, pp. 103–28.

Vane-Wright, R.I., Humphries, C.J. & Williams, P.H. (1991) What to protect? Systematics and the agony of choice. *Biological Conservation* **55**, 235–54.

Veit, R.R. & Lewis, M.A. (1996) Dispersal, population growth, and the Allee effect: dynamics of the house finch invasion of eastern North America. *American Naturalist* **148**, 255–74.

Veldman, M. (1994) *Fantasy, the Bomb and the Greening of Britain: Romantic protest, 1945–80*. Cambridge: Cambrige University Press.

Veltman, C.J., Nee, S. & Crawley, M.J. (1996) Correlates of introduction success in exotic New Zealand birds. *American Naturalist* **147**, 542–57.

Verner, J., McKelvey, K.S., Noon, B.R., Gutierrez, R.J., Gould, G.I. & Beck, T.W. (1992) *The California Spotted Owl: a Technical Assessment of its Current Status*. Albany: US Forest Service, US Department of Agriculture.

Vickery, J., Watkinson, A. & Sutherland, W. (1994) The solutions to the brent geese problem: an economic analysis. *Journal of Applied Ecology* **31**, 371–82.

Victor, P. (1991) Indicators of sustainable development: some lessons from capital theory. *Ecological Economics* **4**, 191–213.

Vincent, P. (1990) *The Biogeography of the British Isles: an Introduction*. London: Routledge.

Vitousek, P. (1988) Diversity and biological invasions of oceanic islands. In: Wilson, E.O. (ed.) *Biodiversity*. Washington, DC: National Academy of Sciences, pp. 181–9.

Vitousek, P.M., Ehrlich, P.R., Ehrlich, A.H. & Matson, P.A. (1986) Human appropriation of the products of photosynthesis, *Bioscience* **36**, 368–73.

Vuilleumier, F., LeCroy, M. & Mayr, E. (1992) New species of birds described from 1981 to 1990. *Bulletin of the British Ornithological Club Centenary* **112A** (Suppl.), 267–309.

Waddington, C.H. (1975) The origin. In:

Worthington, E.B. (ed.) *The Evolution of the IBP.* Cambridge: Cambridge University Press, pp. 4–11.

Walker, B.H. (1992) Biological diversity and ecological redundancy. *Conservation Biology* **6**, 18–23.

Walker, B. (1995) Conserving biological diversity through ecosystem resilience. *Conservation Biology* **9**, 747–52.

Walker, C.H. (1983) Pesticides and birds— mechanisms of selective toxicity. *Agriculture Ecosystems and Environment* **9**, 211–26.

Wallace, G.J., Nickell, W.P. & Bernard, R.F. (1961) Bird mortality in the Dutch elm disease program in Michigan. *Cranbrook Institue of Science Bulletin* **41**, 1–44.

Waller, R.D. (1988) Emutai: crisis and response in Maasailand (1884–1904). In: Johnson, D. & Anderson, D.M. (eds) *The Ecology of Survival.* Lester Crook, pp. 73–112.

Waller, R.D. (1990) Tsetse fly in Western Narok, Kenya. *Journal of African History* **31**, 81–101.

Wallwork, J.A. (1976) *The Distribution and Diversity of Soil Fauna.* London: Academic Press.

Walter, H.S. (1990) Small viable population: the red-tailed hawk of Socorro Island. *Conservation Biology* **4**, 441–3.

Walters, C. & Maguire, J-J. (1996) Lessons for stock assessment from the northern cod collapse. *Reviews in Fish Biology and Fisheries* **6**, 125–37.

Wardle, P. (1991) *Vegetation of New Zealand.* Cambridge: Cambridge University Press.

Warren, M.S. (1987a) The ecology and conservation of the heath fritillary butterfly, *Mellicta athalia* I. Host selection and phenology. *Journal of Applied Ecology* **24**, 467–82.

Warren, M.S. (1987b) The ecology and conservation of the heath fritillary butterfly *Mellicta athalia.* II Adult population structure and mobility. *Journal of Applied Ecology* **24**, 483–98.

Warren, M.S. (1987c) The ecology and conservation of the heath fritillary butterfly *Mellicta athalia.* III Population dynamics and the effect of habitat management. *Journal of Applied Ecology* **24**, 499–514.

Watkins, M.H. (1963) A staple theory of economic growth. *Canadian Journal of Economics and Political Science* **29**, 141–58.

Watkinson, A.R. and Sutherland, W.J. (1995) Sources, sinks and pseudo-sinks. *Journal of Animal Ecology* **64**, 126–30.

Watts, M.J. (1995) 'A new deal in emotions': theory and practice and the crisis of development. In: Crush, J. (ed.) *Power of Development.* London: Routledge, pp. 44–62.

Wayne, R.K. & Jenks, S.M. (1991) Mitochondrial DNA analysis supports a hybrid origin for the endangered red wolf (*Canis rufus*). *Nature* **351**, 565–8.

Wayne, R.K., Benveniste, R.E., Janczewski, D.N. & O'Brien, S.J. (1989) Molecular and biochemical evolution of the carnivora. In: Gittleman, J.L. (ed.) *Carnivore Behavior, Ecology, and Evolution.* Ithaca: Cornell University Press, pp. 465–94.

Wayne, R.K., Van Valkenburgh, B. & O'Brien, S.J. (1991) Molecular distance and divergence time in carnivores and primates. *Molecular Biological Evolution* **8**, 297–310.

WCMC (1992) *Global Biodiversity: Status of the Earth's Living Resources.* London: Chapman & Hall for World Conservation Monitoring Centre.

Webb, N.R. & Thomas, J.A. (1994) Conserving insect habitats in heathland biotopes: a question of scale. In: Edwards, P.J., May, R.M. & Webb, N.R. (eds) *Large-scale Ecology and Conservation Biology.* Oxford: Blackwell Science, pp. 129–51.

Webb, T. (1992) Past changes in vegetation and climate: lessons for the future. In: Peters, R.L. & Lovejoy, T.E. (eds) *Global Warming and Biological Diversity.* New Haven: Yale University Press, pp. 59–75.

Weitzman, M.L. (1995) Diversity functions. In: Perrings, C., Mäler, K-G., Folke, C., Holling, C.S. & Jansson, B-O. (eds) *Biodiversity Loss: Economic and Ecological Issues.* Cambridge: Cambridge University Press, pp. 21–43.

Wells, M. & Brandon, K. (1992) *People and Parks: Linking Protected Area Management with Local Communities.* Washington, DC: World Bank.

Wells, M., Brandon, K. & Hannah, L. (1992) *People and Parks: Linking Protected Area Management and Local Communities.* Washington, DC: World Bank, Worldwide Fund for Nature and the United States Agency for International Development.

Western, D. (1982) The environment and ecology of pastoralists in arid savannas. *Development and Change* **13**, 183–211.

Western, D. (1992) The biodiversity crisis: a challenge for biology. *Oikos* **63**, 29–38.

Western, D. & Wright, R.M. (1994) The background into community-based conservation. In: Western, D., White, R.M. & Strum, S.C. (eds) *Natural Connections: Perspectives in Community-based Conservation.* Washington, DC: Island Press, pp. 1–14.

Whitehead B. (1992) Management brief: food for thought. *Economist* **29 August**, 62–4.

Whitehead, J.C. (1993) Total economic values for

coastal and marine wildlife. *Marine Resource Economics* **8**, 119–32.

Whittaker, R.H. (1972) Evolution and measurement of species diversity. *Taxon* **21**, 213–51.

Wiegert, R.G. (1988) Holism and reductionism in ecology: hypotheses, scale and systems. *Oikos* **53**, 267–9.

Wilcove, D.S., McMillan, M. & Winston, K.C. (1993) What exactly is an endangered species? An analysis of the endangered species list, 1985–1991. *Conservation Biology* **7**, 87–93.

Wilkie, D.S., Sidle, J.S. & Boundzange, G.C. (1992) Mechanised logging, market hunting and a bank loan in Congo. *Conservation Biology* **6**, 570–80.

Wilkinson, C.F. (1992) *Crossing the Next Meridian: Land, Water and the Future of the West.* Washington, DC: Island Press.

Williams, C.B. (1960) The range and pattern of insect abundance. *American Naturalist* **94**, 137–51.

Williams, G., Holmes, J. & Kirby, J.A. (1995) Action plans for United Kingdon and European rare, threatened and internationally important birds. *Ibis* **137**, S209–S213.

Williams, J.D., Warren Jr, M.L., Cummings, K.S., Harris, J.L. & Neves, R.J. (1992) Conservation status of freshwater mussels of the United States and Canada. *Fisheries* **18**, 6–22.

Williams, M. (1989a) *Americans and their Forests: a Historical Geography.* Cambridge: Cambridge University Press.

Williams, M. (1989b) Deforestation: past and present. *Progress in Human Geography* **13**, 176–208.

Williams, P., Gibbons, D., Margules, C., Rebelo, A., Humphries, C. & Pressey, R. (1996) A comparison of richness hotspots, rarity hotspots, and complementary areas for conserving diversity of British birds. *Conservation Biology* **10**, 155–74.

Williams, P.H. (1992) WORLDMAP *priority areas for biodiversity*, Version 3. (London: privately distributed computer program and manual.)

Williams, P.H. (1993) Choosing conservation areas: using taxonomy to measure more of biodiversity. In: Moon, T-Y. (ed.) *International Symposium on Biodiversity and Conservation.* Seoul: Korean Entomological Institute, pp. 194–227.

Williams, P.H., Gaston, K.J. & Humphries, C.J. (1994a) Do conservationists and molecular biologists value differences between organisms in the same way? *Biodiversity Letters* **2**, 67–78.

Williams, P.H. & Humphries, C.J. (1996) Comparing character diversity among biotas. In: Gaston, K.J. (ed.) *Biodiversity: a Biology of Numbers and Difference.* Oxford: Blackwell Science, pp. 54–76.

Williams, P.H., Humphries, C.J. & Gaston, K.J. (1994b) Centres of seed-plant diversity: the family way. *Proceedings of the Royal Society London Series B* **256**, 67–70.

Williamson, M.H. (1981) *Island Populations.* Oxford: Oxford University Press.

Williamson, M.H. (1992) Environmental risks from the release of genetically modified organisms (GMOs)—the need for molecular ecology. *Molecular Ecology* **1**, 3–8.

Williamson, M.H. (1993) Invaders, weeds and the risk from genetically modified organisms. *Experientia* **49**, 219–24.

Williamson, M.H. (1994) Community response to transgenic plant release: predictions from the British experience of invasive plants and feral crop plants. *Molecular Ecology* **3**, 75–9.

Williamson, M.H. (1996) *Biological Invasions.* London: Chapman & Hall.

Williamson, M.H. & Brown, K.C. (1986) The analysis and modelling of British invasions. *Philosophical Transactions of the Royal Society Series B* **314**, 505–22.

Willis, E.O. (1974) Populations and local extinctions on Barro Colorado Island, Panama. *Ecological Monographs* **44**, 153–69.

Willis, K. (1991) The recreational value of the Forestry Commission estate. *Scottish Journal of Political Economy* **38**(1), 58–75.

Willis, K. & Garrod, G. (1995) The benefits of alleviating low flows in rivers. *Water Resources Development* **11**(3), 243–60.

Wilson, E.O. (1988a) *Biodiversity.* Washington, DC: National Academy Press.

Wilson, E.O. (1988b) The current state of biological diversity. In: Wilson, E.O. & Peter, F.M. (eds) *Biodiversity.* Washington, DC: National Academy Press, pp. 3–18.

Wilson, E.O. (1989) Threats to biodiversity. *Scientific American* **261**, 108–16.

Wilson, E.O. (1992) *The Diversity of Life.* London: Penguin Books.

Wilson, E.O. & Peter, F.M. (eds) (1988) *Biodiversity.* Washington, DC: National Academy Press.

Wilson, J.W. (1974) Analytical zoogeography of North American mammals. *Evolution* **28**, 124–40.

Witte, F., Goldschmidt, T., Wanink, J.H. *et al.* (1992) The destruction of an endemic species flock: quantitative data on the decline of the haplochromine cichlids of Lake Victoria. *Environmental Biology of Fishes* **34**, 1–28.

Wootton, T.J. & Bell, D.A. (1992) A metapopulation model of the peregrine falcon in California:

viability and management strategies. *Ecological Applications* **2**, 307–21.

World Bank (1989) *Sub-Saharan Africa: from Crisis to Sustainable Growth; a Long-term Perspective Study.* Washington, DC: World Bank.

World Bank (1992) *World Development Report 1992: Development and the Environment.* New York: Oxford University Press for the World Bank.

World Conservation Union (1990) *1990 United Nations List of National Parks and Protected Areas.* Gland, Switzerland: International Union for Conservation of Nature and Natural Resources.

Worster, D. (1985) *Nature's Economy: a History of Ecological Ideas.* Cambridge: Cambridge University Press.

Wright, D.H., Currie, D.J. & Maurer, B.A. (1993) Energy supply and patterns of species richness on local and regional scales. In: Ricklefs, R.E. & Schluter, D. (eds) *Species Diversity in Ecological Communities: Historical and Geographical Perspectives.* Chicago: University of Chicago Press, pp. 66–74.

Wynne, G., Avery, M., Campbell, L. *et al.* (1995) *Biodiversity Challenge: an Agenda for Conservation in the UK*, 2nd edn. Sandy, UK: Butterfly Conservation, Friends of the Earth, Plantlife, The Wildlife Trusts Partnership, the Royal Society for the Protection of Birds and the World Wide Fund.

Yablokov, A. (1994) Validity of whaling data. *Nature* **367**, 108.

Zemsky, V.A., Berzin, A.A., Mikhaliev, Y.A. & Tormosov, D.D. (1995) Report of the scientific committee, Annexe E. Report of the subcommittee on southern hemisphere baleen whales. Appendix 3. Soviet Antarctic pelagic whaling after W.W. 2: review of actual catch data. *Report International Whaling Commission* **45**, 131–5.

Zerner, C. (1994) Transforming customary law and coastal management practices in the Makalu islands, Indonesia, 1870–1992. In: Western, D., White, R.M. & Strum, S.C. (eds) *Natural Connections: Perspectives in Community-based Conservation.* Washington, DC: Island Press, pp. 80–112.

Zube, E.H. & Busch, M. (1990) park–people relationships: an international review. *Landscape and Urban planning* **19**: 117–31.

Index